Nonequilibrium and Irreversibility

Theoretical and Mathematical Physics

The series founded in 1975 and formerly (until 2005) entitled *Texts and Monographs in Physics* (TMP) publishes high-level monographs in theoretical and mathematical physics. The change of title to *Theoretical and Mathematical Physics* (TMP) signals that the series is a suitable publication platform for both the mathematical and the theoretical physicist. The wider scope of the series is reflected by the composition of the editorial board, comprising both physicists and mathematicians.

The books, written in a didactic style and containing a certain amount of elementary background material, bridge the gap between advanced textbooks and research monographs. They can thus serve as basis for advanced studies, not only for lectures and seminars at graduate level, but also for scientists entering a field of research.

For further volumes:
http://www.springer.com/series/720

Giovanni Gallavotti

Nonequilibrium
and Irreversibility

 Springer

Giovanni Gallavotti
Department of Physics
University of Rome "La Sapienza"
Rome
Italy

ISSN 1864-5879   ISSN 1864-5887   (electronic)
ISBN 978-3-319-38327-9   ISBN 978-3-319-06758-2   (eBook)
DOI 10.1007/978-3-319-06758-2
Springer Cham Heidelberg New York Dordrecht London

Printed on acid-free paper

Springer is part of Springer Science+Business Media (www.springer.com)

*A. Daniela and Barbara
and Camomilla and Olim*

Preface

Every hypothesis must derive indubitable results from mechanically well-defined assumptions by mathematically correct methods. If the results agree with a large series of facts, we must be content, even if the true nature of facts is not revealed in every respect. No one hypothesis has hitherto attained this last end, the Theory of Gases not excepted.

Boltzmann, [1, p. 536, #112]

In recent years renewed interest grew about the problems of nonequilibrium statistical mechanics. I think that this has been stimulated by the new research made possible by the availability of simple and efficient computers and of the simulations they make possible.

The possibility and need of performing systematic studies has naturally led to concentrate efforts in understanding the properties of states that are in stationary nonequilibrium: thus establishing a clear separation between properties of evolution towards stationarity (or equilibrium) and properties of the stationary states themselves: a distinction which until the 1970s was rather blurred.

A system is out of equilibrium if the microscopic evolution involves non-conservatives forces or interactions with external particles that can be modeled by or identified with dissipative phenomena which forbid indefinite growth of the system energy. The result is that nonzero currents are generated in the system with matter or energy flowing and dissipation being generated. In essentially all problems, the regulating action of the external particles can be reliably modeled by non-Hamiltonian forces.

Just as in equilibrium statistical mechanics the stationary states are identified by the time averages of the observables. As is familiar in measure theory, the collections of averages of any kind (time average, phase space average, counting average, etc.) are in general identified with probability distributions on the space of the possible configurations of a system; thus such probability distributions yield the natural formal setting for the discussions with which we shall be concerned here. Stationary states are identified with probability distributions on the microscopic configurations, i.e. on phase space, which of course have to be invariant under time evolution.

A first problem is that in general there will be a very large number of invariant distributions: which ones correspond to stationary states of a given assembly of atoms and molecules? That is, which ones lead to averages of observables that can be identified with time averages under the time evolution of the system?

This has been a key question already in equilibrium: Clausius, Boltzmann, Maxwell (and others) considered it reasonable to think that the microscopic evolution had the property that, in course of time, every configuration was reached from motions starting from any other.

Analyzing this question has led to many developments since the early 1980s: the purpose of this monograph is to illustrate a point of view about them. My interest on the subject started from my curiosity to understand the chain of achievements that led to the birth of Statistical Mechanics: many original works are in German language; hence I thought it of some interest to present and comment on the English translation of large parts of a few papers by Boltzmann and Clausius that I found inspiring at the beginning of my studies. Chapter 6 contains the translations: I have tried to present them as faithfully as possible, adding a few personal comments inserted in form of footnotes or, if within the text, in slanted characters; original footnotes are marked with "NdA."

I have not included the celebrated 1872 paper of Boltzmann, [2, #22], on the Boltzmann's equation, which is widely commented and translated in the literature; I have also included comments on Maxwell's work of 1866, [3, 4], where he derives and amply uses a form of the Boltzmann's equation which we would call today a "weak Boltzmann's equation": this Maxwell's work was known to Boltzmann (who quotes it in [5, #5]) and is useful to single out the important contribution of Boltzmann (the "strong" equation for the one particle distribution and the H-theorem).

Together with the many cross references, Chap. 6 makes, hopefully, clear aspects, relevant for the present book, of the interplay between the three founders of modern statistical mechanics, Boltzmann, Clausius, and Maxwell (it is only possible to quote them in alphabetical order) and their influence on the recent developments.

I start, in Chap. 1, with a review of equilibrium statistical mechanics (Chap. 1) mostly of historical nature. The mechanical interpretation of the second law of thermodynamics (referred here as "the heat theorem") via the ergodic hypothesis and the least action principle is discussed. Boltzmann's equation and the irreversibility problem are briefly analyzed. Together with the partial reproduction of the original works in Chap. 6, I hope to have given a rather detailed account of the birth and role (and eventual "irrelevance") of the ergodic hypothesis from the original "monocyclic" view of Boltzmann, to the "polycyclic" view of Clausius, to the more physical view of Maxwell[1] and to the modern definition of ergodicity and its roots in the discrete conception of space time.

In Chap. 2 thermostats, whose role is to permit the establishment of stationary nonequilibria, are introduced. Ideally, interactions are conservative and therefore thermostats should ideally be infinite systems that can indefinitely absorb the energy introduced in a system by the action of non-conservative external forces. Therefore, models of infinitely extended thermostats are discussed and some of their properties are illustrated. However, great progress has been achieved since the 1980s by studying systems kept in a stationary state thanks to the action of finite thermostats: such systems have great advantage of being often well suited for simulations. The disadvantage is that the forces driving them are not purely Hamiltonian: however, one is (or should be) always careful that at least they respect the fundamental symmetry of Physics, which is time reversal.

This is certainly very important particularly because typically in nonequilibrium we are interested in irreversible phenomena. For instance, the Hoover's thermostats are time reversible and led to new discoveries (works of Hoover, Evans, Morriss, Cohen, and many more). This opened the way to establishing a link with another development in the theory of chaotic system, particularly with the theory of Sinai, Ruelle, Bowen, and Ruelle's theory of turbulence. It achieved a major result of identifying the probability distribution that in a given context would be singled out among the great variety of stationary distributions that it had become clear would be generically associated with any mildly chaotic dynamical system.

It seems that this fact is not (yet) universally recognized and the SRB distribution is often shrugged away as a mathematical nicety.[2] I dedicate a large part of Chap. 2 to trying to illustrate the physical meaning of the SRB distribution relating it to what has been called (by Cohen and me) "chaotic hypothesis". It is also an assumption that requires understanding and some open mindedness: personally I have been influenced by the ergodic hypothesis (of which it is an extension to nonequilibrium phenomena) in the original form of Boltzmann, and for this reason I have proposed here rather large portions of the original papers by Boltzmann and Clausius, see Chap. 6. The reader who is perplexed about the chaotic hypothesis can find some relief in reading the mentioned classics and their even more radical treatment of what today would be chaotic motions, via periodic motions. Finally, the role of dissipation (in time reversible systems) is discussed and its remarkable physical meaning of entropy production rate is illustrated (another key discovery

---

[1] "The only assumption which is necessary for the direct proof is that the system, if left to itself in its actual state of motion, will, sooner or later pass through every phase that is consistent with the equation of energy. Now it is manifest that there are cases in which this does not take place... But if we suppose that the material particles, or some of them, occasionally encounter a fixed obstacle such as the sides of a vessel containing the particles, then, except for special forms of the surface of this obstacle, each encounter will introduce a disturbance into the motion of the system, so that it will pass from one undisturbed path into another...", [6, Vol. 2, p. 714].

[2] It is possible to find in the literature heroic efforts to avoid dealing with the SRB distributions by essentially attempting to do what is actually done (and better) in the original works.

due to the numerical simulations with finite reversible thermostats mentioned above).

In Chap. 3, theoretical consequences of the chaotic hypothesis are discussed: the leading ideas are drawn again from the classic works of Boltzmann see Sects. 6.2 and 6.12: the SRB distribution properties can conveniently be made visible if the Boltzmann viewpoint of discreteness of phase space is adopted. It leads to a combinatorial interpretation of the SRB distribution, which unifies equilibrium and nonequilibrium relating them through the coarse graining of phase space made possible by the chaotic hypothesis. The key question of whether it is possible to define entropy of a stationary nonequilibrium state is discussed in some detail making use of the coarse grained phase space: concluding that while it may be impossible to define a nonequilibrium entropy it is possible to define the entropy production rate and a function that in equilibrium is the classical entropy while out of equilibrium is "just" a Lyapunov function (one of many) maximal at the SRB distribution.

In Chap. 4 several general theoretical consequences of the chaotic hypothesis are enumerated and illustrated: particular attention is dedicated to the role of the time reversal symmetry and its implications on the universal (i.e., widely model independent) theory of large fluctuations: the fluctuation theorem by Cohen and myself, Onsager reciprocity and Green–Kubo formula, the extension of the On-sager–Machlup theory of patterns fluctuations, and an attempt to study the cor-responding problems in a quantum context. Universality is, of course, important because it partly frees us from the nonphysical nature of the finite thermostats.

In Chap. 5 I try to discuss some special concrete applications, just as a modest incentive for further research. Among them, however, there is still a general question that I propose and to which I attempt a solution: it is to give a quantitative criterion for measuring the degree of irreversibility of a process, i.e., to give a measure of the quasistatic nature of a process.

In general, I have avoided technical material preferring heuristic arguments to mathematical proofs: however, when possible references have been given for readers who find some interest in the topics treated and want to master the (important) details. The same applies to the Appendices (A–K).

In Chap. 6 several classic papers are presented, all but one in partial translation from the original German language. These papers illustrate my personal route to studying the birth of ergodic theory and its relevance for statistical mechanics, [7–11] and, implicitly, provide motivation for the choices (admittedly very per-sonal) made in the first five chapters and in the Appendices.

Appendices A–K contain a few complements, and the remaining appendices deal with technical problems that are still unsolved. Appendix M (with more details in appendices N, O, P) gives an example of the work that may be necessary in actual constructions of stationary states in the apparently simple case of a forced pendulum in presence of noise. Appendices Q–T discuss an attempt (*work in progress*) at studying a stationary case of BBGKY hierarchy with no random forces but out of equilibrium. I present this case because I think it is instructive although the results are deeply unsatisfactory: it is part unpublished work in strict collaboration with G. Gentile and A. Giuliani.

The booklet represents a viewpoint, my personal, and does not pretend to be exhaustive: many important topics have been left out (like [12–15], just to mention a few works that have led to further exciting developments). I have tried to present a consistent theory including some of its unsatisfactory aspects.

The Collected papers of Boltzmann, Clausius, Maxwell are freely available: about Boltzmann I am grateful (and all of us are) to Wolfgang Reiter, in Vienna, for actively working to obtain that *Österreichische Zentralbibliothek für Physik* undertook and accomplished the task of digitizing the "Wissenschaftliche Abhandlungen" and the "Populäre Schriften" at

https://phaidra.univie.ac.at/detail_object/o:63668
https://phaidra.univie.ac.at/detail_object/o:63638

respectively, making them freely available.
This is *Version* **2.0**$^{©}$: 20 March 2014
Version **1.0**$^{©}$: 31 December 2008

Acknowledgments

I am indebted to D. Ruelle for his teaching and examples. I am indebted to E. G. D. Cohen for his constant encouragement and stimulation as well as, of course, for his collaboration in the developments in our common works and for supplying many ideas and problems. To Guido Gentile and Alessandro Giuliani for their close collaboration in an attempt to study heat conduction in a gas of hard spheres. Finally, a special thank you to Prof. Wolf Beiglböck for his constant interest and encouragement.

Rome, October 2013   Giovanni Gallavotti

References

1. Boltzmann, L.: Wissenschaftliche Abhandlungen. In: Hasenöhrl, F. (ed.) vol. 1, 2, 3 (http://www.esi.ac.at/further/Boltzmannonline.html). Barth, Leipzig (1909)
2. Boltzmann, L.: Weitere Studien über das Wärmegleichgewicht unter Gasmolek ülen, vol. 1, #22. In: Hasenöhrl, F. Hasenöhrl (ed.) Wissenschaftliche Abhandlungen, Chelsea, New York (1968)
3. Maxwell, J.C.: On the dynamical theory of gases. Philos. Trans. **157**, 49–88 (1867)
4. Maxwell, J.C.: On the dynamical theory of gases. In: Maxwell, J.C., Niven, W.D. (eds.) The Scientific Papers, vol. 2. Cambridge University Press, Cambridge (1964)

5. Boltzmann, L.: Studien Über die mechanische Bedeutung des zweiten Hauptsatzes der Wärmetheorie. In: Hasenöhrl, F. (ed.) Wissenschaftliche Abhandlungen. vol. 1, #5. Chelsea, New York (1968)

6. Maxwell, J.C.: The Scientific Papers of J.C. Maxwell. In: Niven, W.D. (ed.), vol. 1, 2. Cambridge University Press, Cambridge (1964)

7. Gallavotti, G.: L' hypothèse ergodique et Boltzmann. In: Chemla, K. (ed.) Dictionnaire Philosophique, Presses Universitaires, Paris, pp. 1081–1086 (1989)

8. Gallavotti, G.: Trattatello di Meccanica Statistica. In: Quaderni del CNR-GNFM, vol. 50. Firenze (1995)

9. Gallavotti, G.: Ergodicity, ensembles, irreversibility in Boltzmann and beyond. J. Stat. Phys. **78**, 1571–1589 (1995)

10. Gallavotti, G.: Statistical Mechanics: A Short Treatise. Springer, Berlin (2000)

11. Gallavotti, G.: Equilibrium statistical mechanics. In: Francoise, J.P., Naber, G.L., Tsun, T.S. (eds.) Encyclopedia of Mathematical Physics, vol. 1, pp. 51–87. Elsevier, Amsterdam (2006)

12. Bertini, L., De Sole, A., Gabrielli, D., Jona-Lasinio, G., Landim, C.: Fluctuations in stationary nonequilibrium states of irreversible processes. Phys. Rev. Lett. **87**, 040601 (2001)

13. Derrida, B., Lebowitz, J.L., Speer, E.R.: Exact free energy functional for a driven diffusive open stationary nonequilibrium system. Phys. Rev. Lett. **89**, 030601 (2002)

14. Gerschenfeld, A., Derrida, B., Lebowitz, J.L.: Anomalous Fouriers law and long range correlations in a 1D non-momentum conserving mechanical model. J. Stat. Phys. **141**, 757–766 (2010)

15. Bricmont, J., Kupiainen, A.: Diffusion in energy conserving coupled maps. Commun. Math. Phys. **321**, 311–369 (2013)

Contents

Chapter 1
Equilibrium

1.1 Many Particles Systems: Kinematics, Timing

Mechanical systems in interaction with thermostats will be modeled by evolution equations describing the time evolution of the point $x = (X, \dot{X}) = (x_1, \ldots, x_N, \dot{x}_1, \ldots, \dot{x}_N) \in R^{6N}$ representing positions and velocities of all particles in the ambient space R^3.

It will be often useful to distinguish between the positions and velocities of the N_0 particles in the "system proper" (or "test system" as in [1]), represented by $(X^{(0)}, \dot{X}^{(0)}) = (x_1^{(0)}, \ldots, x_{N_0}^{(0)}, \dot{x}_1^{(0)}, \ldots, \dot{x}_{N_0}^{(0)})$ and by the positions and velocities of the N_j particles in the various thermostats $(X^{(j)}, \dot{X}^{(j)}) = (x_1^{(j)}, \ldots, x_{N_j}^{(j)}, \dot{x}_1^{(j)}, \ldots, \dot{x}_{N_j}^{(j)})$, $j = 1, \ldots, m$: for a total of $N = \sum_{j=0}^{m} N_j$ particles.

Time evolution is traditionally described by *differential equations*

$$\dot{x} = F(x) \tag{1.1.1}$$

whose solutions (given initial data $x(0) = (X(0), \dot{X}(0))$) yield a trajectory $t \to x(t) = (X(t), \dot{X}(t))$ representing motions developing in continuous time t in "phase space" (i.e. the space where the coordinates of x dwell).

A better description is in terms of *maps* whose n-th iterate represents motions developing at discrete times t_n. The point representing the state of the system at time t is denoted $S_t x$ in the continuous time models or, at the n-th observation, $S^n \xi$ in the discrete time models.

The connection between the two representations of motions is illustrated by means of the following notion of *timing event*.

Physical observations are always performed at discrete times: i.e. when some special, prefixed, *timing* event occurs, typically when the state of the system is in a set $\Xi \subset R^{6N}$ and can possibly trigger the action of a "measurement apparatus", e.g. shooting a picture after noticing that a chosen observable assumes a prefixed value. If Ξ comprises the collection of the timing events, i.e. of the states ξ of the

G. Gallavotti, *Nonequilibrium and Irreversibility*,
Theoretical and Mathematical Physics, DOI: 10.1007/978-3-319-06758-2_1,
© Springer International Publishing Switzerland 2014

system which can induce the act of measurement, motion of the system can also be represented as a map $\xi \to S\xi$ defined on Ξ.[1]

For this reason mathematical models are often maps which associate with a timing event ξ, i.e. a point ξ in the manifold Ξ of the measurement inducing events, the next timing event $S\xi$.

Here x, ξ will not be necessarily points in R^{6N} because it is possible, and sometimes convenient, to use other coordinates: therefore, more generally, x, ξ will be points on a manifold M or Ξ of dimension $6N$ or $6N - 1$, respectively, called the *phase space*, or the space of the states. The dimension of the space Ξ of the timing events is one unit less than that of M: because, by definition, timing events correspond to a prefixed value of some observable $f(x)$. Furthermore sometimes the system admits conservation laws allowing a description of the motions in terms of fewer than $6N$ coordinates.

Of course the "*section*" Ξ of the timing events has to be chosen so that every trajectory, or at least all trajectories but a set of 0 probability with respect to the random choices that are supposed to generate the initial data, crosses infinitely many times the set Ξ, which in this case is also called a *Poincaré's section* and has to be thought of as a codimension 1 surface drawn on phase space.

There is a simple relation between the evolution in continuous time $x \to S_t x$ and the discrete representation $\xi \to S^n \xi$ in discrete integer times n, between successive timing events: namely $S\xi \equiv S_{\tau(\xi)}\xi$, if $\tau(\xi)$ is the time elapsing between the timing event ξ and the subsequent one $S\xi$.

Timing observations with the realization of special or "intrinsic" events (i.e. $x \in \Xi$), rather than at "extrinsic" events like at regularly spaced time intervals, is for good reasons: namely to discard information that is of little relevance.

It is clear that, fixed $\tau > 0$, two events $x \in M$ and $S_\tau x$ will evolve in a strongly correlated way. It will forever be that the event $S_\tau^n x$ will be followed τ later by the next event; which often is an information of little interest[2]: which is discarded if observations are timed upon the occurrence of dynamical events $x \in \Xi$ which (usually) occur at "random" times, i.e. such that the time $\tau(x)$ between an event $x \in \Xi$ and the successive one $S_{\tau(x)}x$ has a nontrivial distribution when x is randomly selected by the process that prepares the initial data. This is quite generally so when Ξ is a codimension 1 surface in phase space M which is crossed transversely by the continuous time trajectories.

The discrete time representation, timed on the occurrence of intrinsic dynamical events, can be particularly useful (physically and mathematically) in cases in which the continuous time evolution shows singularities: the latter can be avoided by choosing timing events which occur when the point representing the system is

---

[1] Sometimes the observations can be triggered by a clock arm indicating a chosen position on the dial: in this case the phase space will be R^{6N+1} and the space Ξ will coincide with R^{6N}. But in what follows we shall consider measurements triggered by some observable taking a prefixed value, unless otherwise stated.

[2] In the case of systems described in continuous time the data show always a 0-Lyapunov exponent and this remains true it the observations are made at fixed time intervals

neither singular nor too close to a singularity (i.e. avoiding situations in which the physical measurements become difficult or impossible).

Very often, in fact, models idealize the interactions as due to potentials which become infinite at some *exceptional* configurations, (see also [2]). For instance the Lennard-Jones interparticle potential, for the pair interactions between molecules of a gas, diverges as r^{-12} as the pair distance r tends to 0; or the model of a gas representing atoms as elastic hard spheres, with a potential becoming $+\infty$ at their contacts.

An important, paradigmatic, example of timed observations and of their power to disentangle sets of data arising without any apparent order has been given by Lorenz [3].

A first aim of the Physicist is to find relations, which are general and model independent, between time averages of a few (*very few*) observables. Time average of an observable F on the motion starting at $x \in M$, or in the discrete time case starting at $\xi \in \Xi$, is defined as

$$\langle F \rangle = \lim_{T \to \infty} \frac{1}{T} \int_0^T F(S_t x) dt \quad \text{or} \quad \langle F \rangle = \lim_{n \to \infty} \frac{1}{n} \sum_{j=0}^{n-1} F(S^j \xi) \qquad (1.1.2)$$

and in principle the averages might depend on the starting point (and might even not exist). There is a simple relation between timed averages and continuous time averages, provided the observable $\tau(\xi)$ admits an average $\overline{\tau}$, namely if $\widetilde{F}(\xi) \overset{def}{=} \int_0^{\tau(\xi)} F(S_t \xi) dt$

$$\langle F \rangle = \lim_{n \to \infty} \frac{1}{n\overline{\tau}} \sum_{j=0}^{n-1} \widetilde{F}(S^n \xi) \equiv \frac{1}{\overline{\tau}} \langle \widetilde{F} \rangle \qquad (1.1.3)$$

if the limits involved exist.

Only later, as a further and more interesting problem, the properties specific of given systems or classes of systems become the object of quantitative investigations.

1.2 Birth of Kinetic Theory

The classical example of general, model independent, results is offered by Thermodynamics: its laws establish general relations which are completely independent of the detailed microscopic interactions or structures (to the point that it is not even necessary to suppose that bodies consist of atoms).

It has been the first task of Statistical Mechanics to show that Thermodynamics, under suitable assumptions, is or can be regarded as a consequence of simple, but very general, mechanical models of the elementary constituents of matter motions.

The beginnings go back to the classical age [4]: however in modern times *atomism* can be traced to the discovery of Boyle's law (1660) for gases [5, p. 43], which could be explained by imagining the gas as consisting of individual particles linked by elastic springs. This was a static theory in which particles moved only when their container underwent compression or dilation. The same static view was to be found in Newton (postulating nearest neighbor interactions) and later in Laplace (postulating existence and somewhat peculiar properties of *caloric* to explain the nature of the nearest neighbor molecular interactions). The correct view, assigning to the atoms the possibility of free motion was heralded by Bernoulli [5, p. 57], (1738). In his theory molecules move and exercise pressure through their collisions with the walls of the container: the pressure is not only proportional to the density but also to the average of the square of the velocities of the atoms, as long as their size can be neglected. Very remarkably he introduces the definition of temperature via the gas law for air: the following discovery of Avogadro (1811), "law of equivalent volumes", on the equality of the number of molecules in equal volumes of rarefied gases in the same conditions of temperature and pressure [6], made the definition of temperature independent of the particular gas employed allowing a macroscopic definition of absolute temperature.

The work of Bernoulli was not noticed until much later, and the same fate befell on the work of Herapath (1821), who was "unhappy" about Laplace's caloric hypotheses and proposed a kinetic theory of the pressure deriving it as proportional to the average velocity rather than to its square, but *that was not* the reason why it was rejected by the *Philosophical Transactions of the Royal Society* [7], and sent to temporary oblivion.

A little later Waterston (1843), proposed a kinetic theory of the equation of state of a rarefied gas in which temperature was proportional to the average squared velocity. His work on gases was first published inside a book devoted to biology questions (on the physiology of the central nervous system) and later written also in a paper submitted to the *Philosophical Transactions of the Royal Society*, but rejected [7], and therefore it went unnoticed. Waterston went further by adopting at least in principle a model of interaction between the molecules proposed by Mossotti [8], holding the view that it should be possible to formulate a unified theory of the forces that govern microscopic as well as macroscopic matter.

The understanding of heat as a form of energy transfer rather than as a substance, by Mayer (1841), and Joule (1847), provided the first law of Thermodynamics on *internal energy* and soon after Clausius [9] formulated the second law, see p. 19, on the impossibility of cyclic processes whose only outcome would be the transfer of heat from a colder thermostat to a warmer one and showed Carnot's efficiency theorem [10], to be a consequence and the basis for the definition of the *entropy*.[3]

And at this point the necessity of finding a connection between mechanics and thermodynamics had become clear and urgent.

---

[3] The meaning of the word was explained by Clausius himself [11, p. 390]: "I propose to name the quantity S the entropy of the system, after the Greek word ἡ τροπή "the transformation" [12] , [*in German Verwandlung*]. I have deliberately chosen the word entropy to be as similar as possible to the word energy: the two quantities to be named by these words are so closely related in physical significance that a certain similarity in their names appears to be appropriate."

The kinetic interpretation of absolute temperature as proportional to the average kinetic energy by Krönig [13], was the real beginning of Statistical Mechanics (because the earlier work of Bernoulli and Waterstone had gone unnoticed). The speed of the molecules in a gas was linked to the speed of sound: therefore too fast to be compatible with the known properties of diffusion of gases. The mean square velocity u (in a rarefied gas) could nevertheless be more reliably computed via Krönig proposal and from the knowledge of the gas constant R (with no need of knowing Avogadro's number): because $pV = nRT$, with n the number of moles, and $\frac{3}{2}nRT = \frac{3}{2}Nmu^2 = \frac{3}{2}Mu^2$ with M being the mass of the gas enclosed in the volume V at pressure p. This gives speeds of the order of 500 m/sec, as estimated by Clausius [14, p. 124].

Clausius noted that compatibility could be restored by taking into account the collisions between molecules: he introduced, (1858), the *mean free path* λ given in terms of the atomic radius a as $\lambda = \frac{1}{n\pi a^2}$ with n the numerical density of the gas. Since Avogadro's number and the atomic sizes were not yet known this could only show that λ could be expected to be much smaller than the containers size L (being, therefore, proportional to $\frac{L}{N}(\frac{L}{a})^2$): however it opened the way to explaining why breaking an ampulla of ammonia in a corner of a room does not give rise to the sensation of its smell to an observer located in another corner, not in a time as short as the sensation of the sound of the broken glass.

The size a, estimated as early as 1816 by T. Young and later by Waterston (1859), to be of the order of 10^{-8} cm, can be obtained from the ratio of the volume L^3 of a gas containing N molecules to that of the liquid into which it can be compressed (the ratio being $\varrho = \frac{3}{4\pi}\frac{L^3}{a^3 N}$) and by the mean free path ($\lambda = \frac{L^3}{N4\pi a^2}$) which can be found (Maxwell [15, p. 386]) from the liquid dynamical viscosity ($\eta = \frac{c}{Na^2}\sqrt{3nRT}$ with n number of moles, and c a numerical constant of order 1 [16, Eqs. (8.1.4), (8.1.8)]; thereby *also* expressing N and a in terms of macroscopically measurable quantities ϱ, η, measured carefully by Loschmidt [7, p. 75]

Knowledge of Avogadro's number N and of the molecular radius a allows to compute the diffusion coefficient of a gas with mass density ϱ and average speed V: following Maxwell in [17, Vol. 2, pp. 60, 345] it is $D = \frac{RT}{cNa^2V\varrho}$, with c a constant of order 1. This makes quantitative Clausius explanation of diffusion in a gas.[4]

(Footnote 3 continued)
More precisely the German word really employed by Clausius [11, p. 390], is *Verwandlungsinhalt* or "transformation content".

[4] The value of D depends sensitively on the assumption that the atomic interaction potential is proportional to r^{-4} (hence at constant pressure D varies as T^2). The agreement with the few experimental data available (1866 and 1873) induced Maxwell to believe that the atomic interaction would be proportional to r^{-4} (hard core interaction would lead to D varying as $T^{\frac{3}{2}}$ as in his earlier work [15]).

1.3 Heat Theorem and Ergodic Hypothesis

The new notion of entropy had obviously impressed every physicist and the young Boltzmann attacked immediately the problem of finding its mechanical interpretation. Studying his works is as difficult as it is rewarding. Central in his approach is what I will call here *heat theorem* or *Clausius' theorem* (abridging the original diction "main theorem of the theory of heat").[5]

The interpretation of absolute temperature as average kinetic energy was already spread: inherited from the earlier works of Krönig and Clausius and will play a key role in his subsequent developments. But Boltzmann provides a kinetic theory argument for this.[6]

With this key knowledge Boltzmann published his first attempt at reducing the heat theorem to mechanics: he makes the point that it is a form of the least action principle. Actually he considers an extension of the principle: the latter compares close motions which in a given time interval develop and connect fixed initial and final points. The extension considered by Boltzmann compares close periodic motions.

The reason for considering only periodic motions has to be found in the basic philosophical conception that the motion of a system of points makes it wander in the part of phase space compatible with the constraints, visiting it entirely. It might take a long time to do the travel but eventually it will be repeated. In his first paper on the subject Boltzmann mentions that in fact motion might, sometimes, be not periodic but even so it could possibly be regarded as periodic with infinite period [18, #2, p. 30]: the real meaning of the statement is discussed in p. 139 Sect. 6.1, see also Appendix B.

This is what is still sometimes called the *ergodic hypothesis*: Boltzmann will refine it more and more in his later memoirs but it will remain essentially unchanged. There are obvious mathematical objections to imagine a point representing the system wandering through all phase space points without violating the regularity or uniqueness theorems for differential equations (as a space filling continuous curve cannot be continuously differentiable and must self intersect on a dense set of points): however it becomes soon clear that Boltzmann does not really consider the world (i.e. space and time) continuous: continuity is an approximation; derivatives and integrals are approximations of ratios of increments and of sums. [7]

---

[5] For a precise formulation see p. 18.

[6] Boltzmann [18], see also Sect. 6.1.

[7] From [19, p. 227] *Differential equations require, just as atomism does, an initial idea of a large finite number of numerical values and points ...... Only afterwards it is maintained that the picture never represents phenomena exactly but merely approximates them more and more the greater the number of these points and the smaller the distance between them. Yet here again it seems to me that so far we cannot exclude the possibility that for a certain very large number of points the picture will best represent phenomena and that for greater numbers it will become again less accurate, so that atoms do exist in large but finite number.* For other relevant quotations see Sect. 1.1 and 5.2 in [16].

Thus motion was considered periodic: a view, at the time shared by Boltzmann, Maxwell, Clausius.[8]

The paper [18, #2] has the ambitious title "On the mechanical meaning of the fundamental theorem of heat theory": under the assumption that [18, #2, p. 24], "*an arbitrarily selected atom visits every site of the region occupied by the body in a given time (although very long) of which the times t_1 and t_2 are the beginning and end of the time interval*[9] *when motions velocities and directions return to themselves in the same sites, describing a closed path, thence repeating from then on their motion*",[10] Boltzmann shows that the average of the variation of kinetic energy in two close motions (interpreted as work done on the system) divided by the average kinetic energy is an exact differential.

The two close motions are two periodic motions corresponding to two equilibrium states of the system that are imagined as end products of an infinitesimal step of a quasi static thermodynamic transformation. The external potential does not enter into the discussion: therefore the class of transformations to which the result applies is very restricted (in the case of a gas it would restrict to the isovolumic transformations, as later stressed by Clausius, see Sect. 6.7). The time scale necessary for the recurrence is not taken into account.

The "revolutionary idea" is that the states of the system are identified with a periodic trajectory which is used to compute the average values of physical quantities: this is the concept of state as a stationary distribution. An equilibrium state is identified with the average values that the observables have in it: in the language of modern measure theory this is a probability distribution, with the property of being invariant under time evolution.

The states considered are equilibrium states: and thermodynamics is viewed as the theory of the relations between equilibrium averages of observables; *not as a theory of the transformations from one equilibrium state to another or leading to an equilibrium state* as will be done a few years later with the Boltzmann's equation [20, #22] in the case of gases, relating to Maxwell's theory of the approach to equilibrium (in rarefied gases) and improving it [21], see also Sect. 6.14.

The derivation is however quite obscure and forces the reader to use hindsight to understand it: the Boltzmann versus Clausius controversy on the results in the "Relation between the second fundamental theorem of the theory of heat", *which will be abridged hereafter, except in* Chap. 6, *as heat theorem*,[11] and the general principles of Mechanics" [22], makes all this very clear, for more details see the following Sects. 6.5, 6.7.

---

[8] Today it seems unwelcome because we have adjusted, under social pressure, to think that chaotic motions are non periodic and ubiquitous, and their images fill both scientific and popular magazines. It is interesting however that the ideas and methods developed by the mentioned Authors have been the basis of the chaotic conception of motion and of the possibility of reaching some understating of it. See also Sects. 3.6, 3.7 below.

[9] The recurrence time.

[10] For Clausius' view see p. 8 and for Maxwell's view see footnote p. viii in the Introduction.

[11] The second fundamental theorem is not the second law but a logical consequence of it, see Sect. 6.1.

Receiving from Boltzmann the comment that his results were essentially the same as Boltzmann's own in the paper [18, #2], Clausius reacted politely but firmly. He objected to Boltzmann the obscurity of his derivation obliging the reader to suitably interpret formulae that in reality are not explained (and ambiguous). Even accepting the interpretation which makes the statements correct he stresses that Boltzmann establishes relations between properties of motions under an external potential that does not change: thus limiting the analysis very strongly (and to a case in which in thermodynamics the exactness of the differential $\frac{dQ}{T}$, main result in Boltzmann's paper, would be obvious because the heat exchanged would be function of the temperature).

Boltzmann acknowledged, in a private letter, the point, as Clausius reports, rather than profiting from the fact that his formula remains the same, under further interpretations, even if the external potential changes, as explained by Clausius through his exegesis of Boltzmann's work: and Boltzmann promised to take the critique into account in the later works. A promise that he kept in the impressing series of papers that followed in 1871 [23, #18], [24, #19], [25, #20] referred here as the "trilogy", see Sects. 6.8, 6.9 and 6.10, just before the formulation of the Boltzmann's equation, which will turn him into other directions, although he kept coming back to the more fundamental heat theorem, ergodic hypothesis and ensembles theory in several occasions, and mainly in 1877 and 1884 [26, #42], [27, #73].

The work of Clausius [22], for details see Sects. 1.4, 6.4 and 6.5, is formally perfect from a mathematical viewpoint: no effort of interpretation is necessary and his analysis is clear and convincing with the mathematical concepts (and even notations [28, T.I, p. 337]) of Lagrange's calculus of variations carefully defined and employed. Remarkably he also goes back to the principle of least action, an aspect of the heat theorem which is now forgotten, and furthermore considers the "complete problem" taking into account the variation (if any) of the external forces.

He makes a (weaker) ergodicity assumption: each atom or small group of atoms undergoes a periodic motion and the statistical uniformity follows from the large number of evolving units [22],

...temporarily, for the sake of simplicity we shall assume, as already before, that all points describe closed trajectories. For all considered points, and that move in a similar manner, we suppose, more specifically, that they go through equal paths with equal period, although other points may run through other paths with other periods. If the initial stationary motion is changed into another, hence on a different path and with different period, nevertheless these will be still closed paths each run through by a large number of points.

And later [22],

...in this work we have supposed, until now, that all points move along closed paths. We now want to give up this special hypothesis and concentrate on the assumption that motion is stationary.

For the motions that do not follow closed paths, the notion of recurrence, literally taken, is no longer useful so that it is necessary to analyze it in another sense. Consider, therefore, first the motions which have a given component in a given di-

rection, for instance the x direction in our coordinate system. Then it is clear that motions proceed back and forth, for the elongation, speed and return time. The time interval within which we find again every group of points that behave in the same way, approximately, admits an average value...

He does not worry about the time scales needed to reach statistical equilibrium: which, however, by the latter assumption are strongly reduced. The first systematic treatment of the time scales will appear a short time later [20, #22], in Boltzmann's theory of diluted gases: via the homonym equation time scales are evaluated in terms of the free flight time and become reasonably short and observable compared to the super-astronomical recurrence times.

Clausius' answer to Boltzmann [29] see also Sect. 6.6, is also a nice example on how a scientific discussion about priority and strong critique of various aspects of the work of a fellow scientist can be conducted without transcending out of reasonable bounds: the paper provides an interesting and important clarification of the original work of Boltzmann, which however remains a breakthrough.

Eventually Boltzmann, after having discussed the mechanical derivation of the heat theorem and obtained the theory of ensembles (the ones today called microcanonical and canonical) and Boltzmann's equation, finds it necessary to rederive it via a combinatorial procedure in which every physical quantity is regarded as discrete [26, #42], see also Sect. 6.12, and remarkably showing that the details of the motion (e.g. periodicity) are completely irrelevant for finding that equilibrium statistics implies macroscopic thermodynamics.

1.4 Least Action and Heat Theorem

Boltzmann and Clausius theorems are based on a version of the action principle for periodic motions. If $t \to x(t)$ is a periodic motion developing under the action of forces with potential energy $V(x)$ (in the application V will be the total potential energy, sum of internal and external potentials) and with kinetic energy $K(x)$, then the action of x is defined, if its period is i, by

$$\mathcal{A}(x) = \int_0^i \left(\frac{m}{2}\dot{x}(t)^2 - V\left(x(t)\right) \right) dt \qquad (1.4.1)$$

We are interested in periodic variations δx that we represent as

$$\delta x(t) = x'(\frac{i'}{i}t) - x(t) \overset{def}{=} x'(i'\varphi) - x(i\varphi) \qquad (1.4.2)$$

where $\varphi \in [0, 1]$ is the *phase*, as introduced by Clausius [22], see also Sect. 6.5. The role of φ is simply to establish a correspondence between points on the initial trajectory x and on the varied one x': it is manifestly an arbitrary correspondence (which

could be defined differently without affecting the final result) which is convenient
to follow the algebraic steps. It should be noted that Boltzmann does not care to
introduce the phase and this makes his computations difficult to follow (in the sense
that once realized what is the final result, it is easier to reproduce it rather than to
follow his calculations).

Set $\overline{F}(x) = i^{-1} \int_0^i F(x(t)) dt$ for a generic observable $F(\xi)$, then the new form
of the action principle for periodic motions is

$$\delta(\overline{K} - \overline{V}) = -2\overline{K}\delta \log i - \delta \overline{\widetilde{V}} \qquad (1.4.3)$$

if $\delta \widetilde{V}$ is the variation of the external potential driving the varied motion, yielding the
correction to the expression of the action principle at fixed temporal extremes and
fixed potential, namely $\delta(\overline{K} - \overline{V}) = 0$ [30, Eq. 2.24.41].

The connection with the heat theorem derives from the remark that in the infini-
tesimal variation of the orbit its total energy $\overline{U} \stackrel{def}{=} \overline{K} + \overline{V}$ changes by $\delta(\overline{K} + \overline{V}) + \delta \overline{\widetilde{V}}$
so $\delta \overline{U} - \delta \widetilde{V}$ is interpreted as the heat δQ received by the system and Eq. (1.4.3) can
be rewritten[12] $\frac{\delta Q}{\overline{K}} = 2\delta \log i\overline{K}$, and $\overline{K}^{-1}$ is an integrating factor for δQ: and the
primitive function is the logarithm of the ordinary action $S = 2 \log i\overline{K}$ up to an
additive constant.

In reality it is somewhat strange that both Boltzmann and Clausius call Eq. (1.4.3)
a "generalization of the action principle": the latter principle uniquely determines
a motion, i.e. it determines its equations; Eq. (1.4.3) instead does not determine a
motion but it only establishes a relation between the variation of average kinetic and
potential energies of close periodic motions under the assumption that they satisfy the
equations of motion; and it does not establish a variational property (unless coupled
with the second law of thermodynamics, see footnote at p. 19).

To derive it, as it will appear in Appendix A, one proceeds as in the analysis of the
action principle and this seems to be the only connection between the Eq. (1.4.3) and
the mentioned principle. Boltzmann formulates explicitly [18, #2, Sect. IV], what he
calls a generalization of the action principle and which is the Eq. (1.4.3) (with $\widetilde{V} = 0$
in his case):

*If a system of points under the influence of forces, for which the "vis viva" principle
holds [i.e. the kinetic energy variation equals the work of the acting forces], performs
some motion, and if all points undergo an infinitesimal change of the kinetic energy
and if they are constrained to move on a trajectory close to the preceding one,
then $\delta \sum \frac{m}{2} \int c \, ds$ equals the total variation of the kinetic energy times half the
time during which the motion takes place, provided the sum of the products of the
infinitesimal displacements of the points times their velocities and the cosine of the
angle at each of the extremes are equal, for instance when the new limits are located
on the lines orthogonal to the old trajectory limits.*

---

[12] From Eq. (1.4.3): $-\delta(\overline{K} + \overline{V}) + 2\delta\overline{K} + \delta\overline{\widetilde{V}} = -2\overline{K}\delta \log i$; i.e. $-\delta Q = -2\delta\overline{K} - 2\overline{K} \log i$, hence
$\frac{\delta Q}{\overline{K}} = 2\delta \log(\overline{K}i)$.

It would be, perhaps, more appropriate to say that Eq. (1.4.3) follows from $\mathbf{f} = m\mathbf{a}$ or, also, that it follows from the action principle because the latter is equivalent to $\mathbf{f} = m\mathbf{a}$.[13]

The check of Eq. (1.4.3) is detailed in Appendix A (in Clausius' version and extension): here I prefer to illustrate a simple explicit example, even though it came somewhat later, in [31, #39], see also Sect. 6.11.

The example is built on a case in which all motions are really periodic, namely a one-dimensional system with potential $\varphi(x)$ such that $|\varphi'(x)| > 0$ for $|x| > 0$, $\varphi''(0) > 0$ and $\varphi(x) \xrightarrow[x\to\infty]{} +\infty$. All motions are periodic (systems with this property are called *monocyclic*, see Sect. 6.13). We suppose that the potential $\varphi(x)$ depends on a parameter V.

Define *a state* a motion with given energy U and given V. And:

U = total energy of the system $\equiv K + \varphi$
T = time average of the kinetic energy K
V = the parameter on which φ is supposed to depend
p = $-$ average of $\partial_V \varphi$.

A state is parameterized by U, V and if such parameters change by dU, dV, respectively, let

$$dW = -pdV, \qquad dQ = dU + pdV, \qquad \overline{K} = T. \tag{1.4.4}$$

Then the heat theorem is in this case:

Theorem ([32, #6], [31, #39], [33]): *The differential* $(dU + pdV)/T$ *is exact.*

In fact let $x_\pm(U, V)$ be the extremes of the oscillations of the motion with given U, V and define S as:

$$S = 2\log 2 \int_{x_-(U,V)}^{x_+(U,V)} \sqrt{K(x; U, V)}dx = 2\log \int_{x_-(U,V)}^{x_+(U,V)} 2\sqrt{U - \varphi(x)}dx \tag{1.4.5}$$

so that $dS = \dfrac{\int \left(dU - \partial_V \varphi(x)dV\right)\frac{dx}{\sqrt{K}}}{\int K \frac{dx}{\sqrt{K}}} \equiv \dfrac{dQ}{T}$, and $S = 2\log i\overline{K}$ if $\dfrac{dx}{\sqrt{K}} = \sqrt{\dfrac{2}{m}}dt$ is used

to express the period i and the time averages via integrations with respect to $\dfrac{dx}{\sqrt{K}}$.

Therefore Eq. (1.4.3) is checked in this case. This completes the discussion of the case in which motions are periodic. In Appendix C an interpretation of the proof of

---

[13] This is an important point: the condition Eq. (1.4.3) does not give to the periodic orbits describing the state of the system any variational property (of minimum or maximum): the consequence is that it does not imply $\int \frac{\delta Q}{T} \le 0$ in the general case of a cycle but only $\int \frac{\delta Q}{T} = 0$ in the (considered) reversible cases of cycles. This comment also applies to Clausius' derivation. The inequality seems to be derivable only by applying the second law in Clausius formulation. It proves existence of entropy, however, see comment at p. 137.

the above theorem in a general monocyclic system is analyzed. See Appendix D for the extension to Keplerian motion [27].

Both Boltzmann and Clausius were not completely comfortable with the periodicity. As mentioned, Boltzmann imagines that each point follows the same periodic trajectory which, if not periodic, "can be regarded as periodic with infinite period" [18, p. 30, #2], see also Appendix B below: a statement not always properly interpreted which, however, will evolve, thanks to his own essential contributions, into the ergodic hypothesis of the 20th century (for the correct meaning see comment at p. 139 in Sect. 6.1, see also Appendix B).

Clausius worries about such a restriction more than Boltzmann does; but he is led to think the system as consisting of many groups of points which closely follow an essentially periodic motion. This is a conception close to the Ptolemaic conception of motion via cycles and epicycles [34].

1.5 Heat Theorem and Ensembles

The identification of a thermodynamic equilibrium state with the collection of time averages of observables was made, almost without explicit comments i.e *as if it needed neither discussion nor justification*, in the Boltzmann's paper of 1866 [18, #2], see also Sect. 6.1.

As stressed above the analysis relied on the assumption that motions are periodic.

A first attempt to eliminate his hypothesis is already in the work of 1868 [35, #5], see also Sect. 6.2, where the Maxwellian distribution is derived first for a gas of hard disks and then for a gas of atoms interacting via very short range potentials.

In this remarkable paper the canonical distribution for the velocity distribution of a single atom is obtained from the microcanonical distribution of the entire gas. The ergodic hypothesis appears initially in the form: the molecule goes through all possible [internal] states because of the collisions with the others. However the previous hypothesis (i.e. periodic motion covering the energy surface) appears again to establish as a starting point the microcanonical distribution for the entire gas.

The argument is based on the fact that the collisions, assumed of negligible duration in a rarefied gas, see p. 144, change the coordinates via a transformation with Jacobian determinant 1 (because it is a canonical map) and furthermore since the two colliding atoms are in arbitrary configurations then the distribution function, being invariant under time evolution, must be a function of the only conserved quantity for the two atoms, i.e. the sum of their energies.

Also remarkable is that the derivation of the Maxwellian distribution for a single particle from the uniform distribution on the N-particles energy surface (microcanonical, i.e. just the uniform distribution of the kinetic energy as the interactions are assumed instantaneous) is performed

(1) by decomposing the possible values of the kinetic energy of the system into a sum of values of individual particle kinetic energies

(2) each of which susceptible of taking finitely many values (with degeneracies, dimension dependent, accounted in the 2 and 3 dimensional space cases)
(3) solving the combinatorial problem of counting the number of ways to realize the given values of the total kinetic energy and particles number
(4) taking the limit in which the energy levels become dense and integrating over all particles velocities but one letting the total number increase to ∞.

The combinatorial analysis has attracted a lot of attention particularly, [36], if confronted with the later similar (but *different*) analysis in [26, #42], see the comments in Sect. 6.12.

The idea that simple perturbations can lead to ergodicity (in the sense of uniformly dense covering of the energy surface by a single orbit) is illustrated in an example in a subsequent paper [32, #6], see also Sect. 6.3. But the example, chosen because all calculations could be done explicitly and therefore should show how ergodicity implies the microcanonical distribution, is a mechanical problem with 2 degrees of freedom which however is *not ergodic* on *all* energy surfaces: see comments in Sect. 6.3.

It is an example similar to the example on the two body problem examined in [31, #39], and deeply discussed in the concluding paper [27, #73].

In 1871 Clausius also made an attempt to eliminate the assumption, as discussed in [22, 29], see also Sect. 6.5. In the same year Boltzmann [23, #18], considered a gas of polyatomic molecules and related the detailed structure of the dynamics to the determination of the invariant probability distributions on phase space that would yield the time averages of observables in a given stationary state without relying on the periodicity.

Under the assumption that *"the different molecules take all possible states of motion"* Boltzmann undertakes again [23, #18], see also Sect. 6.8, the task of determining the number of atoms of the $N = \varrho V$ (ϱ = density, V = volume of the container) molecules which have given momenta $\mathbf{p}$ and positions $\mathbf{q}$ ($\mathbf{p}$ are the momenta of the r atoms in a molecule and $\mathbf{q}$ their positions) determined within $d\mathbf{p}, d\mathbf{q}$, denoted $f(\mathbf{p}, \mathbf{q})d\mathbf{p}d\mathbf{q}$, greatly extending Maxwell's derivation of

$$f(p, q) = \varrho \frac{e^{-hp^2/2m}}{\sqrt{2m \, \pi^3 h^{-3}}} d^3 p \, d^3 q \qquad (1.5.1)$$

for monoatomic gases (and elastic rigid bodies) [15].

The main assumption is no longer that the motion is periodic: only the individual molecules assume in their motion all possible states; and even that is not supposed to happen periodically but it is achieved thanks to the collisions with other molecules; no periodicity any more.

Furthermore, [23, #18, p. 240]:

Since the great regularity shown by the thermal phenomena induces to suppose that f is almost general and that it should be independent from the properties of the special nature of every gas; and even that the general properties depend only weakly from the form of the equations of motion, with the exception of the cases in which the complete integration does not present insuperable difficulties.

and in fact Boltzmann develops an argument that shows that in presence of binary collisions in a rarefied gas the function f has to be $f = Ne^{-hU}$ where U is the total energy of the molecule (kinetic plus potential). This is derived as a consequence of Liouville's theorem and of the conservation of energy in each binary collision.

The binary collisions assumption troubles Boltzmann, [23]:

An argument against is that so far the proof that such distributions are the unique that do not change in presence of collisions is not yet complete. It remains nevertheless established that a gas in the same temperature and density state can be found in many configurations, depending on the initial conditions, a priori improbable and even that will never be experimentally observed.

The analysis is based on the realization that in binary collisions, involving two molecules of n atoms each, with coordinates $(\mathbf{p}_i, \mathbf{q}_i)$, $i = 1, 2$, (here $\mathbf{p}_1 = (p_1^{(1)}, \ldots p_1^{(n)})$, $\mathbf{q}_1 = (q_1^{(1)}, \ldots q_1^{(n)})$, $\mathbf{p}_2 = (p_2^{(1)}, \ldots p_2^{(n)})$, etc., are the momenta and positions of the atoms $1, \ldots, n$ in each of the two molecules), only the total energy and total linear and angular momenta of the pair are constant (by the second and third Newtonian laws) and, furthermore, the volume elements $d\mathbf{p}_1 d\mathbf{q}_1 d\mathbf{p}_2 d\mathbf{q}_2$ do not change (by the Liouville's theorem).

Visibly unhappy with the nonuniqueness Boltzmann resumes the analysis in a more general context: *after all a molecule is a collection of interacting atoms.* Therefore one can consider a gas as a giant molecule and apply to it the above ideas.

Under the assumption that there is only one constant of motion he derives in the subsequent paper [24, #19], see also Sect. 6.9, that the probability distribution has to be what we call today a *microcanonical distribution* and that it implies a canonical distribution for a (small) subset of molecules.

The derivation is the same that we present today to the students. It has been popularized by Gibbs [37], who acknowledges Boltzmann's work but curiously quotes it as [38], i.e. with the title of the first section of Boltzmann's paper [24, #19], see also Sect. 6.9, which refers to a, by now, somewhat mysterious "principle of the last multiplier of Jacobi". The latter is that in changes of variable the integration element is changed by a "last multiplier" that we call now the *Jacobian determinant* of the change. The true title ("*A general theorem on thermal equilibrium*") is less mysterious although quite unassuming given the remarkable achievement in it: this is the first work in which the general theory of the ensembles is discovered simultaneously with the equivalence between canonical and microcanonical distributions for large systems.

Of course Boltzmann does not solve the problem of showing uniqueness of the distribution (we know that this is essentially never true in presence of chaotic dynamics [16, 39]). *Therefore to attribute a physical meaning to the distributions he has to show that they allow to define average values related by the laws of thermodynamics: i.e. he has to go back to the derivation of a result like that of the heat theorem to prove that $\frac{dQ}{T}$ is an exact differential.*

The periodicity assumption is long gone and the result might be not deducible within the new context. He must have felt relieved when he realized, few days later [25, #19], see also Sect. 6.11, that a heat theorem could also be deduced under the same assumption (uniform density on the total energy surface) that the equilibrium distribution is the microcanonical one, without reference to the dynamics.

Defining the heat dQ received by a system as the variation of the total average energy dE plus the work dW performed by the system on the external particles (i.e. the average variation in time of the potential energy due to its variation generated by a change in the external parameters) it is shown that $\frac{dQ}{T}$ is an exact differential if T is proportional to the average kinetic energy, see Sect. 6.13 for the details.

This makes statistical mechanics of equilibrium independent of the ergodic hypothesis and it will be further developed in the 1884 paper, see Sect. 6.13, into a very general theory of statistical ensembles, extended and made popular by Gibbs.

In the arguments used in the "trilogy" [23, #18], [24, #19], [25, #20], dynamics intervenes, as commented above, only through binary collisions (molecular chaos) treated in detail: the analysis will be employed a little later to imply, via the conservation laws of Newtonian mechanics and Liouville's theorem, particularly developed in [23, #18], the new well known Boltzmann's equation, which is presented explicitly immediately after the trilogy [20, #22].

1.6 Boltzmann's Equation, Entropy, Loschmidt's Paradox

Certainly the result of Boltzmann most well known and used in technical applications is the *Boltzmann's equation* [20, #22]: his work is often identified with it (although the theory of ensembles could well be regarded as his main achievement). It is a consequence of the analysis in [23, #18] (and of his familiarity, since his 1868 paper, see Sect. 6.2, with the work of Maxwell [21]). It attacks a completely new problem: namely it no longer deals with determining the relation between properties of different equilibrium states, as done in the analysis of the heat theorem.

The subject is to determine how equilibrium is reached and to show that the evolution of a very diluted N atoms gas[14] from an initial state, which is not time invariant, towards a final equilibrium state admits a "Lyapunov function", if evolution occurs in isolation from the external world: i.e a function $H(f)$ of the *empirical distribution* $Nf(p, q)d^3p d^3q$, giving the number of atoms within the limits $d^3p d^3q$, which evolves monotonically towards a limit H_∞ which is the value achieved by $H(f)$ when f is the canonical distribution.

There are several assumptions and approximations, some hidden. Loosely, the evolution should keep the empirical distribution smooth: this is necessary because in principle the state of the system is a precise configuration of positions and velocities (a "delta function" in the $6N$ dimensional phase space), and therefore the empirical

---

[14] Assume here for simplicity the gas to be monoatomic.

distribution is a reduced description of the microscopic state of the system, which is supposed to be sufficient for the macroscopic description of the evolution; and only binary collisions take place and do so losing memory of the past collisions (the *molecular chaos hypothesis*). For a precise formulation of the conditions under which Boltzmann's equations can be derived for, say, a gas microscopically consisting of elastic hard balls see [16, 40, 41].

The hypotheses are reasonable in the case of a rarefied gas: however the consequence is deeply disturbing (at least to judge from the number of people that have felt disturbed). It might even seem that chaotic motion is against the earlier formulations, adopted or considered seriously not only by Boltzmann, but also by Clausius and Maxwell, linking the heat theorem to the periodicity of the motion and therefore to the recurrence of the microscopic states.

Boltzmann had to clarify the apparent contradiction, first to himself as he initially might have not realized it while under the enthusiasm of the discovery. Once challenged he easily answered the critiques ("sophisms", see Sect. 6.11), although his answers very frequently have been missed [31] and for details see Sect. 6.11.

The answer relies on a more careful consideration of the time scales: already Thomson [42], had realized and stressed quantitatively the deep difference between the time (actually, in his case, the number of observations) needed to see a recurrence in a isolated system and the time to reach equilibrium.

The second is a short time measurable in "human units" (usually from microseconds to hours) and in rarefied gases it is of the order of the average free flight time, as implied by the Boltzmann's equation which therefore also provides an explanation of why approach to equilibrium is observable at all.

The first, that will be called T_∞, is by far longer than the age of the Universe already for a very small sample of matter, like $1 \, cm^3$ of hydrogen in normal conditions which Thomson and later Boltzmann estimated to be of about $10^{10^{19}}$ times the age of the Universe [43, Vol. 2, Sect. 88] (or "equivalently"(!) times the time of an atomic collision, 10^{-12} s).

The above mentioned function $H(f)$ is simply

$$H(f) = -k_B N \int f(p,q) \log(f(p,q)\delta^3) d^3 p d^3 q \qquad (1.6.1)$$

where δ is an arbitrary constant with the dimension of an action and k_B is an arbitrary constant. If f depends on time following the Boltzmann equation then $H(f)$ is monotonic non decreasing and constant if and only if f is the one particle reduced distribution of a canonical distribution.

It is also important that if f has an equilibrium value, given by a canonical distribution, and the system is a rarefied gas so that the potential energy of interaction can be neglected, then $H(f)$ coincides, up to an arbitrary additive constant, with the entropy per mole of the corresponding equilibrium state provided the constant k_B is suitably chosen and given the value $k_B = RN_A^{-1}$, with R the gas constant and N_A Avogadro's number.

This induced Boltzmann to define $H(f)$ also as *the entropy of the evolving state* of the system, thus extending the definition of entropy far beyond its thermodynamic definition (where it is a consequence of the second law of thermodynamics). Such extension is, in a way, arbitrary: because it seems reasonable only if the system is a rarefied gas.

In the cases of dense gases, liquids or solids, the analogue of Boltzmann's equation, when possible, has to be modified as well as the formula in Eq. (1.6.1) and the modification is not obvious as there is no natural analogue of the equation. Nevertheless, after Boltzmann's analysis and proposal, *in equilibrium*, e.g. canonical or microcanonical not restricted to rarefied gases, the entropy can be identified with $k_B \log W$, W being the volume (normalized via a dimensional constant) of the region of phase space consisting of the microscopic configurations with the same empirical distribution $f(p, q) \overset{def}{=} \frac{\varrho(p,q)}{\delta^3}$ and it is shown to be given by the "Gibbs entropy"

$$S = -k_B \int \varrho(\mathbf{p}, \mathbf{q}) \log \varrho(\mathbf{p}, \mathbf{q}) \frac{d^{3N}\mathbf{p}\,d^{3N}\mathbf{q}}{\delta^{3N}} \qquad (1.6.2)$$

where $\varrho(\mathbf{p}, \mathbf{q}) \frac{d^{3N}\mathbf{p}\,d^{3N}\mathbf{q}}{\delta^{3N}}$ is the equilibrium probability for finding the microscopic configuration $(\mathbf{p}, \mathbf{q})$ of the N particles in the volume element $d^{3N}\mathbf{p}\,d^{3N}\mathbf{q}$ (made adimensional by the arbitrary constant δ^{3N}).

This suggests that in general (i.e. not just for rarefied gases) there could also exist a simple Lyapunov function controlling the approach to stationarity, with the property of reaching a maximum value when the system approaches a stationary state.

It has been recently shown in [44, 45] that H, defined as proportional to the logarithm of the volume in phase space, divided by a constant with same dimension as the above δ^{3N}, of the configurations that attribute the same empirical distribution to the few observables relevant for macroscopic Physics, is monotonically increasing, if regarded over time scales short compared to T_∞ and provided the initial configuration is not extremely special.[15] The so defined function H is called "Boltzmann's entropy".

However there may be several such functions besides the just defined Boltzmann's entropy. Any of them would play a fundamental role similar to that of entropy in equilibrium thermodynamics if it could be shown to be independent of the arbitrary choices made to define it: like the value of δ, the shape of the volume elements $d^{3N}\mathbf{p}\,d^{3N}\mathbf{q}$ or the metric used to measure volume in phase space: this however does not seem to be the case [46], except in equilibrium and this point deserves further analysis, see Sect. 3.12.

The analysis of the physical meaning of Boltzmann's equation has led to substantial progress in understanding the phenomena of irreversibility of the macroscopic

---

[15] For there will always exist configurations for which $H(f)$ or any other extension of it decreases, although this can possibly happen only for a very short time (of "human size") to start again increasing forever approaching a constant (until a time T_∞ is elapsed and in the unlikely event that the system is still enclosed in its container where it has remained undisturbed and if there is still anyone around to care).

evolution controlled by a reversible microscopic dynamics and it has given rise to a host of mathematical problems, most of which are still open, starting with controlled algorithms of solution of Boltzmann's equation or, what amounts to the same, theorems of uniqueness of the solutions: for a discussion of some of these aspects see [16].

The key conceptual question is, however, how is it possible that a microscopically reversible motion could lead to an evolution described by an irreversible equation (i.e. an evolution in which there is a monotonically evolving function of the state).

One of the first to point out the incompatibility of a monotonic approach to equilibrium, as foreseen by the Boltzmann's equation, was Loschmidt. And Boltzmann started to reply very convincingly in [31], for details see Sect. 6.11, where Sect. II is dedicated to the so called *Loschmidt's paradox*: in it is remarked that if there are microscopic configurations in which the $H(f)$, no matter how it is defined, is increasing at a certain instant there must also be others in which at a certain instant $H(f)$ is decreasing and they are obtained by reversing all velocities leaving positions unchanged, i.e. by applying the *time reversal* operation.

This is inexorably so, no matter which definition of H is posed. In the paper [31, p. 121, #39] a very interesting analysis of irreversibility is developed: I point out here the following citations:

In reality one can compute the ratio of the numbers of different initial states which determines their probability, which perhaps leads to an interesting method to calculate thermal equilibria. Exactly analogous to the one which leads to the second main theorem. This has been checked in at least some special cases, when a system undergoes a transformation from a non uniform state to a uniform one. Since there are infinitely many more uniform distributions of the states than non uniform ones, the latter case will be exceedingly improbable and one could consider practically impossible that an initial mixture of nitrogen and oxygen will be found after one month with the chemically pure oxygen on the upper half and the nitrogen in the lower, which probability theory only states as not probable but which is not absolutely impossible

To conclude later, at the end of Sec.II of [31, p. 122, #39]:

But perhaps this interpretation relegates to the domain of probability theory the second law, whose universal use appears very questionable, yet precisely thanks to probability theory will be verified in every experiment performed in a laboratory.

The work is partially translated and commented in the following Sect. 6.11.

1.7 Conclusion

Equilibrium statistical mechanics is born out of an attempt to find the mechanical interpretation of the second law of equilibrium thermodynamics.[16] Or at least the mechanical interpretation of heat theorem (which is its logical consequence) consequence.

This leads, via the ergodic hypothesis, to establishing a connection between the second law and the least action principle. The latter suitably extended, first by Boltzmann and then by Clausius, is indeed related to the second law: more precisely to the existence of the entropy function (i.e. to $\oint \frac{dQ}{T} = 0$ in a reversble cycle, although not to $\oint \frac{dQ}{T} \le 0$ in general cycles).

It is striking that all, Boltzmann, Maxwell, Clausius, Helmholtz, ... tried to derive thermodynamics from mechanical relations valid for all mechanical system, whether with few or with many degrees of freedom. This was made possible by more or less strong ergodicity assumptions. And the heat theorem becomes in this way an identity always valid. This is a very ambitious viewpoint and the outcome is the Maxwell–Boltzmann distribution on which, forgetting the details of the atomic motions, the modern equilibrium statistical mechanics is developing.

The mechanical analysis of the heat theorem for equilibrium thermodynamics stands *independently* of the parallel theory for the approach to equilibrium based on the Boltzmann's equation: therefore the many critiques towards the latter do not affect the equilibrium statistical mechanics as a theory of thermodynamics. Furthermore the approach to equilibrium is studied under the much more restrictive assumption that the system is a rarefied gas. Its apparently obvious contradiction with the basic equations assumed for the microscopic evolution was brilliantly resolved by Boltzmann [31, #39], and Thomson [42], ... (but rarely understood at the time) who realized the probabilistic nature of the second law as the dynamical law of entropy increase.

A rather detailed view of the reception that the work of Boltzmann received and is still receiving can be found in [49] where a unified view of several aspects of Boltzmann's work are discussed, not always from the same viewpoint followed here.

The possibility of extending the H function *even when the system is not in equilibrium* and interpreting it as a state function defined on stationary states or as a Lyapunov function, is questionable and will be discussed in what follows. In fact (out of equilibrium) the very existence of a well defined function of the "state" (even if restricted to stationary states) of the system which deserves to be called entropy is

---

[16] "The entropy of the universe is always increasing" *is not a very good statement of the second law* [47, Sect. 44.12] The second law in Kelvin-Planck's version "A process whose *only* net result is to take heat from a reservoir and convert it to work is impossible"; and entropy is defined as a function S such that if heat ΔQ is added reversibly to a system at temperature T, the increase in entropy of the system is $\Delta S = \frac{\Delta Q}{T}$ [47, 48]. The Clausius' formulation of the second law is "It is impossible to construct a device that, operating in a cycle will produce no effect other than the transfer of heat from a cooler to a hotter body" [48, p. 148]. In both cases the existence of entropy follows as a theorem, Clausius' "fundamental theorem of the theory of heat", here called "heat theorem".

a problem: for which no physical basis seems to exist indicating the necessity of a solution one way or another.

The next natural question, and not as ambitious as understanding the approach to stationary states (equilibria or not), is to develop a thermodynamics for the stationary states of systems. These are states which are time independent but in which currents generated by non conservative forces, or other external actions,[17] occur. Is it possible to develop a general theory of the relations between time averages of various relevant quantities, thus extending thermodynamics?

References

1. Feynman, R.P., Vernon, F.L.: The theory of a general quantum system interacting with a linear dissipative system. Ann. Phys. **24**, 118–173 (1963)
2. Bonetto, F., Gallavotti, G., Giuliani, A., Zamponi, F.: Chaotic hypothesis, fluctuation theorem and singularities. J. Stat. Phys. **123**, 39–54 (2006)
3. Lorenz, E.: Deterministic non periodic flow. J Atmos Sci **20**, 130–141 (1963)
4. Lucretius, T.: De Rerum Natura. Rizzoli, Milano (1976)
5. S.G. Brush. History of Modern Physical Sciences, Vol. I: The Kinetic Theory of Gases. Imperial College Press, London (2003)
6. Avogadro, A.: Essai d'une manière de determiner les masses relatives des molecules élémentaires des corps, et les proportions selon lesquelles elles entrent dans ces combinaisons. Journal de Physique, de Chimie et d'Histoire naturelle, translated in http://lem.ch.unito.it/chemistry/essai.html. **73**, 58–76 (1811)
7. Brush, S.G.: The Kind of Motion that We Call Heat (I, II). North Holland, Amsterdam (1976)
8. Mossotti, O.F.: Sur les forces qui régissent la constitution intérieure des corps. Stamperia Reale, Torino (1836)
9. Clausius, R.: On the motive power of heat, and on the laws which can be deduced from it for the theory of heat. Philos. Mag. **2**, 1–102 (1851)
10. Carnot, S.: Réflections sur la puissance motrice du feu et sur les machines propres a développer cette puissance. Bachelier, reprint Gabay, 1990, Paris (1824)
11. Clausius, R.: Über einige für anwendung bequeme formen der hauptgleichungen der mechanischen wärmetheorie. Annalen der Physik und Chemie **125**, 353–401 (1865)
12. Lidddell, H.G., Scott, R.: A Greek-English Lexicon. Oxford University Press, Oxford (1968)
13. Krönig, A.: Grundzüge einer Theorie der Gase. Annalen der Physik und, Chemie, XCIX:315–322 (1856)
14. Clausius, R.: The nature of the motion which we call heat. Philos. Mag. **14**, 108–127 (1865)
15. Maxwell, J.C.: Illustrations of the dynamical theory of gases. In: Niven, W.D. (ed.) The Scientific Papers of J.C. Maxwell, vol. 1. Cambridge University Press, Cambridge (1964)
16. Gallavotti, G.: Statistical Mechanics: A Short Treatise. Springer, Berlin (2000)
17. Maxwell, J.C.: The Scientific Papers of J.C. Maxwell. In: Niven, W.D. (ed.), vols. 1, 2. Cambridge University Press, Cambridge (1964)
18. Boltzmann, L.: Über die mechanische Bedeutung des zweiten Hauptsatzes der Wärmetheorie, volume 1, #2 of Wissenschaftliche Abhandlungen, ed. F. Hasenöhrl. Chelsea, New York (1968)
19. Boltzmann, L.: Theoretical Physics and Philosophical Writings. In: Mc Guinness, B. (ed.), Reidel, Dordrecht (1974)
20. Boltzmann, L.: Weitere Studien über das Wärmegleichgewicht unter Gasmolekülen.In: Hasenöhrl, F. (ed.) Wissenschaftliche Abhandlungen, vol. 1, #22. Chelsea, New York (1968)

---

[17] Like temperature differences imposed on the boundaries.

21. Maxwell, J.C.: On the dynamical theory of gases. In: Niven, W.D. (ed.) The Scientific Papers of J.C. Maxwell, vol. 2. Cambridge University Press, Cambridge (1964)
22. Clausius, R.: Ueber die zurückführung des zweites hauptsatzes der mechanischen wärmetheorie und allgemeine mechanische prinzipien. Annalen der Physik **142**, 433–461 (1871)
23. Boltzmann, L.: Über das Wärmegleichgewicht zwischen mehratomigen Gasmolekülen. In: Hasenöhrl. F. (ed.) Wissenschaftliche Abhandlungen, vol. 1, #18. Chelsea, New York (1968)
24. Boltzmann, L.: Einige allgemeine sätze über Wärmegleichgewicht. In: Hasenöhrl, F. (ed.) Wissenschaftliche Abhandlungen, vol. 1, #19. Chelsea, New York (1968)
25. Boltzmann, L.: Analytischer Beweis des zweiten Hauptsatzes der mechanischen Wärmetheorie aus den Sätzen über das Gleichgewicht des lebendigen Kraft. In: Hasenöhrl, F. (ed.) Wissenschaftliche Abhandlungen, vol. 1, #20. Chelsea, New York (1968)
26. Boltzmann, L.: Über die Beziehung zwischen dem zweiten Hauptsatze der mechanischen Wärmetheorie und der Wahrscheinlichkeitsrechnung, respektive den Sätzen über das Wärmegleichgewicht. In: Hasenöhrl, F. (ed.) Wissenschaftliche Abhandlungen, vol. 2, #42. Chelsea, New York (1968)
27. Boltzmann, L.: Über die Eigenshaften monozyklischer und anderer damit verwandter Systeme. Wissenschaftliche Abhandlungen, vol. 3, #73. Chelsea, New York, 1968 (1884)
28. Lagrange, J.L.: Oeuvres. Gauthiers-Villars, Paris (1867–1892)
29. Clausius, R.: Bemerkungen zu der prioritätreclamation des hrn. boltzmann. Annalen der Physik **144**, 265–280 (1872)
30. Gallavotti, G.: The Elements of Mechanics (II nd edn). http://ipparco.roma1.infn.it, Roma (I edition was Springer 1984) (2008)
31. Boltzmann, L.: Bemerkungen über einige Probleme der mechanischen Wärmetheorie. In: Hasenöhrl, F. (ed.) Wissenschaftliche Abhandlungen, vol. 2, #39. Chelsea, New York (1877)
32. Boltzmann, L.: Lösung eines mechanischen Problems. In: Hasenöhrl, F. (ed.) Wissenschaftliche Abhandlungen, vol. 1, #6. Chelsea, New York (1968)
33. Helmholtz, H.: Studien zur Statistik monocyklischer Systeme. Wissenschaftliche Abhandlungen, vol. III. Barth, Leipzig (1895)
34. Gallavotti, G.: Quasi periodic motions from Hypparchus to Kolmogorov. Rendiconti Accademia dei Lincei, Matematica e applicazioni, **12**, 125–152, (2001) e chao-dyn/9907004
35. Boltzmann, L.: Studien über das Gleichgewicht der lebendigen Kraft zwischen bewegten materiellen Punkten. In: Hasenöhrl, F. (ed.) Wissenschaftliche Abhandlungen, vol. 1, #5. Chelsea, New York (1968)
36. Bach, A.: Boltzmann's probability distribution of 1877. Arch. Hist. Exact. Sci. **41**, 1–40 (1990)
37. Gibbs, J.: Elementary principles in statistical mechanics. Schribner, Cambridge (1902)
38. Boltzmann, L.: Zusammenhang zwischen den Sätzen über das Verhalten mehratomiger Gasmoleküle mit Jacobi's Prinzip des letzten Multiplicators. In: Hasenöhrl, F. (ed.) Wissenschaftliche Abhandlungen, vol. 1, p. 259. Chelsea, New York (1968)
39. Gallavotti, G., Bonetto, F., Gentile, G.: Aspects of the ergodic, qualitative and statistical theory of motion. Springer, Berlin (2004)
40. Lanford, O.: Time evolution of large classical systems. In: Moser, J. (ed.) Dynamical Systems, Theory and Applications, vol. 38, pp. 1–111. Lecture Notes in Physics, Berlin (1974)
41. Spohn, H.: On the integrated form of the BBGKY hierarchy for hard spheres. arxiv: math-ph/0605068, pp. 1–19, (2006)
42. Thomson, W.: The kinetic theory of dissipation of energy. Proceedings of the Royal Society of Edinburgh **8**, 325–328 (1874)
43. Boltzmann, L.: Lectures on gas theory. English edition annotated by S. Brush. University of California Press, Berkeley (1964)
44. Goldstein, S., Lebowitz, J.L.: On the (boltzmann) entropy of nonequilibrium systems. Physica D **193**, 53–66 (2004)
45. Garrido, P.L., Goldstein, S., Lebowitz, J.L.: Boltzmann entropy for dense fluids not in local equilibrium. Phys. Rev. Lett. **92**, 050602 (+4) (2005)
46. Gallavotti, G.: Counting phase space cells in statistical mechanics. Commun. Math. Phys. **224**, 107–112 (2001)

47. Feynman, R.P., Leighton, R.B., Sands, M.: The Feynman Lectures in Physics, vol. I, II, III. Addison-Wesley, New York (1963)
48. Zemansky, M.W.: Heat and Thermodynamics. McGraw-Hill, New York (1957)
49. Uffink, J.: Boltzmann's work in statistical physics. In: Zalta, E.N. (ed.) The Stanford Encyclopedia of Philosophy. Winter 2008 edition (2008)
50. Gallavotti, G.: Entropy, thermostats and chaotic hypothesis. Chaos **16**, 043114 (+6) (2006)

Chapter 2
Stationary Nonequilibrium

2.1 Thermostats and Infinite Models

The essential difference between equilibrium and nonequilibrium is that in the first case time evolution is conservative and Hamiltonian while in the second case time evolution takes place under the action of external agents which could be, for instance, external nonconservative forces.

Nonconservative forces perform work and tend to increase the kinetic energy of the constituent particles: therefore a system subject only to this kind of forces cannot reach a stationary state. For this reason in nonequilibrium problems there must exist other forces which have the effect of extracting energy from the system balancing, in average, the work done or the energy injected on the system.

This is achieved in experiments as well as in theory by adding thermostats to the system. Empirically a thermostat is a device (consisting also of particles, like atoms or molecules) which maintains its own temperature constant while interacting with the system of interest.

In an experimental apparatus thermostats usually consist of large systems whose particles interact with those of the system of interest: so large that, for the duration of the experiment, the heat that they receive from the system affects negligibly their temperature.

However it is clear that locally near the boundary of separation between system and thermostat there will be variations of temperature which will not increase indefinitely, because heat will flow away towards the far boundaries of the thermostats containers. But eventually the temperature of the thermostats will start changing and the experiment will have to be interrupted: so it is necessary that the system reaches a satisfactorily stationary state before the halt of the experiment. This is a situation that can be achieved by suitably large thermostatting systems.

There are two ways to model thermostats. At first the simplest would seem to imagine the system enclosed in a container C_0 in contact, through separating walls, with other containers $\Theta_1, \Theta_2, \ldots, \Theta_n$ as illustrated in Fig. 2.1.

G. Gallavotti, *Nonequilibrium and Irreversibility*,
Theoretical and Mathematical Physics, DOI: 10.1007/978-3-319-06758-2_2,
© Springer International Publishing Switzerland 2014

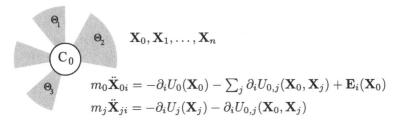

Fig. 2.1 C_0 represents the system container and Θ_j the thermostats containers whose temperatures are denoted by T_j, $j = 1, \ldots, n$. The thermostats are infinite systems of interacting (or free) particles which at all time are supposed to be distributed, far away from C_0, according to a Gibbs' distribution at temperatures T_j. All containers have elastic walls and $U_j(\mathbf{X}_j)$ are the potential energies of the internal forces while $U_{0,j}(\mathbf{X}_0, \mathbf{X}_j)$ is the interaction potential between the particles in C_0 and those in the infinite thermostats

The box C_0 contains the *"system of interest"*, or *"test system"* to follow the terminology of the pioneering work by Feynman and Vernon [1], consisting of N_0 particles while the containers labeled $\Theta_1, \ldots, \Theta_n$ are *infinite* and contain particles with average densities $\varrho_1, \varrho_2, \ldots, \varrho_n$ and temperatures at infinity $T_1, T_2, \ldots, T_n$ which constitute the *"thermostats"*, or *"interaction systems"* to follow [1]. Positions and velocities are denoted $\mathbf{X}_0, \mathbf{X}_1, \ldots, \mathbf{X}_n$, and $\dot{\mathbf{X}}_0, \dot{\mathbf{X}}_1, \ldots, \dot{\mathbf{X}}_n$ respectively, particles masses are $m_0, m_1, \ldots, m_n$. The $\mathbf{E}$ denote external, non conservative, forces.

The temperatures of the thermostats are defined by requiring that initially the particles in each thermostat have an initial distribution which is asymptotically a Gibbs distribution with densities $\rho_1, \ldots, \rho_n$, with inverse temperatures $(k_B T_1)^{-1}, \ldots,$ $(k_B T_n)^{-1}$ and interaction potentials $U_j(\mathbf{X}_j)$ generated by a short range pair potential φ with at least the usual stability properties, [2, Sect. 2.2] i.e. enjoying the lower boundedness property $\sum_{i<j}^{1,n} \varphi(q_i - q_j) \geq -Bn$, $\forall n$, with $B \geq 0$.

Likewise $U_0(\mathbf{X}_0)$ denotes the potential energy of the pair interactions of the particles in the test system and finally $U_{0,j}(\mathbf{X}_0, \mathbf{X}_j)$ denotes the interaction energy between particles in C_0 and particles in the thermostat Θ_j, also assumed to be generated by a pair potential (e.g. the same φ, for simplicity).

The interaction between thermostats and test system are supposed to be *efficient* in the sense that the work done by the external forces and by the thermostats forces will balance, in the average, and keep the test system within a bounded domain in phase space or at least keep its distribution essentially concentrated on bounded phase space domains with a probability which goes to one, as the radius of the phase space domain tends to infinity, at a time independent rate, thus being compatible with the realization of a stationary state.

The above model, first proposed in [1], in a quantum mechanical context, is a typical model that seems to be accepted widely as a physically sound model for thermostats.

However it is quite unsatisfactory; not because infinite systems are unphysical: after all we are used to consider 10^{19} particles in a container of $1\,\text{cm}^3$ as essentially an infinite system; but because it is very difficult to develop a theory of the motion of infinitely many particles distributed with positive density. So far the cases in which the model has been pushed beyond the definition assume that the systems in the thermostats are free systems, as done already in [1], ("free thermostats").

A further problem with this kind of thermostats that will be called "Newtonian" or "conservative" is that, aside from the cases of free thermostats, they are not suited for simulations. And it is a fact that in the last thirty years or so new ideas and progress in nonequilibrium has come from the results of numerical simulations. However the simulations are performed on systems interacting with *finite thermostats*.

Last but not least a realistic thermostat should be able to maintain a temperature gradient because in a stationary state only the temperature at infinity can be exactly constant: in infinite space this is impossible if the space dimension is 1 or 2.[1]

2.2 Finite Thermostats

The simplest finite thermostat models can be illustrated in a similar way to that used in Fig. 2.1:

The difference with respect to the previous model is that the containers $\Theta_1, \ldots, \Theta_n$ are now *finite*, obtained by bounding the thermostats containers at distance ℓ from the origin, by adding a spherical elastic boundary Ω_ℓ of radius ℓ (for definiteness), and contain $N_1, \ldots, N_n$ particles.

The condition that the thermostats temperatures should be fixed is imposed by imagining that there is an extra force $-\alpha_j \dot{\mathbf{X}}_j$ acting on all particles of the j-th thermostat and the multipliers α_j are so defined that the kinetic energies $K_j = \frac{m_j}{2} \dot{\mathbf{X}}_j^2$ are exact constants of motion with values $K_j \overset{def}{=} \frac{3}{2} N_j k_B T_j$, $k_B = $ Boltzmann's constant, $j = 1, \ldots, n$. The multipliers α_j are then found to be[2]:

$$\alpha_j = -\frac{(Q_j + \dot{U}_j)}{3 N_j k_B T_j} \quad \text{with} \quad Q_j \overset{def}{=} -\dot{\mathbf{X}}_j \cdot \partial_{\mathbf{X}_j} U_{0,j}(\mathbf{X}_0, \mathbf{X}_j) \qquad (2.2.1)$$

where Q_j, which is the work per unit time performed by the particles in the test system upon those in the container Θ_j, is naturally interpreted as the *heat* ceded per unit time to the thermostat Θ_j.

The energies $U_0, U_j, U_{0,j}, j > 0$, should be imagined as generated by pair potentials $\varphi_0, \varphi_j, \varphi_{0,j}$ short ranged, stable, smooth, or with a singularity like a hard core or a high power of the inverse distance, and by external potentials generating (or modeling) the containers walls.

---

[1] Because heuristically it is tempting to suppose that temperature should be defined in a stationary state and should tend to the value at infinity following a kind of heat equation: but the heat equation does not have bounded solutions in an infinite domain, like an hyperboloid, with different values at points tending to infinity in different directions *if the dimension of the container is 1 or 2*.

[2] Simply multiplying the both sides of each equation in Fig. 2.2 by $\dot{\mathbf{X}}_j$ and imposing, for each $j = 1, \ldots, n$, that the r.h.s. vanishes.

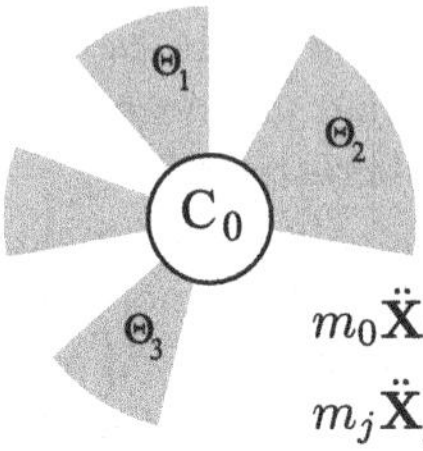

$$m_0\ddot{\mathbf{X}}_{0i} = -\partial_i U_0(\mathbf{X}_0) - \sum_j \partial_i U_{0,j}(\mathbf{X}_0, \mathbf{X}_j) + \mathbf{E}_i(\mathbf{X}_0)$$

$$m_j\ddot{\mathbf{X}}_{ji} = -\partial_i U_j(\mathbf{X}_j) - \partial_i U_{0,j}(\mathbf{X}_0, \mathbf{X}_j) - \alpha_j\dot{\mathbf{X}}_{ji}$$

Fig. 2.2 Finite thermostats model (Gaussian thermostats): the containers Θ_j are finite and contain N_j particles. The thermostatting effect is modeled by an extra force $-\alpha_j\dot{\mathbf{X}}_j$ so defined that the *total* kinetic energies $K_j = m_j/2\dot{\mathbf{X}}_j^2$ are exact constants of motion with values $K_j \overset{def}{=} 3/2N_j k_B T_j$

One can also imagine that thermostat forces act in like manner within the system in C_0: i.e. there is an extra force $-\alpha_0\dot{\mathbf{X}}_0$ which also keeps the kinetic energy K_0 constant ($K_0 \overset{def}{=} N_0\frac{3}{2}k_B T_0$), which could be called an "autothermostat" force on the test system. This is relevant in several physically important problems: for instance in electric conduction models the thermostatting is due to the interaction of the electricity carriers with the oscillations (phonons) of an underlying lattice, and the latter can be modeled (if the masses of the lattice atoms are much larger than those of the carriers) [3] by a force keeping the total kinetic energy (i.e. temperature) of the carriers constant. In this case the multiplier α_0 would be defined by

$$\alpha_0 = \frac{(Q_0 + \dot{U}_0)}{3N_0 k_B T_0} \quad \text{with} \quad Q_0 \overset{def}{=} -\sum_{j>0} \dot{\mathbf{X}}_0 \cdot \partial_{\mathbf{X}_j} U_{0,j}(\mathbf{X}_0, \mathbf{X}_j) \tag{2.2.2}$$

Certainly there are other models of thermostats that can be envisioned: all, including the above, were conceived in order to make possible numerical simulations. The first ones have been the "Nosé-Hoover" thermostats, [4–6]. However they are not really different from the above, or from the similar model in which the multipliers α_j are fixed so that the total energy $K_j + U_j$ in each thermostat is a constant; in the latter, for instance, Q_j is defined as in Eq. (2.2.1)

$$\alpha_j = \frac{Q_j}{3N_j k_B T_j}, \quad k_B T_j \overset{def}{=} \frac{2}{3}\frac{K_j}{N_j} \tag{2.2.3}$$

Such thermostats will be called *Gaussian isokinetic* if $K_j = const$, $j \geq 1$, (hence $\alpha_j = (Q_j + \dot{U}_j)/3N_j k_B T_j$, Eq. (2.2.1) or *Gaussian isoenergetic* if $K_j + U_j = const$ (hence $\alpha_j = Q_j/3N_j k_B T_j$, Eq. (2.2.3).

It is interesting to keep in mind the reason for the attribute "Gaussian" to the models. It is due to the interpretation of the constancy of the kinetic energies K_j or of the total energies $K_j + U_j$, respectively, as a *non holonomic constraint* imposed on the particles. Gauss had proposed to call *ideal* the constraints realized by forces satisfying his principle of *least constraint* and the forces $-\alpha_j\dot{\mathbf{X}}_j$, Eq. (2.2.1) or

(2.2.3), do satisfy the prescription. For completeness the principle is reminded in Appendix E. Here I shall mainly concentrate the attention on the latter Newtonian or Gaussian thermostats.

Remark It has also to be remarked that the Gaussian thermostats generate a *reversible dynamics*: this is *very important* as it shows that Gaussian thermostats do not miss the *essential feature of Newtonian mechanics which is the time reversal symmetry*. Time reversal is a symmetry of nature and any model pretending to be close or equivalent to a faithful representation of nature must enjoy the same symmetry.

Of course it will be important to focus on results and properties which

(1) Have a physical interpretation,
(2) Do not depend on the thermostat model, at least if the numbers of particles $N_0, N_1, \ldots, N_n$ are large.

The above view of the thermostats and the idea that purely Hamiltonian (but infinite) thermostats can be represented equivalently by finite Gaussian termostats external to the system of interest is clearly stated in [7], which precedes the similar [8]

2.3 Examples of Nonequilibrium Problems

Some interesting concrete examples of nonequilibrium systems are illustrated in the following figure (Fig. 2.3).

The multiplier α is $\alpha = \mathbf{E} \cdot \dot{\mathbf{x}}/Nm\dot{\mathbf{x}}^2$ and this is an electric conduction model of N charged particles ($N = 2$ in the figure) in a constant electric field $\mathbf{E}$ and interacting with a lattice of obstacles; it is "autotermostatted" (because the particles in the container C_0 do not have contact with any "external" thermostat). This is a model that appeared since the early days (Drude 1899, [9, Vol. 2, Sect. 35]) in a slightly different form (i.e. in dimension 3, with point particles and with the thermostatting realized by replacing the $-\alpha\dot{\mathbf{x}}$ force with the prescription that after collision of a particle with an obstacle its velocity is rescaled to $|\dot{\mathbf{x}}| = \sqrt{\frac{3}{m}k_B T}$). The thermostat

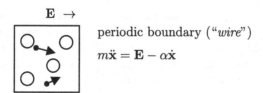

periodic boundary ("*wire*")

$m\ddot{\mathbf{x}} = \mathbf{E} - \alpha\dot{\mathbf{x}}$

Fig. 2.3 A model for electric conduction. The container C_0 is a box with opposite sides identified (periodic boundary). N particles, hard disks ($N = 2$ in the figure), collide elastically with each other and with other fixed hard disks: the mobile particles represent electricity carriers subject also to an electromotive force E; the fixed disks model an underlying lattice whose phonons are phenomenologically represented by the force $-\alpha\dot{\mathbf{x}}$. This is an example of an autothermostatted system in the sense of Sect. 2.2

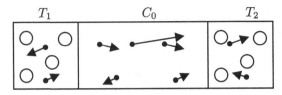

Fig. 2.4 A model for thermal conduction in a gas: particles in the central container C_0 are N_0 hard disks and the particles in the two thermostats are also hard disks; collisions occur whenever the centers of two disks are at distance equal to their diameters. Collisions with the separating walls or bounding walls occur when the disks centers reach them. All collisions are elastic. Inside the two thermostats act thermostatic forces modeled by $-\alpha_j \dot{\mathbf{X}}_j$ with the multipliers α_j, $j = 1, 2$, such that the total kinetic energies in the two boxes are constants of motion $K_j = \frac{N_j}{2} k_B T_j$. If a constant force E acts in the *vertical* direction and the *upper* and *lower* walls of the central container are identified, while the corresponding walls in the *lateral* boxes are reflecting (to break momentum conservation), then this becomes a model for electric and thermal conduction in a gas

forces are a model of the effect of the interactions between the particle (electron) and a background lattice (phonons). This model is remarkable because it is the first nonequilibrium problem that has been treated with real mathematical attention and for which the analog of Ohm's law for electric conduction has been (recently) proved if $N = 1$, [10]

Another example is a model of thermal conduction, Fig. 2.4:

In the model N_0 hard disks interact by elastic collisions with each other and with other hard disks ($N_1 = N_2$ in number) in the containers labeled by their temperatures T_1, T_2: the latter are subject to elastic collisions between themselves and with the disks in the central container C_0; the separations reflect elastically the particles when *their centers* touch them, thus allowing interactions between the thermostats and the main container particles. Interactions with the thermostats take place only near the separating walls.

If one imagines that the upper and lower walls of the *central* container are identified (realizing a periodic boundary condition)[3] and that a constant field of intensity E acts in the vertical direction then two forces conspire to keep it out of equilibrium, and the parameters $\mathbf{F} = (T_2 - T_1, E)$ characterize their strength: matter and heat currents flow.

The case $T_1 = T_2$ has been studied in simulations to check that the thermostats are "efficient" at least in the few cases examined: i.e. that the simple interaction, via collisions taking place across the boundary, is sufficient to allow the systems to reach a stationary state, [11]. A mathematical proof of the above efficiency (at $E \neq 0$), however, seems very difficult (and desirable).

To insure that the system and thermostats can reach a stationary state a further thermostat could be added $-\alpha_0 \dot{\mathbf{X}}_0$ that keeps the total kinetic energy K_0 constant

---

[3] Reflecting boundary conditions on all walls of the side thermostat boxes are imposed to avoid that a current would be induced by the collisions of the "flowing" particles in the central container with the thermostats particles.

and equal to some $\frac{3}{2}N_0k_BT_0$: this would model a situation in which the particles in the central container exchange heat with a background at temperature T_0. This autothermotatted case has been considered in simulations in [3].

2.4 Initial Data

Any set of observations starts with a system in a state x in phase space prepared by some well defined procedure. In nonequilibrium problems systems are always large, because the thermostats and, often, the test system are always supposed to contain many particles: therefore any physically realizable preparation procedure will not produce, upon repetition, the same initial state.

It is a basic assumption that whatever physically realizable preparation procedure is employed it will produce initial data which have a random probability distribution which has a density in the region of phase space allowed by the external constraints. This means, for instance, that in the finite model in Sect. 2.2 the initial data could be selected randomly with a distribution of the form

$$\mu_0(dx) = \varrho(x) \prod_{j=1}^{n} \delta(K_j, T_j) \prod_{j=0}^{n} d\mathbf{X}_j d\dot{\mathbf{X}}_j \qquad (2.4.1)$$

where $x = (\mathbf{X}_0, \mathbf{X}_1, \ldots, \mathbf{X}_n)$ and $\delta(K_j, T_j) = \delta(K_j - \frac{3}{2}N_j k_B T_j)$ and ϱ is a bounded function on phase space.

If observations are performed at timed events, see Sect. 1.1, and are described by a map $S : \Xi \to \Xi$ on a section Ξ of phase space then Eq. (2.4.1) is replaced by

$$\mu_0(dx) = \varrho(x)\,\delta_\Xi(x) \prod_{j=1}^{n} \delta(K_j, T_j) \prod_{j=0}^{n} d\mathbf{X}_j d\dot{\mathbf{X}}_j \qquad (2.4.2)$$

where $\delta_\Xi(x)$ is the delta function imposing that the point x is a timing event, i.e. $x \in \Xi$.

The assumption about the initial data is very important and should not be considered lightly. Mechanical systems as complex as systems of many point particles interacting via short range pair potentials will, *in general*, admit *uncountably many* probability distributions μ which are invariant, hence stationary, under time evolution i.e. such that for all measurable sets $V \subset \Xi$,

$$\mu\left(S^{-1}V\right) = \mu(V) \qquad (2.4.3)$$

where S is the evolution map and "measurable" means any set that can be obtained by a countable number of operations of union, complementation and intersection from the open sets, i.e. *any reasonable set*. In the continuous time representation Ξ is replaced by the full phase space X and the invariance condition becomes $\mu(S_{-t}W) = \mu(W)$ for all $t > 0$ and all measurable sets W.

In the case of infinite Newtonian thermostats the random choice with respect to μ_0 in Eq. (2.4.1) will be with x being chosen with the Gibbs distribution $\mu_{G,0}$ *formally*, [2], given by

$$\mu_{G,0}(dx) = const \, e^{-\sum_{j=0}^{n} \beta_j \left(K_j + U_j \right)} dx \qquad (2.4.4)$$

with $\beta_j^{-1} = k_B T_j$ and some average densities ϱ_j assigned to the particles in the thermostats: satisfying the initial condition of assigning to the configurations in each thermostat the temperature T_j and the densities ϱ_j, but obviously not invariant.[4]

To compare the evolutions in infinite Newtonian thermostats and in large Gaussian thermostats it is natural to choose the initial data in a consistent way (i.e. coincident) in the two cases. Hence in both cases (Newtonian and Gaussian) it will be natural to choose the data with the same distribution $\mu_{G,0}(dx)$, Eq. (2.4.4), and imagine that in the Gaussian case the particles *outside* the finite region, bounded by a reflecting sphere Ω_ℓ of radius ℓ, occupied by the thermostats the particles are "frozen" in the initial positions and velocities of x.

In both cases the initial data can be said to have been chosen respecting the constraints (at given densities and temperatures).

Assuming that physically interesting initial data are generated on phase space M by the above probability distributions $\mu_{G,0}$, Eq. (2.4.4), (or any distribution with density with respect to $\mu_{G,0}$) means that the invariant probability distributions μ that we consider *physically relevant* and that can possibly describe the statistical properties of stationary states are the ones that can be obtained as limits of time averages of iterates of distributions $\mu_{G,0}$. More precisely, in the continuous time cases,

$$\mu(F) = \lim_{\tau \to \infty} \frac{1}{\tau} \int_0^\tau dt \int_M \mu_{G,0}(dx) F(S_t x) \qquad (2.4.5)$$

or, in the discrete time cases:

$$\mu(F) = \lim_{k \to \infty} \frac{1}{k} \sum_{q=0}^{k-1} \int_\Xi \delta_\Xi(x) \mu_{G,0}(dx) F(S^q x) \qquad (2.4.6)$$

---

[4] Not even if $\beta_j = \beta$ for all $j = 0, 1, \ldots, n$ because the interaction between the thermostats and the test system are ignored. In other words the initial data are chosen as independently distributed in the various thermostats and in the test system with a canonical distribution in the finite test system and a Gibbs distribution in the infinite reservoirs case. Of course any distribution with a density with respect to $\mu_{G,0}$ will be equivalent to it, for our purposes.

for all continuous observables F on the test system,[5] where possibly the limits ought to be considered over subsequences (which do not depend on F).

It is convenient to formalize the above analysis, to underline the specificity of the assumption on the initial data, into the following:

Initial data hypothesis *In a finite mechanical system the stationary states correspond to invariant probability distributions μ which are time averages of probability distributions which have a density on the part of phase space compatible with the constraints.*

The assumption, *therefore*, declares "unphysical" the invariant probability distributions that are not generated in the above described way. It puts very severe restriction on which could possibly be the statistical properties of nonequilibrium or equilibrium states.[6]

In general stationary states obtained from initial data chosen with distributions which have a density as above are called SRB distributions. They are not necessarily unique although they are unique in important cases, see Sect. 2.6.

The physical importance of the choice of the initial data in relation to the study of stationary states has been proposed, stressed and formalized by Ruelle, [12–15].

For instance if a system is in equilibrium, i.e. no nonconservative forces act on it and all thermostats are Gaussian and have equal temperatures, then the limits in Eqs. (2.4.5), (2.4.6) are usually supposed to exist, to be ϱ independent and to be equivalent to the Gibbs distribution. Hence the distribution μ has to be

$$\mu(dx) = \frac{1}{Z} e^{-\beta \left(\sum_{j=0}^{n} U_j(\mathbf{X}_j) + \sum_{j=1}^{n} W_j(\mathbf{X}_0, \mathbf{X}_j) \right)} \prod_{j=1}^{n} \delta(K_j, T) \prod_{j=0}^{n} d\mathbf{X}_j \, d\dot{\mathbf{X}}_j \quad (2.4.7)$$

where $\beta = 1/k_B T$, $T_j \equiv T$ and $\delta(K_j, T)$ has been defined after Eq. (2.4.1), provided μ is unique within the class of initial data considered.

In nonequilibrium systems there is the possibility that asymptotically motions are controlled by several attracting sets, typically in a finite number, i.e. closed and disjoint sets $\mathcal{A}$ such that points x close enough to $\mathcal{A}$ evolve at time t into $x(t)$ with distance of $x(t)$ from $\mathcal{A}$ tending to 0 as $t \to \infty$. Then the limits above are not expected to be unique unless the densities ϱ are concentrated close enough to one of the attracting sets.

Finally a warning is necessary: in special cases the preparation of the initial data is, out of purpose or of necessity, such that with probability 1 it produces data which lie in a set of 0 phase space volume, hence of vanishing probability with respect to

---

[5] i.e. depending only on the particles positions and momenta inside $\mathcal{C}_0$, or more generally, within a finite ball centered at a point in $\mathcal{C}_0$.

[6] In the case of Newtonian thermostats, i.e. infinite, the probability distributions to consider for the choice of the initial data are naturally the above $\mu_{G,0}$, Eq. (2.4.5) or distributions with density with respect to them.

μ_0, Eq. (2.4.2), or to any probability distribution with density with respect to volume of phase space. In this case, *of course*, the initial data hypothesis above does not apply: the averages will still exist quite generally but the corresponding stationary state will be different from the one associated with data chosen with a distribution with density with respect to the volume. Examples are easy to construct as it will be discussed in Sect. 3.9 below.

2.5 Finite or Infinite Thermostats? Equivalence?

In the following we shall choose to study *finite* thermostats.

It is clear that this can be of any interest only if the results can, in some convincing way, be related to thermostats in which particles interact via Newtonian forces.

As said in Sect. 2.1 the only way to obtain thermostats of this type is to make them infinite: because the work Q that the test system performs per unit time over the thermostats (heat ceded to the thermostats) will change the kinetic energy of the thermostats and the only way to avoid indefinite heating (or cooling) is that the heat flows away towards (or from) infinity, hence the necessity of infinite thermostats. Newtonian forces and finite thermostats will result eventually in an equilibrium state in which all thermostats temperatures have become equal.

Therefore it becomes important to establish a relation between infinite Newtonian thermostats with only conservative, short range and stable pair forces and finite Gaussian thermostats with additional *ad hoc* forces, as the cases illustrated in Sect. 2.2.

Probably the first objection is that a relation seems doubtful because the equations of motion, and therefore the motions, are different in the two cases. Hence a first step would be to show that *instead* in the two cases the motions of the particles are very close at least if the particles are in, or close to, the test system and the finite thermostats are large enough.

A heuristic argument is that the non Newtonian forces $-\alpha_i \dot{\mathbf{X}}_j$, Eq. (2.2.1), are proportional to the inverse of the number of particles N_j while the other factors (i.e Q_j and $\dot{U}_j$) are expected to be of order $O(1)$ being proportional to the number of particles present in a layer of size twice the interparticle interaction range: hence in large systems their effect should be small (and zero in the limits $N_j \to \infty$ of infinite thermostats). This has been discussed, in the case of a single self-thermostatted test system, in [16], and more generally in [7], accompanied by simulations.

It is possible to go quite beyond a theoretical heuristic argument. However this requires first establishing existence and properties of the dynamics of systems of infinitely many particles. This can be done as described below.

The best that can be hoped is that initial data $\dot{\mathbf{X}}, \mathbf{X}$ chosen randomly with a distribution $\mu_{0,G}$, Eq. (2.4.5), which is a Gibbs distribution with given temperatures and density for the infinitely many particles in each thermostat and with any density for the finitely many particles in the test system, will generate a solution of the equations in Fig. 2.1 with the added prescription of elastic reflection by the boundaries (of the test system and of the thermostats), i.e. a $\dot{\mathbf{X}}(t), \mathbf{X}(t)$ for which both sides of the equation make sense and are equal for all times $t \geq 0$, *with the exception of a set of initial data which has 0 $\mu_{0,G}$-probability.*

At least in the case in which the interaction potentials are smooth, repulsive and short range such a result can be proved, [17–19] in the geometry of Fig. 2.1 in space dimension 2 and in at least one special case of the same geometry in space dimension 3.

If initial data $x = (\dot{\mathbf{X}}(0), \mathbf{X}(0))$ are chosen randomly with the probability $\mu_{G,0}$ the equation in Fig. 2.1 admits a solution $x(t)$, with coordinates of each particle smooth functions of t.

Furthermore, in the same references considered, the finite Gaussian thermostats model, *either isokinetic or isoenergetic*, is realized in the geometry of Fig. 2.2 by terminating the thermostats containers within a spherical surface Ω_ℓ of radius $\ell = 2^k R$, with R being the linear size of the test system and $k \geq 1$ integer.

Imagining the particles external to the ball Ω_ℓ to keep positions and velocities "frozen" in time, the evolution of the particles inside Ω_ℓ will be defined adding to the interparticle forces elastic reflections on the spherical boundaries of Ω_ℓ and the other boundaries of the thermostats and of the test system. It will therefore follow a finite number of ordinary differential equations and at time t the initial data $x = (\dot{\mathbf{X}}(0), \mathbf{X}(0))$, if chosen randomly with respect to the distribution $\mu_{G,0}$ in Eq. (2.4.5), will be transformed into $\dot{\mathbf{X}}^{[k]}(t)$, $\mathbf{X}^{[k]}(t)$ (depending on the regularization parameter $\ell = 2^k R$ and on the isokinetic or isoenergetic nature of the thermostatting forces). Then it is possible to prove the property:

Theorem *Fixed arbitrarily a time $t_0 > 0$ there exist two constants $C, c > 0$ (t_0 –dependent) such that the isokinetic or isoenergetic motions $x_j^{[k]}(t)$ are related as:*

$$|x_j(t) - x_j^{[k]}(t)| \leq C e^{-c 2^k}, \quad \text{if } |x_j(0)| < 2^{k-1} R \qquad (2.5.1)$$

for all $t \leq t_0$, j, with μ_0-probability 1 with respect to the choice of the initial data.

In other words the Newtonian motion and the Gaussian thermostatted motions become rapidly indistinguishable, up to a prefixed time t_0, if the thermostats are large (k large) and if we look at particles initially located within a ball half the size of the confining sphere of radius $\ell = 2^k R$, where the spherical thermostats boundaries are located, i.e. within the ball of radius $2^{k-1} R$.

This theorem is only a beginning, although in the right direction, as one would really like to prove that the evolution of the initial distribution μ_0 lead to a stationary distribution in both cases and that the stationary distributions for the Newtonian and the Gaussian thermostats *coincide in the "thermodynamic limit"* $k \to \infty$.

At this point a key observation has to be made: it is to be expected that in the thermodynamic limit once a stationary state is reached starting from $\mu_{G,0}$ the thermostats temperature (to be suitably defined) should appear varying smoothly toward a value at infinity, in each thermostat Θ_j, equal to the initially prescribed temperature (appearing in the random selection of the initial data with the given distribution $\mu_{G,0}$, Eq. 2.4.5).

Hence the temperature variation should be described, at least approximately, by a solution of the heat equation $\Delta T(q) = 0$ and $T(q)$ not constant, bounded, with Neumann's boundary condition $\partial_n T = 0$ on the lateral boundary of the container $\Omega_\infty = \lim_{\ell \to \infty} \Omega_\ell$ and tending to T_j as $q \in \Theta_j$, $q \to \infty$. However if the space dimension is 1 or 2 there is no such harmonic function.

Therefore the systems considered should be expected to behave as our three dimensional intuition commands only if the space dimension is 3 (or more): it can be expected that the stationary states of the two thermostats models become equal in the thermodynamic limit only if the space dimension is 3.

It is interesting that if really equivalence between the Newtonian and Gaussian thermostats could be shown then the average of the mechanical observable $\sum_{j=1}^{n} 3N_j \alpha_j$, naturally interpreted in the Gaussian case in Eq. (2.2.1), (2.2.3) as entropy production rate, would make sense as an observable also in the Newtonian case[7] with no reference to the thermostats and will have the same average: so that the equivalence makes clear that it *is possible that a Newtonian evolution produces entropy*. I.e. entropy production is compatible with the time reversibility of Newton's equations [7].

2.6 SRB Distributions

The limit probability distributions in Eqs. (2.4.5), (2.4.6) are called *SRB distributions*, from Sinai, Ruelle, Bowen who investigated, and solved in important cases, [20–22], the more difficult question of finding conditions under which, for motions in continuous time on a manifold M, the following limits

$$\lim_{\tau \to \infty} \frac{1}{\tau} \int_0^t F(S_t x)\,dt = \int_X F(y)\mu(dy) \qquad (2.6.1)$$

exist for all continuous observables F, and *for all $x \in M$* chosen randomly according to the initial data hypothesis (Sect. 2.4).[8] A question that in timed observations becomes finding conditions under which, for all continuous observables F, the following limits

$$\lim_{\tau \to \infty} \frac{1}{k} \sum_{q=0}^{k-1} F(S^k x)\,dt = \int_\Xi F(y)\mu(dy) \qquad (2.6.2)$$

---

[7] Because the *r.h.s.* in the quoted formulae are expressed in terms of mechanical quantities Q_j, $\dot{U}_j$ and the temperatures at infinity T_j.

[8] i.e. except possibly for a set V_0 of data x which have zero probability in a distribution with density with respect to the volume and concentrated close enough to an attracting set.

exist *for all* $x \in \Xi$ chosen randomly according to the initial data hypothesis and close enough to an attracting set.

The Eqs. (2.6.1), (2.6.2) express properties stronger than those in the above Eqs. (2.4.5), (2.4.6): no subsequences and no average over the initial data.

Existence of the limits above, outside a set of 0 volume, can be established for systems which are *smooth, hyperbolic and transitive*, also called *Anosov systems* or systems with the *Anosov property*. In the case of *discrete time evolution map* the property is:

Definition (*Anosov map*) Phase space Ξ is a smooth bounded ("compact") Riemannian manifold and evolution is given by a smooth map S with the properties that an infinitesimal displacement dx of a point $x \in \Xi$

(1) Can be decomposed as sum $dx^s + dx^u$ of its components along two transverse planes $V^s(x)$ and $V^u(x)$ which depend continuously on x
(2) $V^\alpha(x)$, $\alpha = u, s$, are covariant under time evolution, in the sense that $(\partial S)(x)V^\alpha(x) = V^\alpha(Sx)$, with $\partial S(x)$ the linearization at x of S ("Jacobian matrix")
(3) Under iteration of the evolution map the vectors dx^s contract exponentially fast in time while the vectors dx^u expand exponentially: in the sense that $|\partial S^k(x)dx^s| \leq Ce^{-\lambda k}|dx^s|$ and $|\partial S^{-k}(x)dx^u| \leq Ce^{-\lambda k}|dx^u|$, $k \geq 0$, for some x-independent $C, \lambda > 0$.
(4) There is a point x with a dense trajectory ("transitivity").

Here ∂S^k denoted the Jacobian matrix $\frac{\partial S^k(x)_i}{\partial x_j}$ of the map S^k at x. Thus $\partial S^k(x)dx$ is an infinitesimal displacement of $S^n x$ and the lengths $|dx^\alpha|$ and $|\partial S^k(x)dx^\alpha|$, $\alpha = s, u$, are evaluated through the metric of the manifold Ξ at the points x and $S^k x$ respectively.

Anosov maps have many properties which will be discussed in the following and that make the evolutions associated with such maps a paradigm of chaotic motions. For the moment we just mention a remarkable property, namely

Theorem (SRB)[9] *If S is a Anosov map on a manifold then there exists a unique probability distribution μ on phase space Ξ such that for all choices of the density $\varrho(x)$ defined on Ξ the limits in Eq. (2.6.2) exist for all continuous observables F and for all x outside a zero volume set.*

Given the assumption on the initial data it follows that in Anosov systems the probability distributions that give the statistical properties of the stationary states are uniquely determined as functions of the parameters on which S depends.

For evolutions on a smooth bounded manifold M *developing in continuous time* there is an analogous definition of "Anosov flow". For obvious reasons the infinitesimal displacements dx pointing in the flow direction cannot expand nor contract with time: hence the generic dx will be covariantly decomposed as a sum $dx^s + dx^u + dx^0$ with dx^s, dx^u exponentially contracting under S_t: in the sense that for some C, λ

---

[9] SRB stands for Sinai-Ruelle-Bowen [14].

it is $|\partial S_t dx^s| \leq C e^{-\lambda t} |dx^s|$ and $|\partial S_{-t} dx^u| \leq C e^{-\lambda t} |dx^u|$ as $t \to +\infty$, while (of course) $|\partial S_t dx^0| \leq C |dx^0|$ as $t \to \pm\infty$; furthermore there is a dense orbit and there is no τ such that the map S_τ^n admits a non trivial constant of motion.[10] Then the above theorem holds without change replacing in its text Eq. (2.6.2) by Eq. (2.6.1), [22–24].

2.7 Chaotic Hypothesis

The latter mathematical results on Anosov maps and flows suggest a daring assumption inspired by the certainly daring assumption that all motions are periodic, used by Boltzmann and Clausius to discover the relation between the action principle and the second principle of thermodynamics, see Sect. 1.3.

The assumption is an interpretation of a similar proposal advanced by Ruelle, [12], in the context of the theory of turbulence. It has been proposed in [25] and called *"chaotic hypothesis"*. For empirically chaotic evolutions, given by a map S on a phase space Ξ, or for continuous time flows S_t on a manifold M, it can be formulated as

Chaotic hypothesis *The evolution map S restricted to an attracting set $A \subset \Xi$ can be regarded as an Anosov map for the purpose of studying statistical properties of the evolution in the stationary states.*

This means that attracting sets A can be considered "for practical purposes" as smooth surfaces on which the evolution map S or flow S_t has the properties that characterize the Anosov maps. It follows that

Theorem *Under the chaotic hypothesis initial data chosen with a probability distribution with a density ϱ on phase space concentrated near an attracting set A evolve so that the limit in Eqs. (2.6.1) or (2.6.2) exists for all initial data x aside a set of zero probability and for all smooth F and are given by the integrals of F with respect to a unique invariant probability distribution μ defined on A.*

This still holds under much weaker assumptions which, however, will not be discussed given the purely heuristic role that will be played by the chaotic hypothesis.[11]

As the ergodic hypothesis is used to justify using the distributions of the microcanonical ensemble to compute the statistical properties of the equilibrium states and to realize the mechanical interpretation of the heat theorem (i.e. existence of the

---

[10] The last condition excludes evolutions like $S_t(x, \varphi) = (Sx, \varphi + t)$, or reducible to this form after a change of variables, with S an Anosov map and $\varphi \in [0, 2\pi]$ and angle, i.e. the most naive flows for which the condition does not hold are also the only cases in which the theorem statement would fail.

[11] For instance if the attracting set satisfies the property "Axiom A", [14, 26], the above theorem holds as well as the key results, presented in the following, on existence of Markov partitions, coarse graining and fluctuation theorem which are what is really wanted for our purposes, see Sects. 3.3, 3.7, 4.6. The heroic efforts mentioned in the footnote[2] of the preface reflect a misunderstanding of the physical meaning of the chaotic hypothesis.

entropy function), likewise the chaotic hypothesis will be used to infer the nature
of the probability distributions that describe the statistical properties of the more
general stationary nonequilibrium states.

This is a nontrivial task as it will be soon realized, see next section, that in general
in nonequilibrium the probability distribution μ will have to be concentrated on a
set of zero volume in phase space, *even when the attracting sets coincide with the
whole phase space.*

In the case in which the volume is conserved, e.g. in the Hamiltonian Anosov case,
the chaotic hypothesis implies the ergodic hypothesis: which is important because
this shows that assuming the new hypothesis cannot lead to a contradiction between
equilibrium and nonequilibrium statistical mechanics. The hypothesis name has been
chosen precisely because of its assonance with the ergodic hypothesis of which it is
regarded here as an extension.

2.8 Phase Space Contraction in Continuous Time

Understanding why the stationary distributions for systems in nonequilibrium are
concentrated on sets of zero volume is the same as realizing that the volume (generically) contracts under non Hamiltonian time evolution.

If we consider the measure $dx = \prod_{j=0}^{n} d\mathbf{X}_j \, d\dot{\mathbf{X}}_j$ on phase space then, under
the time evolution in continuous time, the volume element dx is changed into $S_t dx$
and the rate of change at $t = 0$ of the volume dx per unit time is given by the
divergence of the equations of motion, which we denote $-\sigma(x)$. Given the equations
of motion the divergence can be computed: for instance in the model in Fig. 2.2, i.e.
an isoenergetic Gaussian thermostats model, and $K_j \overset{def}{=} 1/2\dot{\mathbf{X}}_j^2$ is the total kinetic
energy in the j-th thermostat, it is (Eq. 2.2.3)[12]:

$$\sigma(x) = \sum_{j>0} \frac{Q_j}{k_B T_j}, \quad Q_j = -\partial_{\mathbf{X}_j} W_j(\mathbf{X}_0, \mathbf{X}_j) \cdot \dot{\mathbf{X}}_j, \quad N_j k_B T_j \overset{def}{=} \frac{2}{3} K_j \quad (2.8.1)$$

The expression of σ, that will be called the *phase space contraction rate* of the
Liouville volume, has the interesting feature that Q_j can be naturally interpreted as
the heat that the reservoirs receive per unit time, therefore the phase space contraction
contains a contribution that can be identified as the entropy production per unit time.[13]

---

[12] Here a factor $(1 - 2/N_j)$ is dropped from each addend. Keeping it would cause only notational
difficulties and eventually it would have to be dropped on the grounds that the number of particles
N_j is very large.

[13] In the Gaussian isokinetic thermostats Q_j has to be replaced by $Q_j + \dot{U}_j$, Eq. (2.2.1). Notice
that this is true (always neglecting a factor $O(1/N)$ as in the previous footnote) in spite of the fact
that the kinetic energy K_j is not constant in this case: this can be checked by direct calculation or
by remarking that α_j is a homogeneous function of degree -1 in the velocities.

Note that the name is justified *without any need to extend the notion of entropy to nonequilibrium situations*: the thermostats keep the same temperature all along and are regarded as systems in equilibrium (in which entropy is a well defined notion).

In the isokinetic thermostat case σ may contain a further term equal to $\frac{d}{dt}\sum_{j>0} U_j/k_B T_j$ which forbids us to give the naive interpretation of entropy production rate to the phase space contraction. To proceed it has to be remarked that the above σ *is not really unambiguously defined*.

In fact the notion of phase space contraction depends on what we call volume: for instance if we use as volume element

$$\mu_0(dx) = e^{-\beta\left(K_0+U_0+\sum_{j=1}^{n}\left(U_j+W_j(\mathbf{X}_0,\mathbf{X}_j)\right)\right)} \prod_{j=1}^{n} \delta(K_j, T_j) \prod_{j=0}^{n} d\mathbf{X}_j d\dot{\mathbf{X}}_j \quad (2.8.2)$$

with $\beta = 1/k_B T > 0$, arbitrary, the variation rate $-\sigma'(x)$ of a volume element is different; if we call $-\beta H_0(x)$ the argument of the exponential, the new contraction rate is $\sigma'(x) = \sigma(x) + \beta \dot{H}_0(x)$ where $\dot{H}_0$ has to be evaluated via the equations of motion so that $\dot{H}_0 = -\sum \alpha_j \dot{\mathbf{X}}_0^2 + E(\mathbf{X}_0) \cdot \dot{\mathbf{X}}_0$ and therefore

$$\sigma'(x) = \sum_{j>0} \frac{Q_j}{k_B T_j} + \frac{d}{dt} D(x) \quad (2.8.3)$$

where D is a suitable observable (in the example $D = \beta H_0(x)$).

The example shows a special case of the *general property* that if the volume is measured using a different density or a different Riemannian metric on phase space the new volume contracts at a rate differing form the original one by a *time derivative* of some function on phase space.

In other words $\sum_{j>0} Q_j/k_B T_j$ does not depend, in the cases considered, on the system of coordinates while D does *but it has* 0 *time average*.

An immediate consequence is that σ should be considered as defined *up to a time derivative* and therefore only its time averages over long times can possibly have a physical meaning: the limit as $\tau \to \infty$ of

$$\langle \sigma \rangle_\tau \overset{def}{=} \frac{1}{\tau} \int_0^\tau \sigma(S_t x) dt \quad (2.8.4)$$

is independent of the metric and the density used to define the measure of the volume elements; it might still depend on x.

In the timed evolution the time $\tau(x)$ between successive timing events x and $S_{\tau(x)}x$ will have, under the chaotic hypothesis on $S = S_{\tau(x)}$, an average value $\overline{\tau}$ (x-independent except for a set of data x enclosed in a 0 volume set) and the phase space contraction between two successive timing events will be $\exp - \int_0^{\tau(x)} \sigma(S_t x) dt \equiv (\det \partial S(x)/\partial x)^{-1}$ so that

$$\sigma_+ = \lim_{n\to+\infty} \frac{-1}{n\bar{\tau}} \log\left(\det\frac{\partial S^n(x)}{\partial x}\right) = \lim_{n\to+\infty} \frac{-1}{n\bar{\tau}} \sum_{j=1}^{n} \log\det\frac{\partial S}{\partial x}\left(S^j x\right) \quad (2.8.5)$$

which will be a constant σ_+ for all points x close to an attracting set for S and outside a set of zero volume. It has to be remarked that the value of the constant is a well known quantity in the theory of dynamical systems being equal to

$$\sigma_+ = \frac{1}{\bar{\tau}}\sum_i \lambda_i \quad (2.8.6)$$

with λ_i being the SRB Lyapunov exponents of S on the attracting set for S.[14]

In the nonequilibrium models considered in Sect. 2.2 the value of $\sigma(x)$ differs from $\varepsilon(x) = \sum_{j>0} Q_j/k_B T_j$ by a time derivative so that, at least under the chaotic hypothesis, the *average phase space contraction equals the entropy production rate of a stationary state*, and $\sigma_+ \equiv \varepsilon_+$.

An important remark is that $\sigma_+ \geq 0$, [27], if the thermostats are efficient in the sense that motions remain confined in phase space, see Sect. 2.1: the intuition is that it is so because $\sigma_+ < 0$ would mean that any volume in phase space will grow larger and larger with time, thus revealing that the thermostats are not efficient ("it is not possible to inflate a balloon inside a (small enough) safe").

Furthermore if $\sigma_+ = 0$ it can be shown, quite generally, that the phase space contraction is the time derivative of an observable, [27, 28] and, by choosing conveniently the measures of the volume elements, a probability distribution will be obtained which admits a density over phase space and which is invariant under time evolution.

A special case is if it is even $\sigma(x) \equiv 0$: in this case the normalized volume measure is an invariant distribution.

A more interesting example is the distribution Eq. (2.8.2) when $T_j \equiv T_i \stackrel{def}{=} T$ and $\mathbf{E} = \mathbf{0}$. It is a distribution which, for the particles in C_0, is a Gibbs distribution with special boundary conditions

$$\mu_0(dx) = e^{-\beta\left(K_0 + U_0 + \sum_{j=1}^{n}\left(U_j + W_j(\mathbf{X}_0, \mathbf{X}_j)\right)\right)} \prod_{j=1}^{n} \delta(K_j, T) \prod_{j=0}^{n} d\mathbf{X}_j d\dot{\mathbf{X}}_j \quad (2.8.7)$$

and therefore it provides an appropriate distribution for an equilibrium state, [6] and [2, 8]. The more so because of the following consistency check, [6]:

Theorem *If N_i is the number of particles in the i-th thermostat and its temperature is $k_B T_i = \beta^{-1}(1 - 1/3N_i)$ then the distribution in Eq. 2.8.7 is stationary.*

---

[14] The Lyapunov exponents are associated with invariant probability distributions and therefore it is necessary to specify that here the exponents considered are the ones associated with the SRB distribution.

To check: notice that a volume element $dx = \prod_{j=0}^{n} d\mathbf{X}_j d\dot{\mathbf{X}}_j$ is reached at time t by a volume element that at time $t - dt$ had size $e^{\sigma(\mathbf{X},\dot{\mathbf{X}})dt} dx$ and had total energy $H(\mathbf{X} - \dot{\mathbf{X}}dt) = H(\mathbf{X}, \dot{\mathbf{X}}) - dH$. Then compute $-\beta dH + \sigma dt$ via the equations of motion in Fig. (2.2) with the isokinetic constraints Eq. 2.2.3 for $k_B T_i = \beta^{-1}(1 - 1/3N_i)$ obtaining $-\beta dH + \sigma dt \equiv 0$, i.e. proving the stationarity of Eq. 2.8.7.

This remarkable result suggests to *define* stationary nonequilibria as invariant probability distributions for which $\sigma_+ > 0$ and to extend the notion of equilibrium states as invariant probability distributions for which $\sigma_+ = 0$. In this way *a state is in stationary nonequilibrium if the entropy production rate $\sigma_+ > 0$*.

It should be remarked that (in systems satisfying the chaotic hypothesis), as a consequence, the SRB probability distributions for nonequilibrium states are concentrated on *attractors*, defined as subsets $\mathcal{B}$ of the attracting sets $\mathcal{A}$ which have full phase space volume, i.e. full area on the surface $\mathcal{A}$, and minimal fractal dimension, although the closure of $\mathcal{B}$ is the whole $\mathcal{A}$ (which in any event, *under the chaotic hypothesis* is a smooth surface).[15]

In systems out of equilibrium it is convenient to introduce the *dimensionless entropy production rate* and *phase space contraction* as $\varepsilon(x)/\varepsilon_+$ and $\sigma(x)/\varepsilon_+$ and, since ε and σ differ by a time derivative of some function $D(x)$, the finite time averages

$$p = \frac{1}{\tau} \int_0^\tau \frac{\varepsilon(S_t x)}{\varepsilon_+} dt \quad \text{and} \quad p' = \frac{1}{\tau} \int_0^\tau \frac{\sigma(S_t x)}{\sigma_+} dt \tag{2.8.8}$$

will differ by $D(S_\tau x) - D(x)/\tau$ which will tend to 0 as $\tau \to \infty$ (in Anosov systems or under the chaotic hypothesis). Therefore for large τ the statistics of p and p' in the stationary state will be close, at least if the function D is bounded (as in Anosov systems).

2.9 Phase Space Contraction in Timed Observations

In the case of discrete time systems (not necessarily arising from timed observations of a continuous time evolution) the phase space contraction (per timing interval) can be naturally defined as

$$\sigma_+ = \lim_{n \to +\infty} -\frac{1}{n} \sum_{j=1}^{n} \log |\det \frac{\partial S}{\partial x}(S^j x)| \tag{2.9.1}$$

as suggested by Eq. (2.8.5).

---

[15] An attracting set $\mathcal{A}$ is a closed set such that all data x close enough to $\mathcal{A}$ evolve so that the distance of $S^n x$ to $\mathcal{A}$ tends to 0 as $n \to \infty$. A set $\mathcal{B} \subset \mathcal{A}$ with full SRB measure is called an *attractor* if it has minimal Hausdorff dimension, [14]. Typically $\mathcal{B}$ is in general a fractal set whether or not $\mathcal{A}$ is a smooth manifold.

There are several interesting interaction models in which the pair potential is unbounded above: like models in which the molecules interact via a Lennard-Jones potential. As mentioned in Sect. 1.1 this is a case in which observations timed to suitable events become particularly useful.

In the case of unbounded potentials (and finite thermostats) a convenient timing could be when the minimum distance between pairs of particles reaches, while decreasing, a prefixed small value r_0; the next event will be when all pairs of particles, after separating from each other by more than r_0, come back again with a minimum distance equal to r_0 and decreasing. This defines a timing events surface Ξ in the phase space M.

An alternative Poincaré's section could be the set $\Xi \subset M$ of configurations in which the total potential energy $W = \sum_{j=1}^{n} W_j(\mathbf{X}_0, \mathbf{X}_j)$ becomes larger than a prefixed bound $\overline{W}$ with a derivative $\dot{W} > 0$.

Let $\tau(x)$, $x \in \Xi$ be the time interval from the realization of the event x to the realization of the next one $x' = S_{\tau(x)}x$. The phase space contraction is then $\exp \int_0^{\tau(x)} \sigma(S_t x)dt$, in the sense that the volume element ds_x in the point $x \in \Xi$ where the phase space velocity component orthogonal to Ξ is v_x becomes in the time $\tau(x)$ a volume element around $x' = S_{\tau(x)}x$ with

$$ds_{x'} = \frac{v_x}{v_{x'}} e^{-\int_0^{\tau(x)} \sigma(S_t x)dt} ds_x \qquad (2.9.2)$$

Therefore if, as in several cases and in most simulations of the models in Sect. 2.2:

(1) v_x is bounded above and below away from infinity and zero
(2) $\sigma(x) = \varepsilon(x) + \dot{D}(x)$ with $D(x)$ *bounded on* Ξ (but possibly unbounded on the full phase space M)
(3) $\tau(x)$ is bounded and (for x outside a zero volume set) has average $\overline{\tau} > 0$

setting $\widetilde{\varepsilon}(x) = \int_0^{\tau(x)} \varepsilon(S_t x)dt$ it follows that the entropy production rate and the phase space contraction have the same average $\varepsilon_+ = \sigma_+$ and likewise

$$p = \frac{1}{m} \sum_{k=0}^{m-1} \frac{\widetilde{\varepsilon}(S^k x)}{\widetilde{\varepsilon}_+} \quad \text{and} \quad p' = \frac{1}{m} \sum_{k=0}^{m-1} \frac{\widetilde{\sigma}(S^k x)}{\widetilde{\sigma}_+} \qquad (2.9.3)$$

will differ by $1/m\big(D(S^m x) - D(x) + \log v_{S^m x} - \log v_x\big) \xrightarrow[m \to \infty]{} 0$.

This shows that in cases in which $D(x)$ is unbounded in phase space but there is a timing section Ξ on which it is bounded and which has the properties (1)–(3) above it is more reasonable to suppose the chaotic hypothesis for the evolution S timed on Ξ rather than trying to extend the chaotic hypothesis to evolutions in continuous time for the evolution S_t on the full phase space.

2.10 Conclusions

Nonequilibrium systems like the ones modeled in Sect. 2.2 undergo, in general, motions which are empirically chaotic at the microscopic level. The chaotic hypothesis means that we may as well assume that the chaos is maximal, i.e. it arises because the (timed) evolution has the Anosov property.

The evolution is studied through timing events and is therefore described by a map S on a "Poincaré's section" Ξ in the phase space M.

It is well known that in systems with few degrees of freedom the attracting sets are in general fractal sets: the chaotic hypothesis implies that instead one can neglect the fractality (at least if the number of degrees of freedom is not very small) and consider the attracting sets as smooth surfaces on which motion is strongly chaotic in the sense of Anosov.

The hypothesis implies (therefore) that the statistical properties of the stationary states are those exhibited by motions;

(1) That follow initial data randomly chosen with a distribution with density over phase space
(2) Strongly chaotic as in the chaotic hypothesis

and the stationary states of the system are described by the SRB distributions μ which are uniquely associated with each attracting set.

Systems in equilibrium (which in our models means that neither nonconservative forces nor thermostats act) satisfying the chaotic hypothesis can have no attracting set other than the whole phase space, which generically is the energy surface,[16] and have as unique SRB distribution the Liouville distribution, i.e. the chaotic hypothesis implies for such systems that the equilibrium states are described by microcanonical distributions. This means that nonequilibrium statistical mechanics based on the chaotic hypothesis cannot enter into conflict with the equilibrium statistical mechanics based on the ergodic hypothesis.

The main difficulty of a theory of nonequilibrium is that whatever model is considered, e.g. any of the models in Sect. 2.2, there will be dissipation which manifests itself through the non vanishing divergence of the equations of motion: this means that volume is not conserved no matter which metric we use for it, unless the time average σ_+ of the phase space contraction vanishes. Introduction of non Newtonian forces can only be avoided by considering infinite thermostats.

Since the average σ_+ cannot be negative in stationary nonequilibrium systems its positivity is identified with the signature of a genuine nonequilibrium, while the cases in which $\sigma_+ = 0$ are equilibrium systems, possibly "in disguise". If $\sigma_+ > 0$ there cannot be any stationary distribution which has a density on phase space: the stationary states give probability 1 to a set of configurations which have 0 volume in phase space (yet they may be dense in phase space, and often are such, if σ_+ is small).

---

[16] Excluding, for instance, specially symmetric cases, like spherical containers with elastic boundary.

Therefore any stationary distribution describing a nonequlibrium state cannot be described by a suitable density on phase space or on the attracting set, thus obliging us to develop methods to study such singular distributions.

If the chaotic hypothesis is found too strong, one has to rethink the foundations: the approach that Boltzmann used in his discretized view of space and time, started in [29, p. 5] and developed in detail in [30, p. 42], could be a guide.

References

1. Feynman, R.P., Vernon, F.L.: The theory of a general quantum system interacting with a linear dissipative system. Ann. Phys. **24**, 118–173 (1963)
2. Gallavotti, G.: Statistical Mechanics: A Short Treatise. Springer, Berlin (2000)
3. Gallavotti, G.: Chaotic hypothesis: onsager reciprocity and fluctuation-dissipation theorem. J. Stat. Phys. **84**, 899–926 (1996)
4. Nosé, S.: A unified formulation of the constant temperature molecular dynamics methods. J. Chem. Phys. **81**, 511–519 (1984)
5. Hoover, W.: Canonical equilibrium phase-space distributions. Phys. Rev. A **31**, 1695–1697 (1985)
6. Evans, D.J., Morriss, G.P.: Statistical Mechanics of Nonequilibrium Fluids. Academic Press, New York (1990)
7. Williams, S.R., Searles, D.J., Evans, D.J.: Independence of the transient fluctuation theorem to thermostatting details. Phys. Rev. E **70**, 066113(+6) (2004). doi:10.1103/PhysRevE.70. 066113
8. Gallavotti, G.: Entropy, thermostats and chaotic hypothesis. Chaos **16**, 043114(+6) (2006)
9. Becker, R.: Electromagnetic Fields and Interactions. Blaisdell, New York (1964)
10. Chernov, N.I., Eyink, G.L., Lebowitz, J.L., Sinai, Ya.G.: Steady state electric conductivity in the periodic Lorentz gas. Commun. Math. Phys. **154**, 569–601 (1993)
11. Garrido, P., Gallavotti, G.: Boundary dissipation in a driven hard disk system. J. Stat. Phys. **126**, 1201–1207 (2007)
12. Ruelle, D.: What are the measures describing turbulence. Prog. Theor. Phys. Suppl. **64**, 339–345 (1978)
13. Ruelle, D.: Measures describing a turbulent flow. Ann. NY Acad. Sci. **357**, 1–9 (1980)
14. Eckmann, J.P., Ruelle, D.: Ergodic theory of chaos and strange attractors. Rev. Mod. Phys. **57**, 617–656 (1985)
15. Ruelle, D.: Smooth dynamics and new theoretical ideas in non-equilibrium statistical mechanics. J. Stat. Phys. **95**, 393–468 (1999)
16. Evans, D.J., Sarman, S.: Equivalence of thermostatted nonlinear responses. Phys. Rev. E **48**, 65–70 (1993)
17. Gallavotti, G., Presutti, E.: Nonequilibrium, thermostats and thermodynamic limit. J. Math. Phys. **51**, 015202(+32) (2010)
18. Gallavotti, G., Presutti, E.: Fritionless thermostats and intensive constants of motion. J. Stat. Phys. **139**, 618–629 (2010)
19. Gallavotti, G., Presutti, E.: Thermodynamic limit for isokinetic thermostats. J. Math. Phys. **51**, 032901(+9) (2010)
20. Sinai, Ya.G.: Markov partitions and C-diffeomorphisms. Funct. Anal. Appl. **2**(1), 64–89 (1968)
21. Bowen, R.: Markov partitions for axiom a diffeomorphisms. Am. J. Math. **92**, 725–747 (1970)
22. Bowen, R., Ruelle, D.: The Ergodic theory of axiom a flows. Inv. Math. **29**, 181–205 (1975)
23. Arnold, V., Avez, A.: Ergodic Problems of Classical Mechanics. Benjamin, New York (1966)
24. Gentile, G.: A large deviation theorem for anosov flows. Forum Math. **10**, 89–118 (1998)

25. Gallavotti, G., Cohen, E.G.D.: Dynamical ensembles in nonequilibrium statistical mechanics. Phys. Rev. Lett. **74**, 2694–2697 (1995)
26. Ruelle, D.: Turbulence, Strange Attractors and Chaos. World Scientific, New York (1995)
27. Ruelle, D.: Positivity of entropy production in nonequilibrium statistical mechanics. J. Stat. Phys. **85**, 1–25 (1996)
28. Gallavotti, G., Bonetto, F., Gentile, G.: Aspects of the Ergodic, Qualitative and Statistical Theory of Motion. Springer, Berlin (2004)
29. Boltzmann, L.: Studien über das gleichgewicht der lebendigen kraft zwischen bewegten materiellen Punkten. In: Hasenöhrl, F. (ed.) Wissenschaftliche Abhandlungen, vol. 1(5). Chelsea, New York (1968)
30. Boltzmann, L.: Über die beziehung zwischen dem zweiten hauptsatze der mechanischen wärmetheorie und der wahrscheinlichkeitsrechnung, respektive den Sätzen über das Wärmegleichgewicht. In: Hasenöhrl. F. (ed.) Wissenschaftliche Abhandlungen vol. 2(42). Chelsea, New York (1968)

Chapter 3
Discrete Phase Space

3.1 Recurrence

Simulations have played a key role in the recent studies on nonequilibrium. And simulations operate on computers to obtain solutions of equations in phase space: therefore phase space points are given a digital representation which might be very precise but rarely goes beyond 32 bits per coordinate. If the system contains a total of N particles each of which needs four coordinates to be identified (in the simplest 2-dimensional models, six otherwise) this gives a phase space (virtually) containing $\mathcal{N}_{tot} = (2^{32})^{4N}$ points which cover a phase space region of desired size V in velocity and L in position with a lattice of mesh $2^{-32} V$ or $2^{-32} L$ respectively.

Therefore the "fiction" of a discrete phase space, used by Boltzmann in his foundational works, [1, #42, p. 167], has been taken extremely seriously in modern times with the peculiarity that it is seldom even mentioned in the numerical simulations.

A simulation is a code that operates on discrete phase space points transforming them into other points. In other words it is a map $\overline{S}$ which associates with any point on phase space a new one with a precise deterministic rule that can be called a *program*.

All programs for simulating solutions of ordinary differential equations have some serious drawbacks: for instance it is very likely that the map $\overline{S}$ defined by a program is not invertible, unlike the true solution to a differential equation of motion, which obeys a uniqueness theorem: different initial data might be mapped by $\overline{S}$ into the same point.

Since the number $\mathcal{N}_{tot}$ is finite, all points will undergo a motion, as prescribed by the program $\overline{S}$, which will become *recurrent*, i.e. will become eventually a permutation of a subset of the phase space points, hence *periodic*.

The ergodic hypothesis was born out of the natural idea that the permutation would be a *one cycle* permutation: every microscopic state would recur and continue in a cycle, [2]. In simulations, even if dealing with time reversible systems, it would not be reasonable to assume that all the phase space points are part of a permutation, because of the mentioned non invertibility of essentially any program. It is nevertheless possible that, once the *transient* states (i.e. the ones that never recur, being out

G. Gallavotti, *Nonequilibrium and Irreversibility*,
Theoretical and Mathematical Physics, DOI: 10.1007/978-3-319-06758-2_3,
© Springer International Publishing Switzerland 2014

of the permutation cycles) are discarded and motion reduces to a permutation, then the permutation is just a single cycle one.

So in simulations of motions of isolated systems the ergodic hypothesis really consists in two parts: first, the non recurrent phase space points should be "negligible" and, second, the evolution of the recurrent points consists in a single cycle permutation. Two comments:

(a) Periodicity is not in contrast with chaotic behavior: this is a point that Boltzmann and others (e.g. Thomson) clarified in several papers ([3, #39], [4], see also Sect. 6.11) for the benefit of the few that at the time listened.

(b) The recurrence times are beyond any observable span of time (as soon as the particles number N is larger than a few units), [5, Sect. 88].

In presence of dissipation, motions develop approaching a subset of phase space, the attracting set $\mathcal{A}$ and on it the attractor $\mathcal{B}$, p. 40, which has therefore zero volume because volume is not invariant and is asymptotically, hence forever, decreasing.

Nevertheless in the above discrete form the ergodic hypothesis can be formulated also for general nonconservative motions, like the ones in Sect. 2.2. With the difference that, in this case, the *nonrecurrent points will be "most" points*: because in presence of dissipation the attractor set will have 0 volume, see Sects. 2.7 and 2.8 (even in the cases in which the attracting set $\mathcal{A}$ is the entire phase space, like in the small perturbations of conservative Anosov systems)

Therefore it can be formulated by requiring that *on the attracting set* non recurrent points are negligible and the recurrent points form a one cycle permutation. In other words, in this form,

the ergodic hypothesis is the same for conservative and dissipative systems provided phase space is identified with the attracting set.

Of course in chaotic motions the periodicity of motion is not observable as the time scale for the recurrence will remain (in equilibrium as well as out of equilibrium) out of reach.

The chaotic nature of the motions is therefore hidden inside a very regular (and somewhat uninteresting) periodic motion, [2].

In the latter situation the statistics of the motions will be uniquely determined by assigning a probability $\mathcal{N}^{-1}$ to each of the $\mathcal{N}$ configurations on the (discrete version of the) attractor: and this will be the *unique* stationary distribution, see Sect. 3.8.

Remark (1) The uniqueness of the stationary distribution is by no means obvious and, as well, it is not obvious that the motion can be described by a permutation of the points of a regularly discretized phase space. Not even in equilibrium.

(2) Occasionally an argument is found whereby, in equilibrium, motion can be regarded as a permutation "because of the volume conservation because of Liouville's theorem". But this cannot be a sensible argument because of the chaoticity of motion: any volume element will be deformed under evolution and stretched along certain directions while it will be compressed along others. Therefore the points of the discretized phase space should not be thought as small volume elements, with positive volume, but precisely as individual (0 volume) points which the evolution permutes.

(3) Boltzmann argued, in modern terms, that after all we are interested in very few observables, in their averages and in their fluctuations. Therefore we do not have to follow the details of the microscopic motions and all we have to consider are the time averages of a few physically important observables $F_1, F_2, \ldots, F_q, q$ small. This means that we have to understand what is now called a *coarse grained* representation of the motion, collecting together all points on which the observables $F_1, F_2, \ldots, F_q$ assume the same values. Such collection of microscopic states is called a *macrostate*, [6, 7].

The reason why motion appears to reach stationarity and to stay in that situation is that for the overwhelming majority of the microscopic states, i.e. points of a discretized phase space, the interesting observables have the same values. The deviations from the averages are observable over time scales that are most often of human size and have nothing to do with the recurrence times. Boltzmann gave a very clear and inspiring view of this mechanism by developing the Boltzmann's equation, [8, #22]: perhaps realizing its full implications only a few years later when he had to face the conceptual objections of Loschmidt and others, [3, #39], and Sect. 6.11.

3.2 Hyperbolicity: Stable and Unstable Manifolds

If the dynamics is chaotic, i.e. the system is an Anosov system, then points $x + dx$ infinitesimally close to a given $x \in \Xi$ will become $S^n(x + dx)$ and will separate from $S^n x$ exponentially fast as the time $n \to \infty$ with the exception of points $x + dx$ with dx on a tangent plane $V^s(x)$ through x which, instead, approach exponentially fast $S^n x$: this means that points infinitesimally close to x and lying on $V^s(x)$ evolve with the matrix $\partial S^n(x)$ of the derivatives of S^n so that

$$|\partial S^n(x)\, dx| \leq C e^{-\lambda n} |dx|, \qquad n \geq 0, \ dx \in V^s(x) \qquad (3.2.1)$$

for some $C, \lambda > 0$ and for $dx \in V^s(x)$.

Likewise points $x + dx$ infinitesimally close to a given $x \in \Xi$ will also separate from $S^{-n}x$ exponentially fast with the exception of points $x + dx$ with dx on a tangent plane $V^u(x)$ through x which, instead, approach exponentially fast $S^{-n}x$:

$$|\partial S^{-n}(x)\, dx| \leq C e^{-\lambda n} |dx|, \qquad n \geq 0, \ dx \in V^u(x) \qquad (3.2.2)$$

Furthermore if the planes $V^u(x), V^s(x)$ depend continuously on x and if there is a point with dense orbit they will form "integrable" families of planes, i.e. $V^u(x)$ and $V^s(x)$ will be everywhere tangent to smooth surfaces, $W^u(x)$ and $W^s(x)$, without boundary, continuously dependent on x and, for all $x \in \Xi$, dense on Ξ, [9, p. 267], [10, Sect. 4.2].

The surfaces $W^\alpha(x)$, $\alpha = u, s$, are called stable and unstable manifolds through x and their existence as smooth, dense, surfaces without boundaries will be taken here as a property characterizing the kind of chaotic motions in the system. Mathematically a system admitting such surfaces is called *smooth and uniformly hyperbolic* (if furthermore have a dense orbit then they are Anosov systems).

A simple but at first unintuitive property of the invariant manifolds is that *although* they are locally smooth surfaces (if S is smooth) their tangent planes $V^\alpha(x)$ are not very smoothly dependent on x if x is moved out of the corresponding $W^\alpha(x)$: if the $V^\alpha(x)$ are defined by assigning their unit normal vectors $n^\alpha(x)$ (in a coordinate system) and $|x - y|$ is the distance between x and y, it is in general $|n^\alpha(x) - n^\alpha(y)| \leq L_\beta |x - y|^\beta$ where β can be prefixed arbitrarily close to 1 at the expenses of a suitably large choice of L_β, [11]: see Appendix F for an argument explaining why even very smooth Anosov maps only enjoy Hölder continuity of the planes $V^\alpha(x)$.

This implies that although the Jacobian $\partial S(x)$ of the map is a smooth function of x nevertheless the restriction of the $\partial S(x)$ to vectors tangent to the manifolds $W^\alpha(x)$ or to functions of them, like the logarithms of the determinants $\lambda_\alpha(x) \overset{def}{=} \log|\det \partial S(x)|_{V^\alpha(x)}|$, $\alpha = s, u$, depend on x only Hölder continuously: namely the exist constants L_β^α such that

$$\begin{cases} |\partial S(x)|_{V^\alpha(x)} - \partial S(y)|_{V^\alpha(y)}| \leq L_\beta^\alpha |x - y|^\beta \\ |\lambda_\alpha(x) - \lambda_\alpha(y)| \leq L_\beta^\alpha |x - y|^\beta \end{cases} \tag{3.2.3}$$

for $\alpha = u, s$, $\beta < 1$ and $|x - y|$ equal to the distance between x and y.

The Hölder continuity property will play an important role in the following. The reason for this apparent anomaly[1] is that it is possible to give a formal expression for the derivatives of $n^\alpha(x)$ which gives a formally finite value for the derivatives of $n^\alpha(x)$ along $W_\alpha(x)$ but a value of the derivatives along $W_{\alpha'}(x)$ formally undefined if $\alpha' \neq \alpha$, see Sect. 10.1 and Problem 10.1.5 in [10], see also Appendix F.

Often there is no point with a dense orbit because the system admits a finite number of invariant, closed, attracting sets $\mathcal{A}_i \subset \Xi$ which are not dense in Ξ: motions starting on $\mathcal{A}_i$ stay there and those starting close enough to $\mathcal{A}_i$ evolve approaching $\mathcal{A}_i$ exponentially fast; furthermore each of the $\mathcal{A}_i$ admits a motion dense in $\mathcal{A}_i$ and cannot be further decomposed.

Beyond the chaotic hypothesis a more general assumption could be that the surfaces $W^\alpha(x)$, $\alpha = u, s$, exist outside a set of zero volume in Ξ and may have boundary points. Such systems are called simply *hyperbolic*: however the basic proposal, *and the ratio behind the chaotic hypothesis*, is to build the intuition about chaotic motions upon smooth and uniformly hyperbolic dynamics $x \to Sx$.

Therefore the attracting sets $\mathcal{A}_i$ that will be considered should be visualized as smooth surfaces which attract exponentially fast the nearby points: the interesting properties of the dynamics will be related to the motions of points on such surfaces. In this way the motion is attracted by a smooth surface on which motions are uniformly hyperbolic and on which there is a dense orbit (i.e. the restriction of the evolution to

---

[1] Naive expectation would have been that if the manifold Ξ and the map S are smooth, say C^∞, also $V^\alpha(x)$ and $W^\alpha(x)$ depend as smoothly on x.

$\mathcal{A}_i$ is an Anosov system for each i). *This means that S satisfies the chaotic hypothesis,* Sect. 2.7 and gives a clearer intuitive interpretation of it.

3.3 Geometric Aspects of Hyperbolicity: Rectangles

Perhaps the deep meaning of hyperbolicity is that it leads to a natural definition of coarse grained partitions, whose elements will be called *rectangles*, as it will be discussed in the next sections, after setting up some geometrical definitions.

A geometric consequence of the hyperbolicity implied by the chaotic hypothesis is that it is possible to give a natural definition of sets E which have a boundary ∂E which consists of two parts one of which, $\partial^s E$, is compressed by the action of the evolution S and stretched under the action of S^{-1} and the other, $\partial^u E$ to which the opposite fate is reserved. Such sets are called *rectangles* and their construction is discussed in this section.

Inside the ball $B_\gamma(x)$ of radius γ centered at x the surface elements $W^s_\gamma(x) \subset W^s(x) \cap B_\gamma(x)$, $W^u_\gamma(x) \subset W^u(x) \cap B_\gamma(x)$ *containing x and connected*[2] have the geometric property that, if γ is small enough (independently of x), near two close points ξ, η on an attracting set $\mathcal{A} \subseteq \Xi$ there will be a point defined as $\zeta \overset{def}{=} [\xi, \eta] \in \mathcal{A} \subset \Xi$ whose n-th iterate in the past will be exponentially approaching $S^{-n}\xi$ while its n-th iterate in the future will be exponentially approaching $S^n\eta$ as $n \to +\infty$ (Fig. 3.1).

This can be used to define special sets E that will be called *rectangles* because they can be drawn by giving two "axes around a point x", $C \subset W^u_\gamma(x)$, $D \subset W^s_\gamma(x)$ (γ small); the axes around x will be connected surface elements with a boundary which has zero measure relative to the area measure on $W^s_\gamma(x)$ or $W^u_\gamma(x)$ and which

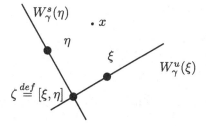

Fig. 3.1 Representation of the operation that associates $[\xi, \eta]$ with the two points ξ and η as the intersection of a short connected part $W^u_\gamma(\xi)$ of the unstable manifold of ξ and of a short connected part $W^s_\gamma(\eta)$ of the stable manifold of η. The size γ is short "enough", compared to the diameter of Ξ, and it is represented by the segments to the *right* and *left* of η and ξ. The ball $B_\gamma(x)$ is not drawn

---

[2] Notice that under the chaotic hypothesis motions on the attracting sets $\mathcal{A}$ are Anosov systems so that the stable and unstable manifolds of every point are dense: therefore $W^s(x) \cap B_\gamma(x)$ is a dense family of layers in $B_\gamma(x)$, but only one is connected and contains x, if γ is small enough.

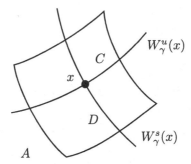

Fig. 3.2 A *rectangle* A with axes C, D crossing at a center x

are the closures of their internal points (relative to $W_\gamma^s(x)$ and $W_\gamma^u(x)$); an example could be the connected parts, containing x, of the intersections $C = W^u(x) \cap B_\gamma(x)$ and $D = W^s(x) \cap B_\gamma(x)$.

The boundaries of the surface elements will be either 2 points, if the dimension of $W^s(x)$, $W^u(x)$ is 1 as in the above figures or, more generally, continuous connected surfaces each of dimension one unit lower than that of $W^s(x)$, $W^u(x)$.

Then define E as the set

$$E = C \times D \equiv [C, D] \overset{def}{=} \bigcup_{y \in C, \, z \in D} \{[y, z]\}, \tag{3.3.1}$$

and call x the *center* of E with respect to the *pair of axes*, C and D. This is illustrated in Fig. 3.2. We shall say that C is an unstable axis and D a stable one.

If C, D and C', D' are two pairs of axes for the same rectangle we say that C and C', or D and D', are "parallel"; one has either $C \equiv C'$ or $C \cap C' = \emptyset$.

A given rectangle E can be constructed as having any internal point $y \in E$ as center, by choosing an appropriate pair of axes.

The boundary of $E = C \times D$ is composed by sides $\partial^s E$, $\partial^u E$, see Fig. 3.3, each not necessarily connected as a set. The first are parallel to the stable axis C and the other two to the unstable axis D, and they can be defined in terms of the boundaries ∂C and ∂D of C and D considered as subsets of the unstable and stable lines that contain them.

The stable and unstable parts of the boundary are defined as

$$\partial^s E = [\partial C, D], \qquad \partial^u E = [C, \partial D]. \tag{3.3.2}$$

which in the 2-dimensional case consist of two pairs of parallel lines, as shown in Fig. 3.3.

Remark As mentioned any point x' in E is the intersection of unstable and stable surfaces C', D' so that E can be written also as $C' \times D'$: hence any of its points can be a center for E. It is also true that if $C \times D = C' \times D'$ then $C \times D' = C' \times D$.

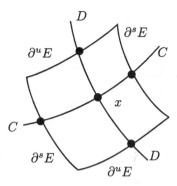

Fig. 3.3 The stable and unstable boundaries of a *rectangle* $E = C \times D$ in the simple 2-dimensional case in which the boundary really consists of two pairs of parallel axes

For this reason given a rectangle any such C' will be called an *unstable axis* of the rectangle and any D' will be called a *stable axis* and the intersection $C' \cap D'$ will be a point x' called the center of the rectangle for the axes C', D'.

If C, C' are two parallel *stable* axes of a rectangle E and a map $\theta : C \to C'$ is established by defining $\xi' = \theta(\xi)$ if ξ, ξ' are on the *same unstable* axis through $\xi \in C$: then it can be shown that the map θ maps sets of positive relative area on C to sets of positive area on C'. The corresponding property holds for the correspondence established between two parallel unstable axes D, D' by their intersections with the stable axes of E.

This property is called *absolute continuity* of the foliations $W^s(x)$ with respect to $W^u(x)$ and of $W^u(x)$ with respect to $W^s(x)$.

3.4 Symbolic Dynamics and Chaos

To visualize and take advantage of the chaoticity of motion we imagine that phase space can be divided into *cells* E_j, with pairwise disjoint interiors, determined by the dynamics. They consist of rectangles $E_j = C_j \times D_j$ as in Fig. 3.2 with the axes C_j, D_j crossing at a "center" $x_j = C_j \cap D_j$, and the size of their diameters can be supposed smaller than a prefixed $\delta > 0$.

The basic property of hyperbolicity and transitivity is that the cells $E_1, E_2, \ldots, E_k$ can be so adapted to enjoy of the two properties below:
(1) the "stable part of the boundary" of E_j, denoted $\partial^s E_j$ under the action of the evolution map S ends up as a subset of the union of all the stable boundaries of the rectangles and likewise the unstable boundary of E_i is mapped into the union of all the unstable boundaries of the rectangles under S^{-1}. In formulae

$$S\partial^s E_j \subset \cup_k \partial^s E_k, \qquad S^{-1}\partial^u E_j \subset \cup_k \partial^u E_k, \qquad (3.4.1)$$

$$S\partial^s E_j \subset \cup_k \partial^s E_k, \qquad S^{-1}\partial^u E_j \subset \cup_k \partial^u E_k,$$

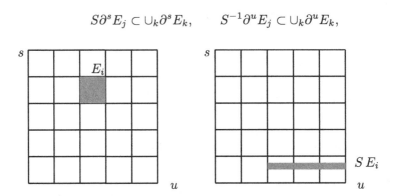

Fig. 3.4 The figures illustrate very symbolically, as 2-dimensional *squares*, a few elements of a Markovian pavement (or Markov partition). An element E_i of it is transformed by S into SE_i in such a way that the part of the boundary that contracts ends up exactly on a boundary of some elements among $E_1, E_2, \ldots, E_n$

In other words no new stable boundaries are created if the cells E_j are evolved towards the future and no new unstable boundaries are created if the cells E_j are evolved towards the past as visualized in the idealized figure, Fig. 3.4 (idealization due to the dimension 2 and to the straight and parallel boundaries of the rectangles). (2) Furthermore the intersections $E_i \cap SE_j$ with internal points have to be connected sets.

Defining $M_{ij} = 1$ if the interior of SE_i intersects the interior of E_j or $M_{ij} = 0$ otherwise, then for each point x there is one sequence of labels $\boldsymbol{\xi} = \{q_j\}_{j=-\infty}^{\infty}$ such that $M_{q_k q_{k+1}} \equiv 1$, for all $k \in Z$, and with $S^k x \in E_{q_k}$: the sequence $\boldsymbol{\xi}$ is called a *compatible history* of x.

Viceversa if the diameter of the cells is small enough then there is only one compatible history for a point x with the exception of points of a set with zero volume.[3]

The requirement of small enough diameter is necessary to imply that the image of the interior of any element E_j intersects E_i (i.e. if $M_{ji} = 1$) in a connected set: true only if the diameter is small enough (compared to the minimum curvature of the stable and unstable manifolds).

The one to one correspondence (aside for a set of zero volume) between points and compatible histories is a key property of hyperbolic smooth evolutions: it converts the evolution $x \to Sx$ into the trivial translation of the history of x which becomes $\boldsymbol{\xi}' \equiv S\boldsymbol{\xi} \overset{def}{=} \{q_{k+1}\}_{k=-\infty}^{\infty}$. This follows from the hyperbolicity definition once it is accepted that it implies existence of Markovian pavements with elements of small enough diameter.

---

[3] The exception is associated with points x which are on the boundaries of the rectangles or on their iterates. In such cases it is possible to assign the symbol ξ_0 arbitrarily among the labels of the rectangles to which x belongs: once made this choice a compatible history $\boldsymbol{\xi}$ determining x exists and is unique.

This means that sequences ξ can be used to identify points of $\varXi$ just as decimal digits are used to identify the coordinates of points (where exceptions occur as well, and for the same reasons, as ambiguities arise in deciding, for instance, whether to use $0.9999\ldots$ or $1.0000\ldots$).

The matrix M will be called a "compatibility matrix". Transitivity (p. 35) implies that the matrix M admits an iterate M^h which, for some $h > 0$, has *no vanishing entry*.

Therefore the points $x \in \varXi$ can be thought as the *possible* outputs of a Markovian process with transition matrix M: for this reason the partitions $\{E_j\}$ of $\varXi$ are called *Markovian*.

Remark (1) The Markovian property has a geometrical meaning (seen from Fig. 3.4 above): imagine each E_i as the "stack" formed by all the connected unstable axes $\delta(x)$, intersections of E_i with the unstable manifolds of its points x, which can also be called unstable "layers" in E_i.

Then if $M_{i,j} = 1$, the expanding layers in each E_i expand under the action of S and their images *fully cover* the layers of E_j with which they overlap.[4]

A corresponding property holds for the stable layers.

(2) It is important to notice that once a Markovian pavement $\mathcal{E} = (E_1, \ldots, E_q)$ with elements with diameter $\leq d$ has been constructed then it is possible to construct a new Markovian pavement $\mathcal{E}_\tau$ whose elements have diameter smaller than a prefixed quantity. It suffices to consider the pavement $\mathcal{E}_\tau$ whose elements are, for instance, the sets which have interior points and have the form

$$E_{\mathbf{q}} \overset{def}{=} E_{q_{-\tau},\ldots,q_\tau} \overset{def}{=} \cap_{i=-\tau}^{\tau} S^{-i} E_{q_i}. \tag{3.4.2}$$

Their diameters will be $\leq 2Ce^{-\tau\lambda}$ if C, λ are the hyperbolicity constants, see Eqs. (3.2.1), (3.2.2). The sets of the above form with non empty interior are precisely the sets $E(\mathbf{q})$ for which $\prod_{j=-\tau}^{\tau-1} M_{q_j q_{j+1}} = 1$.

(3) If $x \in E(\mathbf{q})$, $\mathbf{q} = (q_{-\tau}, \ldots, q_\tau)$ then the symbolic history $\xi = (\xi_i)_{i=-\infty}^\infty$, with $S^{-j}x \in E_{\xi_j}$, of x coincides at times $j \in [-\tau, \tau]$ with $\mathbf{q}$, i.e. $\xi_j = q_j$ for $j \in [-\tau, \tau]$ (except for a set of 0 volume of x's).

(4) A rectangle $E(\mathbf{q})$ can be imagined as the stack of the portions of unstable manifolds of the points on its stable axis $\delta(\mathbf{q}, x) = [x, W_\gamma^u(x)] \cap E(\mathbf{q})$: i.e. $E(\mathbf{q}) = \cup_{x \in D} \delta^u(\mathbf{q}, x)$ (or as the stack $E(\mathbf{q}) = \cup_{x \in C} \delta^s(\mathbf{q}, x)$ with δ^s defined similarly).

The symbolic representation of the portion of unstable manifold $[x, W_\gamma^u(x)] \cap E(\mathbf{q})$ simply consists of the compatible sequences ξ with $\xi_i = q_i$, $i \in [-\tau, \tau]$ and which continue to $i < -\tau$ into the sequence of symbols of x with labels $i < -\tau$ while for $i > \tau$ are arbitrary. The portion of stable manifold has a corresponding representation.

---

[4] Formally let $E_i \in \mathcal{P}$, $x \in E_i$ and $\delta(x) = E_i \cap W_u(x)$: then if $M_{i,j} = 1$, i.e. if the interior of SE_i visits the interior of E_j, it is $\delta(Sx) \subset S\delta(x)$.

(5) The smallest m with the property that $M_{ij}^m > 0$ will be called the *symbolic mixing time*; it gives the minimum time needed to be sure that any symbol i can be followed by any other j in a compatible sequence with compatibility matrix given by the transitive matrix of the pavement $\mathcal{E}$.

Simple examples will be discussed in Sect. 3.5.

3.5 Examples of Hyperbolic Symbolic Dynamics

The paradigmatic example is the simple evolution on the 2-dimensional torus $T^2 = [0, 2\pi]^2$ defined by the transformation.[5]

$$S\varphi = S\begin{pmatrix} \varphi_1 \\ \varphi_2 \end{pmatrix} = \begin{pmatrix} 1 & 1 \\ 1 & 0 \end{pmatrix}\begin{pmatrix} \varphi_1 \\ \varphi_2 \end{pmatrix} \overset{def}{=} \begin{pmatrix} \varphi_1 + \varphi_2 \\ \varphi_1 \end{pmatrix} \mod 2\pi \qquad (3.5.1)$$

It is possible to construct simple examples of Markovian partitions of T^2 because the stable and unstable directions through a point φ are everywhere the directions of the two eigenvectors of the matrix $\begin{pmatrix} 1 & 1 \\ 1 & 0 \end{pmatrix}$.

Hence in the coordinates φ they are straight lines (wrapping densely over T^2 because the slope of the eigenvectors is irrational). For instance Fig. 3.5 gives an example of a partition satisfying the property (1) in Sect. 3.4.

This is seen by remarking that in Fig. 3.5 the union of the stable boundaries of the rectangles, i.e. the lines with negative slope (irrational and equal to $(-\sqrt{5} - 1)/2$)

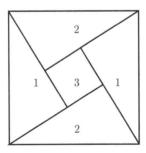

Fig. 3.5 The pavement with three rectangles (E_1, E_2, E_3) of the torus T^2 whose sides lie on two *connected* portions of stable and unstable manifold of the fixed point at the origin. It satisfies the property in Eq. (3.4.1) *but it is not Markovian* because the correspondence between histories and points is not 1–1 even if we allow for exceptions on a set of zero area: the three sets are too large. But the partition whose elements are $E_j \cap SE_j$ has the desired properties, as it follows from the next figure

---

[5] The map is not obtainable as a Poincaré's section of the orbits of a 3-dimensional manifold simply because its Jacobian determinant is not $+1$.

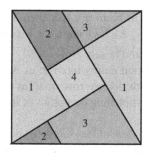 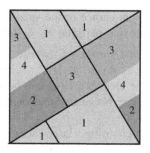

Fig. 3.6 A Markovian pavement (*left*) for S ("Arnold's cat map"). The images under S of the pavement *rectangles* are shown in the *right figure*: corresponding *rectangles* are marked by the corresponding *colors* and numbers

is, because of the periodicity of T^2, a *connected part* of the stable manifold exiting on either side from the origin; likewise the union of the unstable boundaries, i.e. the lines with positive slope (equal to $(\sqrt{5} - 1)/2$) are a *connected part* of the unstable manifold exiting from the origin.

Therefore under action of S the union of the stable boundaries will be still a part of the stable manifold through the origin *shorter by a factor* $(\sqrt{5} - 1)/2 < 1$ hence it will be part of itself, so that the first Eq. (3.4.1) holds. For the unstable boundaries the same argument can be repeated using S^{-1} instead of S to obtain the second Eq. (3.4.1).

One checks that the partition in Fig. 3.5 $\mathcal{E} = (E_1, E_2, E_3)$ generates ambiguous histories in the sense that E_2 is too large and in general the correspondence between points and their symbolic history is $2 - 1$ (and more to 1 on a set of zero area). However by slightly refining the partition (subdividing the set E_2 in Fig. 3.5) a true Markovian partition is obtained as shown in Fig. 3.6.

The examples above are particularly simple because of the 2-dimensionality of phase space and because the stable and unstable manifolds of each point are straight lines.

If the dimension is higher and the manifolds are not flat or if expansion (or contraction) in different directions is different the Markovian partition still exists but its elements may have an irregular boundary. For this reason the reader is referred to the original papers [12, 13], except in the general 2-dimensional case particularly simple and discussed in Appendix G.

3.6 Coarse Graining and Discrete Phase Space

Given the observables $F_1, \ldots, F_r$ the phase space is imagined subdivided in small regions in which the observables have a constant value, for the purpose of their measurements.

A convenient choice of the small regions will be to imagine them constructed from a Markovian partition $\mathcal{E}_0 = (E_1, E_2, \ldots, E_m)$. Given $\tau \geq 0$ consider the *finer Markovian partition* $\mathcal{E}_\tau$ whose elements are the sets $E(\mathbf{q}) \overset{def}{=} \cap_{j=-\tau}^{\tau} S^{-j} E_{q_j}$, $\mathbf{q} = (q_{-\tau}, \ldots, q_\tau)$, $q_j \in \{1, 2, \ldots, m\}$, with non empty interior: as discussed in Eq. (3.4.2) the elements $E(\mathbf{q})$ can be made with diameter as small as pleased by choosing τ large, because of the contraction or stretching properties of their boundaries.

Therefore, choosing τ large enough so that in each of the *cells* $E(\mathbf{q})$ the "interesting" observables $F_1, \ldots, F_r$ have a constant value, we shall call $\mathcal{E}_\tau$ a *coarse grained partition* of phase space into "coarse cells" relative to the observables $F_1, \ldots, F_r$. Should we decide that higher precision is necessary we shall have simply to increase the value of τ. But, given the precision chosen, the time average of any observable F of interest will be of the form

$$\langle F \rangle = \frac{\sum_{\mathbf{q}} F(x_{\mathbf{q}}) w(\mathbf{q})}{\sum_{\mathbf{q}} w(\mathbf{q})} \tag{3.6.1}$$

where $\mathbf{q} = (q_{-\tau}, \ldots, q_\tau)$ are $2\tau + 1$ labels among the labels $1, 2, \ldots, q$ of the Markovian pavement used to construct the coarse cells, $x_{\mathbf{q}}$ denotes a point of the cell $E(\mathbf{q})$, and $w(\mathbf{q})$ are suitable *weights*.

If the system admits a SRB distribution μ_{SRB} then the weights $w(\mathbf{q})$ will, within the precision, be given by

$$\frac{w(\mathbf{q})}{\sum_{\mathbf{q}'} w(\mathbf{q}')} = \mu_{SRB}(E(\mathbf{q})) \tag{3.6.2}$$

and the problem is to determine, for systems of interest, the weights (hence the SRB distribution).

To understand this point it is convenient to consider a discretization of phase space Ξ into *equally spaced* points x, centers of tiny boxes[6] of sides $\delta p, \delta q$ in the momentum and, respectively, positions coordinates and volume $(\delta p \delta q)^{3N} \overset{def}{=} h^{3N}$, by far smaller than the diameter of the largest coarse cell (as usual in simulations).

This will allow us to discuss time evolution in a way deeply different from the usual: it has to be stressed that such "points" or "microscopic" cells, are not associated with any particular observable; they can be thought of as tiny $6N$ dimensional boxes and represent the highest microscopic resolution and will be called *microcells* or *discrete points* of phase space *and have nothing to do with the above coarse cells E_j which are to be thought as much larger and containing very large numbers of microcells.*

Let $\mathcal{N}_0$ be the total number of microcells *regularly spread* on Ξ. The dynamics will be thought as a map of microcells into themselves: it will then be eventually

---

[6] The name is chosen to mark the distinction with respect to the parallelepipeds of the coarse partition.

periodic. The recurrent points will be in general $\mathcal{N} \ll \mathcal{N}_0$, i.e. much less than the number $\mathcal{N}_0$ of points in the discretization of $\varXi$.

No matter how small coarse cells $E(\mathbf{q})$ are chosen, as long as the number of discrete points inside them is very large, it will be impossible to represent the motion as a permutation: not even in the conservative case in which the volume of the cells remains constant. Simply because the cells are deformed by the evolution, being stretched in some direction and compressed in others, if the motion has nonzero Lyapunov exponents (i.e. is chaotic).

The next section will address the question: how can this be reconciled with the numerical simulations, and with the naive view of motion, as a permutation of cells? The phase space volume will generally contract with time: yet we want to describe the evolution in terms of an evolution permuting microscopic states? And *how to determine the weights* $w(\mathbf{q})$ of the coarse cells?

3.7 Coarse Cells, Phase Space Points and Simulations

The new microcells (introduced in the previous section) should be considered as realizations of objects alike to those arising in computer simulations: in them phase space points x are "digitally represented" with coordinates given by a string of integers and the evolution S becomes a *program*, or *code*, $\overline{S}$ simulating the solution of equations of motion suitable for the model under study. The code $\overline{S}$ operates *exactly* on the coordinates (the deterministic round offs, enforced by the particular computer hardware and software, should be considered part of the program).

Assuming the validity of the chaotic hypothesis, i.e. that the evolution map S on phase space $\varXi$ is smooth hyperbolic and with a dense orbit on the attracting sets, then the general properties analyzed in the previous sections will hold. In particular there will be a partition of the attracting set into rectangles with the Markovian property.

The evolution S considered in the approximation in which it acts on the discretized phase space will produce (for approximations careful enough) a chaotic evolution "for all practical purposes", if attention is directed at

(1) looking only at "macroscopic observables" which are constant on the coarse graining scale $\gamma = Ce^{-\lambda\tau}$, see Eq. (3.4.2)[7]; and
(2) looking only at phenomena accessible on time scales far shorter than the recurrence times (always finite in finite representations of motion, but of size always large enough to make the recurrence phenomenon irrelevant).[8]

---

[7] Here it is essential that the chaotic hypothesis holds, i.e. that the system is hyperbolic, otherwise if the system has long time tails the analysis becomes much more involved and so far it can be dealt, even if only qualitatively, on a case by case basis.

[8] To get an idea of the orders of magnitude consider a rarefied gas of N mass m particles of density ϱ at temperature T: the metric on phase space will be $ds^2 = \sum_i (\frac{d\mathbf{p}_i^2}{mk_BT} + \frac{d\mathbf{q}_i^2}{\varrho^{-2/3}})$; each coarse cell will have size at least $\sim \sqrt{mk_BT}$ in momentum and $\sim \varrho^{-\frac{1}{3}}$ in position; this is the minimum

It has to be realized that:

(a) there has to be a small enough division into microcells that allows us to describe evolution S as a map $\overline{S}$ of the microcells (otherwise numerical simulations would not make sense);
(b) however the map $\overline{S}$ approximating the evolution map S cannot be, in general, a permutation of microcells. As in simulations it will happen, *essentially always*, that it will send distinct microcells into the same one. It does certainly happen in nonequilibrium systems in which phase space contracts in the average[9];
(c) even though the map $\overline{S}$ will not be one-to-one, nevertheless it will be such *eventually*: because any map on a finite space is a *permutation* of the points which are recurrent. If, for simplicity, we suppose that the evolution S has only one attracting set $\mathcal{A}$ then the set of recurrent points for $\overline{S}$ is a *discrete representation* of the attracting set, that we call $\overline{\mathcal{A}}$.

The discrete set $\overline{\mathcal{A}}$ will be imagined as a collection of microcells approximating unstable manifolds of the attracting set $\mathcal{A}$. More precisely once the phase space is discretized its points will move towards the attracting set $\overline{\mathcal{A}}$ which will be a finite approximation of the attracting set $\mathcal{A}$ and will appear as arrays of points located on portions of unstable manifolds (which ones will depend on the details of the program for the simulation).

In each rectangle $E(\mathbf{q})$ of a coarse grained partition such arrays, will approximate the intersections of some of the unstable axes of $E(\mathbf{q})$, see p. 50, that we call $\delta(\mathbf{q})$ and whose union will be called $\Delta(\mathbf{q})$ (Fig. 3.7).

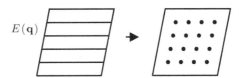

Fig. 3.7 A very schematic and idealized drawing of the intersections $\Delta(\mathbf{q})$ with $E(\mathbf{q})$ of the unstable surfaces which contain the microcells remaining, after a transient time, inside a coarse cell $E(\mathbf{q})$. The second drawing (indicated by the *arrow*) represents schematically the collections of microcells which are on the unstable surfaces which in $E(\mathbf{q})$ give a finite approximation of the attracting set, i.e. of the unstable surface elements $\Delta(\mathbf{q})$

(Footnote 8 continued)
precision required to give a meaning to the particles as separate entities. Each microcell could have coordinates represented with 32 bits will have size of the order of $\sqrt{mk_BT}2^{-32}$ in momentum and $\varrho^{-\frac{1}{3}}2^{-32}$ in position and the number of *theoretically possible* phase space points representable in the computer will be $O((2^{32})^{6N})$ which is obviously far too large to allow anything being close to a recurrence in essentially any simulation of a chaotic system involving more than $N = 1$ particle.

[9] With extreme care it is sometimes, and in equilibrium, possible to represent a chaotic evolution S with a code $\overline{S}$ which is a true permutation: the only example that I know, dealing with a physically relevant model, is in [14] .

(d) The evolution $\overline{S}$ will map the discrete arrays of microcells on the attracting set $\overline{A}$ into themselves. If $t \overset{def}{=} 2\tau$ then any number of unstable axes $\delta(\mathbf{q})$ in $E(\mathbf{q})$ is mapped by S^t to a surface fully covering an equal number of axes in *every other* $E(\mathbf{q}')$ provided $S^t E(\mathbf{q}) \cap E(\mathbf{q}')$ has an interior point (*i.e.* if some point internal to $E(\mathbf{q})$ evolves in time 2τ in a point internal to $E(\mathbf{q}')$).

(e) Every permutation can be decomposed into cycles: assuming that the microcells in the arrays $\cup_{\mathbf{q}} \Delta(\mathbf{q})$ take part in the same *one cycle permutation* is an analogue, and an extension, of the ergodic hypothesis for equilibrium (in the form that every microcell visits all others compatible with the constraints): *however this is not an innocent assumption* and, in the end, it is the reason why the SRB is unique.

Then consistency between expansion of the unstable directions and existence of a cyclic permutation of the microcells in the attracting set $\overline{A}$ puts *severe restrictions* on the number $\mathcal{N}(\mathbf{q})$ of microcells in each coarse grained cell $E(\mathbf{q})$, a fraction of the total number $\mathcal{N}$ of microcells in $\overline{A}$,

$$\mathcal{N}(\mathbf{q}) = \mathcal{N} \frac{w(\mathbf{q})}{\sum_{\mathbf{q}} w(\mathbf{q})} = \mathcal{N} \mu_{SRB}(E_{\mathbf{q}}) \tag{3.7.1}$$

which determine the weights $w(\mathbf{q})$ and, within the precision prefixed, the SRB distribution, as it will be discussed in Sect. 3.8.

The above viewpoint can be found in [2, 15–18].

3.8 The SRB Distribution: Its Physical Meaning

The determination of the weights $w(\mathbf{q})$ can be found through the following *heuristic argument*, [2, 18].

Let $t \overset{def}{=} 2\tau$ and call $n(\mathbf{q}')$ the number of unstable axes $\delta(\mathbf{q}')$ forming the approximate attracting set $\overline{A}$ (see Sect. 3.7) inside $E(\mathbf{q}')$, denoted $\Delta(\mathbf{q}')$ in the previous section.

The numerical density of microcells on the attracting set will be $\frac{N(\mathbf{q}')}{n(\mathbf{q}')|\delta_u(\mathbf{q}')|} = \frac{N(\mathbf{q}')}{\Delta(\mathbf{q}')}$, because the $n(\mathbf{q}')$ unstable axes $\delta_u(\mathbf{q}')$ have (approximately) the same surface $|\delta_u(\mathbf{q}')|$: the expanding action of S^τ will expand the unstable axes by a factor that will be written $e^{A_{u,\tau}(\mathbf{q}')}$, hence the density of microcells will decrease by the same factor. Thus the number of microcells that go from $E(\mathbf{q}')$ to $E(\mathbf{q})$ equals $\frac{N(\mathbf{q}')}{n(\mathbf{q}')|\delta_u(\mathbf{q}')|} e^{-A_{u,\tau}(\mathbf{q}')} |\delta_u(\mathbf{q})| n(\mathbf{q}')$, because the number of axes of $E(\mathbf{q}) \cap \overline{A}$ covered by the images of axes in $E(\mathbf{q}') \cap \overline{A}$, by the Markovian property of the pavement, is $n(\mathbf{q}')$, provided $S^\tau E(\mathbf{q}') \cap E(\mathbf{q})$ has an interior point.

The next remark is that $n(\mathbf{q}) = n(\mathbf{q}')$. The $n(\mathbf{q}')$ axes $\delta(\mathbf{q}')$ will be mapped into $n(\mathbf{q}')$ axes in $E(\mathbf{q})$ (if $S^t E(\mathbf{q}') \cap E(\mathbf{q})$ have an interior point, i.e. if a transition from $E(\mathbf{q}')$ to $E(\mathbf{q})$ is possible at all). The invariance of the approximate attracting set $\overline{A}$ implies that the numbers $n(\mathbf{q})$ are independent of $\mathbf{q}$ otherwise after a number of iterations of $\overline{S}$ greater than the mixing time the number of axes in $E(\mathbf{q})$ will be larger than at the beginning, against the invariance of $\overline{A}$.

Hence the fraction of points initially in $E(\mathbf{q}')$ that ends up in $E(\mathbf{q})$ is $\nu(\mathbf{q}, \mathbf{q}') \overset{def}{=} \frac{1}{|\Delta(\mathbf{q}')|} \frac{1}{e^{\Lambda_{u,\tau}(\mathbf{q}')}} |\Delta(\mathbf{q})|$. Then consistency with evolution as a cyclic permutation is expressed as

$$N(\mathbf{q}) = \sum_{\mathbf{q}'}^{*} \nu_{\mathbf{q},\mathbf{q}'} N(\mathbf{q}') \equiv \sum_{\mathbf{q}'}^{*} \frac{N(\mathbf{q}')}{|\Delta(\mathbf{q}')|} \frac{1}{e^{\Lambda_{u,\tau}(\mathbf{q}')}} |\Delta(\mathbf{q})| \tag{3.8.1}$$

where the $*$ signifies that $S^t E(\mathbf{q}') \cap E(\mathbf{q})$ must have an interior point in common, so that $\sum_{\mathbf{q}}^{*} \nu_{\mathbf{q},\mathbf{q}'} = 1$.

Hence the density $\varrho(q) \overset{def}{=} \frac{N(\mathbf{q})}{\Delta(\mathbf{q})}$ satisfies Eq. (3.8.1), i.e.:

$$\varrho(\mathbf{q}) = \sum_{\mathbf{q}'}^{*} e^{-\Lambda_{u,\tau}(\mathbf{q}')} \varrho(\mathbf{q}') \overset{def}{=} (\mathcal{L}\varrho)(\mathbf{q}) \tag{3.8.2}$$

closely related to the similar equation for invariant densities of Markovian surjective maps of the unit interval, [10].

Remark For later reference it is useful to mention that the expansion per time step at a point $x \in E(\mathbf{q})$ for the map S along the unstable manifold is given by the determinant of the matrix $\partial^u S(x)$, giving the action of the Jacobian matrix $\partial S(x)$ on the vectors of the unstable manifold at x (i.e. it is the restriction of the Jacobian matrix to the space of the unstable vectors); it has a logarithm:

$$\lambda^u(x) \overset{def}{=} \log |\det \partial^u S(x)| \tag{3.8.3}$$

and the expansion at x for the map $S^{2\tau}$ as a map from $S^{-\tau}x$ to $S^\tau x$ is $\Lambda^u(x, \tau) = \log |\det(\partial^u S^{2\tau}(S^{-\tau}x))|$, is

$$\Lambda_{u,\tau}(\mathbf{q}) = \sum_{j=-\tau}^{\tau-1} \lambda_u(S^j x) \tag{3.8.4}$$

by composition of differentiation.

For m larger than the symbolic mixing time the matrix $(\mathcal{L}^m)_{\mathbf{q},\mathbf{q}'}$ has all elements > 0 (because $S^{m\tau} E(\mathbf{q}')$ intersects all $E(\mathbf{q})$ for $\tau > m$, p. 10), and therefore has a simple eigenvector v with positive components to which corresponds the eigenvalue λ with maximum modulus: $v = \lambda \mathcal{L}(v)$ (the "Perron-Frobenius theorem", [10, Problem 4.1.17]) with $\lambda = 1$ (because $\sum_{\mathbf{q}}^{*} \nu_{\mathbf{q},\mathbf{q}'} = 1$). It follows that the consistency requirement uniquely determines $\varrho(\mathbf{q})$ as proportional to $v_{\mathbf{q}}$ and

$$w(\mathbf{q}) = v_{\mathbf{q}}, \qquad \mu_{SRB}(E(\mathbf{q})) = h_l(\mathbf{q})e^{-\Lambda_{u,\tau}(\mathbf{q})}h_r(\mathbf{q}) \qquad (3.8.5)$$

where h_l, h_r are functions of τ and of the symbols in $\mathbf{q}$, which essentially depend only on the first few symbols in the string $\mathbf{q}$ with label close to $-\tau$ or close to τ, respectively, and are uniformly bounded above and below in τ, as it follows from the general theory of equations like Eq. (3.8.2), [19], which gives an exact expression for $v_{\mathbf{q}}$ and μ_{SRB}, Eq. (3.8.7) below.

It should be noticed that the uniform boundedness of h_l, h_r imply (from Eq. 3.8.5)

$$\frac{1}{\tau}\log\sum_{\mathbf{q}}e^{-\Lambda_{u,\tau}(\mathbf{q})} = O(\frac{1}{\tau}) \qquad (3.8.6)$$

reflecting a further result on the theory of SRB distributions, "Pesin's formula", [20, p. 697], [10, Prop. 6.3.4]. This completes the heuristic theory of SRB distributions.

If more observables need to be considered it is always possible to refine the coarse graining and even take the limit of infinitely fine coarse graining:

$$\langle F\rangle_{SRB} = \lim_{\tau\to\infty}\frac{\sum_{\mathbf{q}}e^{-\Lambda_{u,\tau}(\mathbf{q})}\,F(x_{\mathbf{q}})}{\sum_{\mathbf{q}}e^{-\Lambda_{u,\tau}(\mathbf{q})}} \qquad (3.8.7)$$

which is an *exact formula for* μ_{SRB}: the limit can be shown to exist for all choices of (continuous) F, of the particular Markovian partitions $\mathcal{E}$ used for the coarse graining and of the choice of the center $x_{\mathbf{q}}$ in $E(\mathbf{q})$ (rather arbitrarily picked up), [10].

The above viewpoint can be found in [2, 15, 16], [17, p. 684], [18].

3.9 Other Stationary Distributions

So the SRB distribution arises naturally from assuming that dynamics can be discretized on a regular array of points ("microcells") and becomes a one cycle permutation of the microcells on the attracting set. This is so under the chaotic hypothesis and *holds whether the dynamics is conservative (Hamiltonian) or dissipative*.

It is, however, well known that hyperbolic systems admit (uncountably) many invariant probability distributions, besides the SRB. This can be seen by noting that the space of the configurations is identified with a space of compatible sequences, Sect. 3.4.

On such a space uncountably many stochastic processes can be defined, for instance by assigning an arbitrary short range translation invariant potential, and regarding the corresponding Gibbs state as a probability distribution on phase space, [10, Sect. 5].

Yet the analysis just presented apparently singles out SRB as the unique invariant distribution. This is due to our assumption that, in the discretization, microcells are regularly spaced and centered on a regular discrete lattice and evolution eventually permutes them in a (single, by transitivity) cycle consisting of the microcells located on the attracting set (and therefore locally evenly spaced, as inherited from the regularity of the phase space discretization and from the smoothness of the attracting set and of the unstable manifolds).

Other invariant distributions can be obtained by custom made discretizations of phase space which will not cover the attracting set in a regular way. This is what is done when defining the choice of the initial data if other distributions, "not absolutely continuous with respect to the phase space volume", are studied in simulations.

A paradigmatic example is given by the map $S : x \rightarrow 3x$ mod 1: it has an invariant distribution $\widetilde{\mu}$ attributing zero probability to the points x that, in base 3, contain the digit 2: it can be found in a simulation by writing a program in which data have this property and make sure that the round off errors will not destroy it. Almost any "naive" code that simulates this dynamics using double precision reals represented in base 2 will generate, *instead* (due to round-off truncations), the SRB distribution μ for S and $\mu \neq \widetilde{\mu}$: the latter is simply the Lebesgue measure on the unit interval (which on the symbolic dynamics is the Bernoulli's process attributing equal probability $\frac{1}{3}$ to each digit).

The physical representation of the SRB distribution just obtained, see [2, 21], shows that there is no conceptual difference between stationary states in equilibrium and out of equilibrium. If motions are chaotic, in both cases they are permutations of microcells and the *SRB distribution is simply equidistribution over the recurrent microcells*, provided the microcells are *uniformly spread* in phase space. In equilibrium this gives the Gibbs microcanonical distribution and out of equilibrium it gives the SRB distribution (of which the Gibbs' distribution is a very special case).

The above heuristic argument is an interpretation of the mathematical proofs behind the SRB distribution which can be found in [10, 22]. Once Eq. (3.8.5) is given, the expectation values of the observables in the SRB distributions can be formally written as sums over suitably small coarse cells and symmetry properties inherited from symmetries of the dynamics become transparent and can (and will) be used in the following to derive universal properties of the stationary states (for instance extending to systems in non equilibrium the Onsager's reciprocity derived infinitesimally close to equilibrium from the basic time reversal symmetry).

3.10 Phase Space Cells and Entropy

The discrete representation, in terms of coarse grain cells and microcells leads to the possibility of counting the number $\mathcal{N}$ of the microcells on the attracting set and therefore to define a kind of entropy function: see [16].

Consider a smooth, transitive, hyperbolic system S on a bounded phase space Ξ (i.e. an Anosov system). Let μ_{SRB} be the SRB distribution describing the asymptotic behavior of almost all initial data in phase space (in the sense of the volume measure).

As discussed above the SRB distribution admits a rather simple representation which can be interpreted in terms of *"coarse graining"* of the phase space.

Let $\mathcal{E}$ be a "Markov partition" of phase space $\mathcal{E} = (E_1, \ldots, E_k)$ with sets E_j, see p. 52. Let τ be a time such that the size of the $E(\mathbf{q}) = \cap_{j=-\tau}^{\tau} S^{-j} E_{q_j}$ is so small that the physically interesting observables can be viewed as constant inside $E(\mathbf{q})$, so that $\mathcal{E}_\tau$ can be considered a coarse grained partition of phase space, see Sect. 3.6, p. 55.

Then the SRB probability $\mu(E(\mathbf{q}))$ of $E(\mathbf{q})$ is described in terms of the functions $\lambda^u(x) = \log|\det(\partial S)_u(x)|$, Eq. (3.8.2), and the expansion rates $\Lambda_u(x, \tau)$ in Eq. (3.8.5). Here $(\partial S)_u(x)$ (resp. $(\partial S)_s(x)$) is the Jacobian of the evolution map S restricted to the unstable (stable) manifold through x and mapping it to the unstable (stable) manifold through Sx. Selecting a point $x_\mathbf{q} \in E(\mathbf{q})$ for each $\mathbf{q}$, the SRB distribution is given approximately by Eq. (3.8.5) or exactly by Eq. (3.8.7).

Adopting the discrete viewpoint on the structure of phase space, Sect. 3.7, regard motion as a cyclic permutation of microcells and ask on general grounds the question, [16]:

Can we count the number of ways in which the asymptotic SRB state of the system can be realized microscopically?

This extends the question asked by Boltzmann for the equilibrium case in [3, #39], as

In reality one can compute the ratio of the numbers of different initial states which determines their probability, which perhaps leads to an interesting method to calculate thermal equilibria

and answered in [1, #42, p. 166], see Sects. 1.6 and 6.12.

In equilibrium the (often) accepted answer is simple: the number is $\mathcal{N}_0$, i.e. just the number of microcells ("ergodic hypothesis"). This means that we think that dynamics will generate a one cycle permutation of the $\mathcal{N}_0$ microcells on phase space $\mathcal{E}$ (which in this case is also the attracting set), each of which is therefore, representative of the equilibrium state. And the average values of macroscopic observables are obtained simply as:

$$\langle F \rangle = \mathcal{N}_0^{-1} \sum_{\mathbf{q} \in \mathcal{E}_\tau} F(x_\mathbf{q}) \sim \int_{\mathcal{E}} F(y) \mu_{SRB}(dy) \qquad (3.10.1)$$

If W denotes the volume in phase space of the region consisting in the union of the microcells that overlap with the surface where the sum of kinetic energy K plus the potential energy U has a value E while the positions of the particles are confined within a container of volume V then, imagining the phase space discretized into microcells of phase space volume h^{3N}, according to Boltzmann, see for instance p. 372 in [5], the quantity:

$$S_B \overset{def}{=} k_B \log \frac{W}{h^{3N}} \qquad (3.10.2)$$

i.e. k_B times the logarithm of the total number of microcells is, under the ergodic hypothesis (each microcell visits all the others), proportional to the *physical entropy*

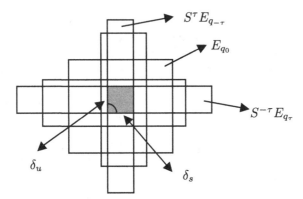

Fig. 3.8 The *shadowed region* represents the intersection $\cap_{-\tau}^{\tau} S^{-j} E_{q_j}$; the angle φ between the stable axis of $S^{\tau} E_{q_{-\tau}}$ and the unstable axis of $S^{-\tau} E_{q_{\tau}}$ is marked in the dashed region (in general it is not 90°) around a corner of the *rectangle* $E(\mathbf{q})$

of the equilibrium state with N particles and total energy E, see [1, #42], (up to an additive constant independent of the state of the system).[10]

A simple extension to systems out of equilibrium is to imagine, as done in the previous sections, that a similar kind of "ergodicity" holds: namely that the microcells that represent the stationary state form a subset of all the microcells, on which evolution acts as a one cycle permutation and that entropy is defined by $k_B \log \mathcal{N}$, with $\mathcal{N}$ being the number of phase space cells *on the attracting set*, which in general will be $\ll \mathcal{N}_0$, if $\mathcal{N}_0$ is the number of regularly spaced microcells in the phase space region compatible with the constraints.

To proceed it is necessary to evaluate the ratio between the fraction of W of the coarse cell $E(\mathbf{q})$, namely $\frac{|E(\mathbf{q})|}{W}$, and the SRB probability $\mu_{SRB}(E(\mathbf{q}))$.

3.11   Counting Phase Space Cells Out of Equilibrium

The ratio between the fraction of available phase space $\frac{|E(\mathbf{q})|}{W}$ and the SRB probability $\mu_{SRB}(E(\mathbf{q}))$ can be estimated heuristically by following the ideas of the previous Sect. 3.8. For this purpose remark that the elements $E(\mathbf{q}) = \cap_{j=-\tau}^{\tau} S^{-j} E_{q_j}$ generating the Markovian partition $\mathcal{E}_\tau$ can be symbolically represented as Fig. 3.8.

The surfaces of the expanding axis of $S^{-\tau} E_{q_\tau}$ and of the stable axis of $S^{\tau} E_{q_{-\tau}}$, indicated with δ_u, δ_s in Fig. 3.8 are (approximately)

$$\delta_u = e^{-\sum_{j=1}^{\tau} \lambda_u(S^j x)} \delta_u(q_\tau), \qquad \delta_s = e^{\sum_{j=1}^{\tau} \lambda_s(S^{-j} x)} \delta_s(q_{-\tau}) \qquad (3.11.1)$$

where $\delta_s(q_{-\tau})$, $\delta_u(q_\tau)$ are the surfaces of the stable axis of $E_{q_{-\tau}}$, and of the unstable axis of E_{q_τ}, respectively.

---

[10] However in [1, #42] the w's denote integers rather than phase space volumes.

From the figure it follows that

$$\frac{|E(\mathbf{q})|}{W} = \frac{\delta_s(q_{-\tau})\delta_u(q_\tau)\sin\varphi}{W} e^{-\sum_{j=0}^{\tau-1}(\lambda_u(S^j x) - \lambda_s(S^{-j}x))} \tag{3.11.2}$$

where φ is the angle at x between $W^s(x)$ and $W^u(x)$ while $\mu_{SRB}(\mathbf{q})$ is given by Eq. (3.8.5).

A nontrivial property which emerges from the above formula is that $\frac{\delta_s(q_{-\tau})\delta_u(q_\tau)\sin\varphi}{W}$ is bounded above and below as soon as it is $\neq 0^{11}$: hence if 2τ is not smaller than the symbolic mixing time. Since $\sum_{\mathbf{q}} \frac{|E(\mathbf{q})|}{W} = 1$ this implies again (see Eq. 3.8.6) a kind of Pesin's formula[12]:

$$\log \sum_{\mathbf{q}} e^{-\sum_{j=0}^{\tau-1}(\lambda_u(S^j x) - \lambda_s(S^{-j}x))} = O(1), \qquad \forall \tau \tag{3.11.3}$$

Then $\frac{|E(\mathbf{q})|}{h^{6N}}$ is larger than the number of microcells in the attractor inside $E(\mathbf{q})$: i.e. $\frac{|E(\mathbf{q})|}{h^{6N}} \geq \mathcal{N}\mu_{SRB}(\mathbf{q})$ where h is the size of the microcells: thus $\mathcal{N}_0 h^{3N} \equiv W$ implies $\mathcal{N} \leq \mathcal{N}_0 \frac{|E(\mathbf{q})|}{W\mu_{SRB}(\mathbf{q})}$.

Therefore $\mathcal{N} \leq \mathcal{N}_0 \times$ the ratio of the r.h.s. of Eq. (3.11.2) to the r.h.s. of Eq. (3.8.5) which, by Eq. 3.8.4, is proportional to $e^{\sum_{j=0}^{\tau-1}(\lambda_s(S^{-j}x) + \lambda_u(S^{-j}x))}$ up to a factor bounded independently of $\mathbf{q}$ away from 0 and ∞ for τ larger than the mixing symbolic time (see p. 53) (because of the remarked consequences of Pesin's formula). Hence

$$\mathcal{N} \leq \mathcal{N}_0 \min_{\mathbf{q}} e^{\sum_{j=0}^{\tau-1}(\lambda_s(S^{-j}x(\mathbf{q})) + \lambda_u(S^{-j}x(\mathbf{q})))} \leq \mathcal{N}_0 e^{-\sigma_+\tau} \tag{3.11.4}$$

because for x on the attractor the quantity $-\sum_{j=0}^{\tau}(\lambda_s(S^j x) + \lambda_u(S^j x))$ has average $-\tau\sigma_+$ with σ_+ = the average phase space contraction.

The picture must hold for all Markovian pavements $\mathcal{E}$ and for all τ's such that the coarse grain cells contain a large number of microcells: i.e. if δ_p, δ_q are the typical sizes in momentum or, respectively, in position of an element of the partition $\mathcal{E}$, for $e^{-\lambda\tau}\delta_p \gg \delta p, e^{-\lambda\tau}\delta_q \gg \delta q$ with λ the maximal contraction of the stable and unstable manifolds under S or, respectively, S^{-1} and $(\delta p\, \delta q)^{3N} = h^{3N}$ is the size of a microcell.

Fix $\tau \leq \bar\tau = \lambda^{-1}\log\theta$ with $\theta = \min(\frac{\delta p}{\delta_p}, \frac{\delta q}{\delta_q})$. So that

---

[11] Simply because q_j have finitely many values, W is fixed and the angle $\varphi = \varphi(x)$ between stable and unstable manifolds at x is bounded away fro 0, π because of the transversality of the manifolds (in Anosov maps).

[12] Informally Pesin's formula is $\sum_{q_0,q_1,\ldots,q_N} e^{-\lambda_u(S^j x)} = O(1)$, see Eq. (3.8.6) and, formally, $s(\mu_{srb}) - \mu_{srb}(\lambda_u) = 0$, where $s(\mu)$ is the Kolmogorov-Sinai entropy, see p. 73, and $\mu(\lambda) \overset{def}{=} \int \lambda\, d\mu$. Furthermore $s(\mu) - \mu(\lambda_u)$ is maximal at $\mu = \mu_{srb}$: "Ruelle's variational principle". See p. 60 and [10, Proposition 6.3.4].

$$S_{cells} = k_B \log \mathcal{N} \le k_B (\log \mathcal{N}_0 - \frac{\sigma_+}{\lambda} \log \theta) \qquad (3.11.5)$$

This inequality does not prove, without extra assumptions, that S_{cells} will depend nontrivially on $\theta, \lambda, \sigma_+$ when $\sigma_+ > 0$. It gives, however, an indication[13] that S_{cells} might not be independent of the precision θ used in defining the microcells and the course grained cells; and the dependence might not be simply an additive constant because σ_+/λ is a dynamical quantity; changing θ to θ' (i.e. our representation of the microscopic motion).

This is in sharp contrast with the equilibrium result that changing the precision changes $\log \mathcal{N}_0$ by a constant independent of the equilibrium state (as in equilibrium the number of microcells changes by $3N \log \frac{h'}{h}$ if the size h^{3N} is changed to h'^{3N}).

Given a precision θ of the microcells, the quantity S_{cells} measures how many "non transient" microcells must be used, in a discretization of phase space, to obtain a *faithful* representation of the attracting set and of its statistical properties on scales $\delta p, \delta q \gg \delta p, \delta q$. Here by "faithful" on scale $\delta p, \delta q$ it is meant that all observables which are constant on such scale will show the correct statistical properties, i.e. that coarse cells of size much larger than θ will be visited with the correct SRB frequency.

3.12 $k_B \log \mathcal{N}$: Entropy or Lyapunov Function?

From the previous sections some conclusions can be drawn.

(1) Although S_{cell} (see Eq. (3.11.5)) gives the cell count it does not seem to deserve to be taken also as a definition of entropy for statistical states of systems out of equilibrium, not even for systems simple enough to admit a transitive Anosov map as a model for their evolution. The reason is that it might not change by a trivial additive constant if the size of the microcells is varied (except in the equilibrium case): the question requires further investigation. It also seems to be a notion distinct from what has become known as the "Boltzmann's entropy", [6, 23].

(2) S_{cell} is also different from the Gibbs' entropy, to which it is equivalent only in equilibrium systems: in nonequilibrium (dissipative) systems the latter can only be defined as $-\infty$ and perpetually decreasing; because *in such systems one can define the rate at which (Gibbs') entropy is "generated" or "ceded to the thermostats" by the system* to be σ_+, i.e. to be the average phase space contraction $\sigma_+ > 0$, see [24, 25].

(3) We also see, from the above analysis, that the SRB distribution appears to be the equal probability distribution among the $\mathcal{N}$ microcells which are not transient.[14] Therefore $S_{cell} = k_B \log \mathcal{N}$ maximizes a natural functional of the probability distributions $(\pi_x)_{x \in \overline{\mathcal{E}}}$ defined on the discretized approximation of the attractor $\overline{\mathcal{E}}$; namely the functional defined by $S(\pi) = -k_B \sum_x \pi_x \log \pi_x$. Even though S_{cells} does not seem interpretable as a function of the stationary state *it can nevertheless be con-*

---

[13] I would say a strong one.

[14] In equilibrium, under the ergodic hypothesis, all microcells are non transient and the SRB distribution coincides with the Liouville distribution.

sidered a Lyapunov function which estimates how far a distribution is from the SRB distribution and reaches it maximum on the SRB distribution.

(4) If we could take $\tau \to \infty$ (hence, correspondingly, $h, \theta \to 0$) the distribution μ which attributes a total weight to $E(\mathbf{q})$ equal to $N(\mathbf{q}) = \mu_{SRB}(E(\mathbf{q})).\mathcal{N}$ would become the exact SRB distribution. However it seems conceptually more satisfactory to suppose that τ will be large but not infinite.

References

1. Boltzmann, L.: Über die Beziehung zwischen dem zweiten Hauptsatze der mechanischen Wärmetheorie und der Wahrscheinlichkeitsrechnung, respektive den Sätzen über das Wärmegleichgewicht, vol. 2, #42 of Wissenschaftliche Abhandlungen, ed. Hasenöhrl, F. Chelsea, New York, (1968)
2. Gallavotti, G.: Ergodicity, ensembles, irreversibility in Boltzmann and beyond. J. Stat. Phys. **78**, 1571–1589 (1995)
3. Boltzmann, L.: Bemerkungen über einige Probleme der mechanischen Wärmetheorie, vol. 2, #39 of Wissenschaftliche Abhandlungen, ed. Hasenöhrl, F. Chelsea, New York (1877)
4. Thomson, W.: The kinetic theory of dissipation of energy. Proc. R. Soc. Edinb. **8**, 325–328 (1874)
5. Boltzmann, L.: Lectures on Gas Theory. English edition annotated by Brush S. University of California Press, Berkeley (1964)
6. Lebowitz, J.L.: Boltzmann's entropy and time's arrow. Phys. Today **46**(9), 32–38, (1993)
7. Garrido, P.L., Goldstein, S., Lebowitz, J.L.: Boltzmann entropy for dense fluids not in local equilibrium. Phys. Rev. Lett. **92**, 050602 (+4) (2005)
8. Boltzmann, L.: Weitere Studien über das Wärmegleichgewicht unter Gasmolekülen, vol. 1, #22 of Wissenschaftliche Abhandlungen, ed. Hasenöhrl, F. Chelsea, New York (1968)
9. Katok, A., Hasselblatt, B.: Introduction to the modern theory of dynamical systems, vol. 54 of Encyclopedia of Mathematics and Its Applications. Cambriidge University Press, Cambridge (1997)
10. Gallavotti, G., Bonetto, F., Gentile, G.: Aspects of the Ergodic, Qualitative and Statistical Theory of Motion. Springer, Berlin (2004)
11. Ruelle, D.: Elements of Differentiable Dynamics and Bifurcation Theory. Academic Press, New York (1989)
12. Sinai, Y.G.: Markov partitions and C-diffeomorphisms. Funct. Anal. Appl. **2**(1), 64–89 (1968)
13. Bowen, R.: Markov partitions for axiom A diffeomorphisms. Am. J. Math. **92**, 725–747 (1970)
14. Levesque, D., Verlet, L.: Molecular dynamics and time reversibility. J. Stat. Phys. **72**, 519–537 (1993)
15. Gallavotti, G.: New methods in nonequilibrium gases and fluids. Open Syst. Inf. Dyn. **6**, 101–136 (1999) (preprint chao-dyn/9610018)
16. Gallavotti, G.: Counting phase space cells in statistical mechanics. Commun. Math. Phys. **224**, 107–112 (2001)
17. Gallavotti, G.: Entropy production in nonequilibrium stationary states: a point of view. Chaos **14**, 680–690 (2004)
18. Gallavotti, G.: Heat and fluctuations from order to chaos. Eur. Phys. J. B **61**, 1–24 (2008)
19. Ruelle, D.: Statistical mechanics of one-dimensional lattice gas. Commun. Math. Phys. **9**, 267–278 (1968)
20. Jiang, M., Pesin, Y.B.: Equilibrium measures for coupled map lattices: existence, uniqueness and finite-dimensional approximations. Commun. Math. Phys. **193**, 675–711 (1998)
21. Gallavotti, G.: Statistical Mechanics. A Short Treatise. Springer, Berlin (2000)

22. Bowen, R.: Equilibrium states and the ergodic theory of Anosov diffeormorphisms. Lecture Notes in Mathematics, vol. 470. Springer, Berlin (1975)
23. Einstein, E.: Zur Theorie des Radiometers. Annalen der Physik **69**, 241–254 (1922)
24. Andrej, L.: The rate of entropy change in non-Hamiltonian systems. Phys. Lett. **111A**, 45–46 (1982)
25. Ruelle, D.: Smooth dynamics and new theoretical ideas in non-equilibrium statistical mechanics. J. Stat. Phys. **95**, 393–468 (1999)

Chapter 4
Fluctuations

4.1 SRB Potentials

Intuition about the SRB distributions, hence about the statistics of chaotic evolutions, requires an understanding of their nature. The physical meaning discussed in Sect. 3.8 is not sufficient because the key notion of *SRB potentials* is still missing. It is therefore time to introduce it.

Given a smooth hyperbolic transitive evolution S on a phase space Ξ (i.e. a Anosov map) consider a Markovian partition $\mathcal{E} = (E_1, \ldots, E_k)$: given a coarse cell $E(\mathbf{q}) = \cap_{i=-\tau}^{\tau} S^{-i} E_{q_i}$, $q_i = 1, 2, \ldots, k$, see Sect. 3.6, the time average of any smooth observable can be computed by the formula Eq. (3.8.7):

$$\langle F \rangle_{SRB} = \lim_{\tau \to \infty} \frac{\sum_{\mathbf{q}} e^{-\Lambda_{u,\tau}(\mathbf{q})} F(x_{\mathbf{q}})}{\sum_{\mathbf{q}} e^{-\Lambda_{u,\tau}(\mathbf{q})}} \tag{4.1.1}$$

which is an exact formula for μ_{SRB}: here $\Lambda_{u,\tau}(\mathbf{q})$ is defined in terms of the function $\lambda_u(x) \overset{def}{=} \log | \det \partial^u S(x)|$ which gives the expansion rate of the area of the unstable manifold through x in one time step and of the function $\Lambda_u(x, \tau) = \sum_{j=-\tau}^{\tau-1} \lambda_u(S^j x)$ which gives the expansion rate of the unstable manifold at $S^{-\tau} x$ when it is transformed into the unstable manifold at $S^{\tau} x$ by the map $S^{2\tau}$. It is, Eq. 4.8.4,

$$\Lambda_{u,\tau}(\mathbf{q}) \overset{def}{=} \Lambda_u(x_{\mathbf{q}}, \tau) = \sum_{j=-\tau}^{\tau-1} \lambda_u(S^j x_{\mathbf{q}}) \tag{4.1.2}$$

where $x_{\mathbf{q}}$ is a point arbitrarily selected in $E(\mathbf{q})$.

Defining the SRB potentials is very natural in the case in which the compatibility matrix M as *no zero entry*, i.e. all transitions are allowed. In this (unrealistic) case the point $x_{\mathbf{q}}$ can be conveniently chosen to be the point whose symbolic representation is the sequence $\ldots, 1, 1, q_{-\tau}, \ldots, q_{\tau}, 1, 1, \ldots$ obtained by extending $\mathbf{q}$ to an infinite sequence by writing the symbol 1 (say) to the right and left of it. More generally given

G. Gallavotti, *Nonequilibrium and Irreversibility*,
Theoretical and Mathematical Physics, DOI: 10.1007/978-3-319-06758-2_4,
© Springer International Publishing Switzerland 2014

a finite sequence $\boldsymbol{\eta} = (\eta_c, \ldots, \eta_{c+d})$ it can be extended to an infinite compatible sequence $\overline{\boldsymbol{\eta}}$ by continuing it, right and left, by the symbol 1 (any other symbol would be equally convenient). Then if

$$\boldsymbol{\xi} = \{\xi_j\}_{j=-\infty}^{j=\infty}, \ \boldsymbol{\xi}_k \overset{def}{=} (\xi_{-k}, \ldots, \xi_k), \ \overline{\boldsymbol{\xi}}_k = (\ldots 1, 1, \xi_{-k}, \ldots, \xi_k, 1, 1, \ldots) \tag{4.1.3}$$

it is (trivially) possible to write $\lambda_u(\boldsymbol{\xi})$ as a sum of "finite range potentials":

$$\lambda_u(x_{\mathbf{q}}) = \sum_{k=0}^{\infty} \Phi(\xi_{-k}, \ldots, \xi_k) = \sum_{k=0}^{\infty} \Phi(\boldsymbol{\xi}_k), \tag{4.1.4}$$

where the potentials $\Phi(\boldsymbol{\xi}_k)$ are defined "telescopically" by

$$\Phi(\xi_0) = \lambda_u(\overline{\boldsymbol{\xi}}_0)$$
$$\Phi(\xi_{-k}, \ldots, \xi_k) = \lambda_u(\overline{\boldsymbol{\xi}}_k) - \lambda_u(\overline{\boldsymbol{\xi}}_{k-1}), \qquad k \geq 1 \tag{4.1.5}$$

Define also $\Phi(\eta_c, \ldots, \eta_{c+d}) = \Phi(\boldsymbol{\eta})$, d even, by setting $\Phi(\boldsymbol{\eta})$ equal to $\Phi(\boldsymbol{\eta}^{tr})$ with $\boldsymbol{\eta}^{tr}$ obtained from $\boldsymbol{\eta}$ by translating it to a string with labels centered at the origin ($\boldsymbol{\eta}^{tr} = (\eta_{-\frac{d}{2}}^{tr}, \ldots, \eta_{\frac{d}{2}}^{tr})$ with $\eta_j^{tr} = \eta_{c+j+\frac{d}{2}}$. For all other finite strings Φ will also be defined, but set $\equiv 0$.

In this way Φ will have been defined as a translation invariant potential which can be $\neq 0$ only for strings $\boldsymbol{\eta} = (\eta_c, \eta_{c+1}, \ldots, \eta_{c+d})$ (with d even). The $\Lambda_{u,\tau}(x_{\mathbf{q}})$ can then be written, from Eqs. (4.1.2), (4.1.4), (4.1.5), simply as

$$\Lambda_{u,\tau}(x_{\mathbf{q}}) = \sum_{j=-\tau}^{\tau} \sum_{k=0}^{\infty} \Phi((S^j \boldsymbol{\xi}_{\mathbf{q}})_k) = \sum_{\eta:\tau} \Phi(\boldsymbol{\eta}) \tag{4.1.6}$$

where $(\boldsymbol{\xi})_k$ denotes the string $(\xi_{-k}, \ldots, \xi_k)$ and $\sum_{\eta:\tau}$ denotes sum over the finite substrings $\boldsymbol{\eta} = (\eta_c, \eta_{c+1}, \ldots, \eta_{c+d})$ of $\boldsymbol{\xi}_{\mathbf{q}}$, i.e. $\boldsymbol{\eta} = (\eta_c, \eta_{c+1}, \ldots, \eta_{c+d})$ with $d+1$ odd and labels $(c, c+1, \ldots, c+d)$, centered in a point $c + \frac{1}{2}(d-1)$ in the interval $[-\tau, \tau]$.

The convergence of the series in Eq. (4.1.6) is implied by the remark that the two strings $\overline{\boldsymbol{\xi}}_k, \overline{\boldsymbol{\xi}}_{k-1}$ coincide at the positions labeled $-(k-1), \ldots, (k-1)$: the hyperbolicity then implies that the points with symbolic representations given by $\overline{\boldsymbol{\xi}}_k, \overline{\boldsymbol{\xi}}_{k-1}$ are close within $Ce^{-\lambda(k-1)}$, see the comment following Eq. (3.4.2).

Therefore, by the smoothness of the map S and the Hölder continuity of $\lambda_u(x)$, $\lambda_s(x)$, Eq. (3.2.3), there exists a constant $B, b > 0$ such that

$$|\Phi(\boldsymbol{\eta})| \leq B \, e^{-\lambda b k} \tag{4.1.7}$$

if k is the length of the string $\boldsymbol{\eta}$; which gives to the SRB averages the expression

$$\langle F \rangle_{SRB} = \lim_{\tau \to \infty} \frac{\sum_{\mathbf{q}} e^{-\sum_{\eta:\tau} \Phi(\eta)} F(x_{\mathbf{q}})}{\sum_{\mathbf{q}} e^{-\sum_{\eta:\tau} \Phi(\eta)}} \tag{4.1.8}$$

where η are the finite substrings of $\overline{\xi}_{\mathbf{q}}$ and $\sum_{\eta:\tau}$ means sum over the $\eta = (\eta_c, \ldots, \eta_{c+d})$ with d even and center of $c, \ldots, c+d$ at a point in $[-\tau, \tau]$.

In the cases in which the compatibility matrix contains some 0 entries (corresponding to "transitions forbidden in one time step") the above representation of the SRB distribution can be still carried out essentially unchanged by taking advantage of the transitivity of the matrix M.

The choice of $x_{\mathbf{q}}$ will be fixed by remarking that, given any compatible string $\mathbf{q}$ (i.e. such that $E(\mathbf{q}) \neq \emptyset$), it can be extended to an infinite compatible sequence $\boldsymbol{\xi}_{\mathbf{q}} = (\ldots, \xi_{-\tau-1}, q_{-\tau}, \ldots, q_\tau, \xi_{\tau+1}, \ldots)$ in a "standard way" as follows.

If m is the symbolic mixing time, see p. 53, for the compatibility matrix, for each symbol q fix a string $\mathbf{a}(q))$ of length m of symbols leading from q to a prefixed (once and for all) symbol $\overline{q}$ and a string $\mathbf{b}(q)$ of length m of symbols leading from $\overline{q}$ to q[1]; then continue the string $q_{-\tau}, \ldots, q_\tau$ by attaching to it $\mathbf{a}(q)$ to the right and $\mathbf{b}(q)$ to the left; and finally continue the so obtained string of length $2m + 2\tau + 1$ to a compatible infinite string in a prefixed way, to the right starting from $\overline{q}$[2] and to the left ending in $\overline{\xi}$. The sequence $\boldsymbol{\xi}_{\mathbf{q}}$ determines uniquely a point $x_{\mathbf{q}}$.

In general given a finite string $\eta = (\eta_c, \eta_{c+1}, \ldots, \eta_{c+d})$ it can be continued to an infinite string by continuing it to the right and to the left in a standard way as above: the standard continuation of η to an infinite string will be denoted $\overline{\eta}$, leaving Eq. (4.1.8) unchanged after reinterpreting in this way the strings $\overline{\xi}_k$ and $\overline{\xi}_{k-1}$ in the definition Eq. (4.1.5).

4.2 Chaos and Markov Processes

In the literature many works can be found which deal with nonequilibrium theory and which are modeled by Markov processes introduced either as fundamental models or as approximations of deterministic models.

Very often this is criticized because of the a priori stochasticity assumption which, to some, sounds as introducing *ex machina* the key property that should, instead, be derived.

The proposal of Ruelle on the theory of turbulence, see [1, 2] but in fact already implicit in his earlier works [3], has been the inspiration of the chaotic hypothesis (Sect. 2.7).

---

[1] This means fixing for each q two strings $\mathbf{a}(q) = \{a_0 = q, a_1, \ldots, a_{m-1}, a_m = \overline{q}\}$ and $\mathbf{b}(q) = \{b_0 = \overline{q}, b_1, \ldots, b_{m-1}, b_m = q\}$ such that $\prod_{i=0}^{m-1} M_{a_i, a_{i+1}} = \prod_{i=0}^{m-1} M_{b_i, b_{i+1}} = 1$.

[2] The continuation has to be done choosing arbitrarily but once and for all a compatible sequence starting with $\overline{q}$: the simplest is to repeat indefinitely a string of length m beginning and ending with $\overline{q}$ to the right and to the left.

The coarse graining theory that follows from the chaotic hypothesis essentially explains why there is little difference between Markov chains evolutions and chaotic evolutions. It is useful to establish a precise and general connection between the two.

From the general expression for the SRB distribution it follows that *if $\Phi(\eta) \equiv 0$ for strings of length $k > k_0$*, i.e. if the potential Φ has finite range rather than an exponential decay to 0 as in Eq. (4.1.6), then the limit in Eq. (4.1.8) would exist (independently of the arbitrariness of the above standard choice of the string representing x_q). And it would be a finite memory, transitive[3] Markov process, therefore equivalent to an ordinary Markov process.

The long range of the potential does not really affect the picture: technically infinite range processes with potentials decaying exponentially fast are in the larger class of stochastic processes known as 1–dimensional *Gibbs distributions*: they have essentially the same properties as Markov chains. In particular the limit in Eq. (4.1.8) does not depend on the arbitrariness of the above standard choice of the string ξ_q), representing x_q), [4].

Furthermore they are translation invariant and exponentially fast mixing in the sense that if $S^t F(x) \overset{def}{=} F(S^t x)$:

$$\mu_{SRB}(S^t F) = \mu_{SRB}(F), \qquad \text{for all } t \in Z$$
$$|\mu_{SRB}(F S^t G) - \mu_{SRB}(F)\mu_{SRB}(G)| \leq \gamma \, ||F|| \, ||G|| \, e^{-\kappa|t|} \qquad (4.2.1)$$

for all $t \in Z$ and for suitable constants γ, κ which depend on the regularity of the functions F, G if they are at least Hölder continuous.

The dimension 1 of the SRB process is remarkable because it is only in dimension 1 that the theory of Gibbs states with exponentially decaying potential is elementary and easy. And nevertheless a rather general deterministic evolution has statistical properties identical to those of a Markov process if, as usual, the initial data are chosen close to an attracting set and outside a zero volume set in phase space.

The result might be at first sight surprising: however it should be stressed that the best sequences of random numbers are precisely generated as symbolic histories of chaotic maps: for a simple, handy and well known although not the best, example see [5, p. 46].

Thus it is seen that the apparently different approaches to nonequilibrium based on Markovian models or on deterministic evolutions are in fact completely equivalent, at least in principle. Under the chaotic hypothesis for any deterministic model a Gibbs process with short range potential Φ which decays exponentially can be constructed equivalent to it, via an algorithm which, in principle, is constructive (because the Markov partitions can be constructed, in principle).

Furthermore truncating the potential Φ to its values of range $<k_0$ with k_0 large enough will approximate the Gibbs process with a finite memory Markov process.

---

[3] Because of transitivity of the compatibility matrix.

The approximation can be pushed as far as wished in may senses, for instance in the sense of "distribution and entropy", see [6],[4] which implies that the process is a "Bernoulli's process".

That is, the symbolic sequences ξ could even be coded, outside a set of μ_{SRB} probability 0, into new sequences of symbols in which the symbols appear without any compatibility restriction and with independent probabilities, [7, 8].

It is also remarkable that the expression for the SRB distribution is the same for systems in equilibrium or in stationary nonequilibrium: it has the form of a Gibbs' distribution of a one dimensional lattice system.

4.3 Symmetries and Time Reversal

In chaotic systems the symbolic dynamics inherits naturally symmetry properties enjoyed by the time evolution (if any). An interesting property is, as an example in particle systems, the standard time reversal antisymmetry (velocity reversal with positions unchanged): it is important because it is a symmetry of nature and it is present also in the models considered in Sect. 3.2, which is the reason why they attracted so much interest.

Time reversal is, in general, defined as a smooth *isometric* map I of phase space which "anticommutes" with the evolution S, namely $IS = S^{-1}I$, and which squares to the identity $I^2 = 1$.

If $\mathcal{E}_0$ is a Markovian partition then also $I\mathcal{E}_0$ has the same property because $IW^s(x) = W^u(Ix)$ and $IW^u(x) = W^s(Ix)$ so that Eq. (3.4.1) hold. Since a time reversal anticommutes with evolution it will be called a "reverse" symmetry.

It is then possible to consider the new Markovian partition $\mathcal{E} = \mathcal{E}_0 \cap I\mathcal{E}_0$ whose elements have the form $E_{ij} \overset{def}{=} E_i \cap IE_j$. Then $IE_{ij} = E_{ji}$. In each set E_{ij} with $i \leq j$ let x_{ij} be a selected center for E_{ij} and choose for E_{ji} the point $x_{ji} = Ix_{ij}$.[5] Define $I(i, j) = (j, i)$.

The partition $\mathcal{E}$ will be called "time reversal symmetric". Then the compatibility matrix will enjoy the property $M_{\alpha, \beta} = M_{I\beta, I\alpha}$, for all pairs $\alpha = (i, j)$ and $\beta = (i', j')$; and the map I will act on the symbolic representation ξ of x transforming it into the representation $I\xi$ of Ix as

---

[4] This means that given $\varepsilon, n > 0$ there is k large enough so that the Gibbs' distribution μ_{SRB} with potential Φ and the Markov process μ_k with potential $\Phi^{[\leq k]}$, truncation of Φ at range k, will be such that $\sum_{\mathbf{q}=(q_0,\dots,q_n)} |\mu_{SRB}(E(\mathbf{q})) - \mu_k(E(\mathbf{q}))| < \varepsilon$ and the Kolmogorov-Sinai's entropy (i.e. $s(\mu) \overset{def}{=} \lim_{n\to\infty} -\frac{1}{n} \sum_{\mathbf{q}=(q_0,\dots,q_n)} \mu(E(\mathbf{q})) \log \mu(E(\mathbf{q}))$ of μ_{SRB} and μ_k are close within ε.

[5] Hence $x_{ii} = Ix_{ii}$ if $E_i \cap IE_i$ is a non empty rectangle. Such a fixed point exists because the rectangles are homeomorphic to a ball. The fixed point theorem can be avoided by associating to $E_i \cap IE_i$ two centers x_{ii}^1 and $x_{ii}^2 = Ix_{ii}^1$: we do not do so to simplify the formulae.

$$(I\,\boldsymbol{\xi})_k = (I\xi_{-k}), \qquad \text{for all } k \in Z$$
$$\lambda_{u,\tau}(x_\alpha) = -\lambda_{s,\tau}(x_{I\alpha}), \tag{4.3.1}$$

A symmetry of this type can be called a "reverse symmetry". The second relation relies on the assumed isometric property of I.

Likewise if P is a symmetry, i.e. it is a smooth *isometric* map of phase space which squares to the identity $P^2 = 1$ and *commutes* with the evolution $PS = SP$, then if $\mathcal{E}_0$ is a Markovian pavement the pavement $\mathcal{E}_0 \cap I\mathcal{E}_0$ is P-symmetric, i.e. $PE_{i,j} = E_{j,i}$ and if $x_{i,j}$ is chosen so that $Px_{i,j} = x_{j,i}$, as in the previous case, then defining $P(i, j) = (j, i)$:

$$(P\,\boldsymbol{\xi})_k = (P\xi_k), \qquad \text{for all } k \in Z$$
$$\lambda_{u,\tau}(x_\alpha) = \lambda_{u,\tau}(x_{P\alpha}), \tag{4.3.2}$$

A symmetry of this type can be called a "direct symmetry".

If a system admits two *commuting* symmetries one direct, P, and one reversed, I, then PI is a reversed symmetry, i.e. a *new* time reversal.

This is interesting in cases in which a time evolution has the symmetry I on the full phase space but not on the attracting set $\mathcal{A}$ and maps the latter into a disjoint set $I\mathcal{A} \neq \mathcal{A}$ (a "repelling set"). If the evolution admits also a direct symmetry P mapping the repelling set back onto the attracting one ($PI\mathcal{A} = \mathcal{A}$), then the map PI maps the attracting set into itself and is a time reversal symmetry for the motions *on the attracting set* $\mathcal{A}$.

For the latter property to hold it, actually, suffices that P, I be just defined on the set $\mathcal{A} \cup I\mathcal{A}$ and commute on it.

The natural question is when it can be expected that I maps the attracting set into itself.

If the system is in equilibrium (e.g. the map S is canonical being generated via timing on a continuous time Hamiltonian flow) then the attracting set is, according to the chaotic hypothesis, the full phase space (of given energy). Furthermore often the velocity inversion is a time reversal symmetry (as in the models of Sect. 3.2). At small forcing an Anosov system can be mapped via a change of coordinates back to the not forced system ("structural stability of Anosov maps") and therefore the existence of a time reversal can be expected also out of equilibrium at small forcing.[6]

The situation becomes very different when the forcing increases: the attracting set $\mathcal{A}$ can become strictly smaller than the full phase space and the velocity reversal I will map it into a disjoint repelling set $I\mathcal{A}$ and I is no longer a time reversal for the interesting motions, i.e. the ones that take place on an attracting set.

Nevertheless it is possible to formulate a property for the evolution S, introduced in [9] and called *Axiom C*, which has the following features

---

[6] The transformation Φ of a perturbed Anosov map into the unperturbed one is in general not a smooth change of coordinates but just Hölder continuous. So that the image of I in the new coordinates might be hard to use.

(1) there is a time reversal symmetry I on the full phase space but the attracting set $\mathcal{A}$ is not invariant and $I\mathcal{A} \neq \mathcal{A}$ is, therefore, a repelling set.
(2) the attracting set $\mathcal{A}$ is mapped onto the repelling set by an isometric map P which commutes with I and S and squares to the identity $P^2 = 1$.
(3) it is structurally stable: i.e. if the evolution S is perturbed then, for small enough perturbations, the properties (1), (2) remain valid for the new evolution S'.

Then for such systems the map PI is a time reversal symmetry for the motions on the attracting set. A precise definition of the Axiom C property and a simple example are in Appendix H.

The interest of the Axiom C is that, expressing a structurally stable property, it might hold quite generally thus ensuring that the original global time reversal, associated with the velocity reversal operation, is a symmetry that "cannot be destroyed". Initially, when there is no forcing, it is a natural symmetry of the motions; increasing the forcing eventually the time reversal symmetry may be *spontaneously broken* because the attracting set is no longer dense on phase space (although it remains a symmetry for the motion in the sense that on the full phase space $IS = S^{-1}I$).

However if the system satisfies Axiom C a new symmetry P is spawned (by virtue of the constraint posed by the geometric Axiom C) which maps the attracting set onto the repelling set and commutes with I and S. Therefore the map $I^* = PI$ maps the attracting set into itself and is a time reversal for the evolution S restricted to the attracting set.

This scenario can repeat itself, as long as the system remains an Axiom C system: the attracting set can split into a pair of smaller attracting and repelling sets and so on until, increasing further and further the forcing, an attracting set may be reached on which motion is no longer chaotic (e.g. it is a periodic motion), [10].

The above is to suggest that time reversal symmetry might be quite generally a symmetry of the motions on the attracting sets and play an important role.

4.4 Pairing Rule and Axiom C

The analysis of Sect. 4.3 would be of greater interest if general relations could be established between phase space properties and attracting set properties. For instance the entropy production is related to the phase space attracting set contraction of the phase space volume but it is not in general related to the area contraction on the attracting surfaces.

It will be seen that when an attracting surface is not the full phase space interesting properties can be derived for the contractions of its area elements: however attracting sets, even when they are smooth surfaces (as under the chaotic hypothesis) are difficult to study and the area contraction is certainly difficult to access.

Therefore it is important to remark the existence of a rather general class of systems for which there is a simple relation between the phase space contraction and the area contraction on an attracting smooth surface.

This is based on another important relation, that will be discussed first, called *pairing rule* for even dimensional systems and the only known class of systems for which it can be proved is in Appendix I, where the corresponding proof is reported, following the original work in [11].

Let $2D$ be the number of the Lyapunov exponents, excluding possibly an even number of vanishing ones, and order the first D exponents, $\lambda_0^+, \ldots, \lambda_{D-1}^+$, in decreasing order while the next D, $\lambda_0^-, \ldots, \lambda_{D-1}^-$, are ordered in increasing order then the *pairing rule* is:

$$\frac{\lambda_j^+ + \lambda_j^-}{2} = const \qquad \text{for all } j = 0, \ldots, D-1; \qquad (4.4.1)$$

the constant will be called *"pairing level"* or *"pairing constant"*: the constant then must be $\frac{1}{2D}\langle\sigma\rangle_+$.[7]

In the cases in which Eq. (4.4.1) has been proved, [11], it holds also in a far *stronger* sense: the *local Lyapunov exponents* i.e. the non trivial eigenvalues[8] of the matrix $\frac{1}{2t}\log(\partial S_t(x)^T \partial S_t(x))$, of which the Lyapunov exponents are the averages, are paired to a j-independent constant but, of course, dependent on the point in phase space and on t. This property will be called the *strong pairing rule*.

Remarks (1) It should be kept in mind that while a pairing rule of the Lyapunov exponents is independent of the metric used on phase space, the strong pairing rule can only hold, if at all, for special metrics.
(2) Consider the systems of Appendix I, described in continuous time on by motions on a manifold M of dimensions $2(D+1)$. and satisfying the pairing rule. Then if S describes the same system on a Poincaré's section Ξ the pairing rule is transformed into a pairing of the $2D$ numbers obtained by removing removing λ_{D+1}^- from the set of $2D+1$ Lyapunov exponents of S.
(3) Among the examples are the evolutions in continuous time for the equations of the reversible model (1) in Sect. 2.3, Fig. 2.3. In the latter systems and more generally in systems in which there is an integral of motion, like in thermostatted systems (see footnote 8) with a isokinetic or isoenergetic Gaussian constraint (see Chap. 2) there will be a second vanishing Lyapunov exponents associated with the variations of the integral: as discussed in Appendix I, in the general theory of the pairing, the two vanishing exponents have to be excluded in checking Eq. 4.1.1.[8]

A *tentative* interpretation of the strong pairing, [12], could be that pairs with elements of opposite signs describe expansion *on the manifold* on which the attractor lies. While the $M \leq D$ pairs consisting of two negative exponents describe contraction of phase space *transversely to the manifold* on which the attractor lies.

---

[7] Because the version of Eq. (2.8.6) for evolutions in continuous time is $\lim_{\tau\to\infty} \frac{1}{\tau}\log\det(\partial S_\tau(x)) = \sum_{i=0}^{2D-1}\lambda_i$.
[8] Since the model is defined in continuous time the matrix $(\partial S_t)^*\partial S_t$ will always have a trivial eigenvalue with average 0.

Then since all pairs are "paired" at the same value $\sigma_{pair}(x) = \frac{\lambda_j^+(x)+\lambda_j^-(x)}{2}$ we would have $\sigma_{\mathcal{A}}(x) = 2(D-M)\sigma_{pair}(x)$ while the full phase space contraction would be $\sigma(x) = 2D\sigma_{pair}(x)$ and we should have proportionality between the phase space contraction $\sigma(x)$ and the area contraction $\sigma_{\mathcal{A}}(x)$ on the attracting set, i.e. (accepting the above heuristic and tentative argument, taken from [12, Eq. (6.4)]):

$$\sigma_{\mathcal{A},\tau}(x) = \frac{(D-M)}{D}\frac{1}{\tau}\sum_{j=0}^{\tau-1}\sigma(S^j x), \qquad \tau \geq 1 \qquad (4.4.2)$$

that is, the "known" phase space contraction $\sigma(x)$, differing from the entropy production $\varepsilon(x)$ by a time derivative, and the "unknown" area contraction $\sigma_{\mathcal{A}}$ on the attracting surface $\mathcal{A}$ (also of difficult access) are proportional via a factor simply related to the loss of dimensionality of the attracting surface compared to the phase space dimensionality.

In a system for which the chaotic hypothesis and the pairing rule hold and there are pairs of Lyapunov exponents consisting of two *negative* exponents, we conclude that a not unreasonable scenario would be that the closure of the attractor is a *smooth lowerdimensional surface*,[9] and if the system is reversible and satisfies the axiom C then on such lower dimensional attracting manifold the motion will still be *reversible* in the sense that there will be a map I^* of the attracting manifold into itself (*certainly different from* the global time reversal map I) which *inverts* the time on the attractor and that can be naturally called a *local time reversal* [9, 12].

The appearance of a non time reversal invariant attracting manifold in a time reversible system can be regarded as a *spontaneous symmetry breaking*: the existence of I^* means that in some sense time reversal symmetry of the system cannot be broken: if it does spontaneously break then it is replaced by a lower symmetry (I^*) which "restores it". The analogy with the symmetries T (broken) and TCP (valid) of Fundamental Physics would be remarkable.

The difficulty of the scenario is that there is no a priori reason to think that attractors should have the above structure: i.e. fractal sets lying on smooth surfaces on phase space *on which motion is reversible*. But the picture is suggestive and it might be applicable to more general situations in which reversibility holds only *on the attracting set* and not in the whole space (like in "strongly dissipative systems") [12].

For an application of the above *tentative* proposal in the cases in which it is coupled with the Axiom C property, see Sect. 5.7 and Appendix H.

---

[9] This does not preclude the possibility that the attractor has a fractal dimension (smoothness of the closure of an attractor has nothing to do with its fractal dimensionality, see [13–15]). The motion on this lower dimensional surface (whose dimension is smaller than that of phase space by an amount equal to the number of paired negative exponents) will still have an attractor (see p. 39) with dimension *lower* than the dimension of the surface itself, as suggested by the Kaplan–Yorke formula, [13].

4.5 Large Deviations

An interesting property of the SRB distribution for Anosov maps is that a *large deviation* law governs the fluctuations of finite time averages of observables. It is an immediate consequence of the property that if F is a smooth observable and S an evolution satisfying the chaotic hypothesis (i.e. S is hyperbolic, regular, transitive) the finite time averages

$$\widetilde{g} = \langle G \rangle_\tau = \frac{1}{\tau} \sum_{j=0}^{\tau-1} G(S^j x) \tag{4.5.1}$$

satisfy a *large deviations law*, i.e. fluctuations off the average $\langle G \rangle_\infty$ as large as τ itself, are controlled by a function $\zeta(\widetilde{g})$ convex and analytic in a (finite) interval $(\widetilde{g}_1, \widetilde{g}_2)$, maximal at $\langle G \rangle_\infty$, [16–18]. This means that the probability that $\widetilde{g} \in [a, b]$ satisfies

$$P_\tau(\widetilde{g} \in [a, b]) \simeq_{\tau \to \infty} e^{\tau \, \max_{\widetilde{g} \in [a,b]} \zeta(\widetilde{g})}, \qquad \forall a, b \in (\widetilde{g}_1, \widetilde{g}_2) \tag{4.5.2}$$

where $\simeq$ means that τ^{-1} times the logarithm of the *l.h.s.* converges to $\max_{[a,b]} \zeta(\widetilde{g})$ as $\tau \to \infty$, and the interval $(\widetilde{g}_1, \widetilde{g}_2)$ is non trivial if $\langle G^2 \rangle_\infty - \langle G \rangle_\infty^2 > 0$, [17, 19] and [4, Appendix 6.4].

If $\zeta(\widetilde{g})$ is quadratic at its maximum (i.e. at $\langle G \rangle_\infty$) then this implies a central limit theorem for the fluctuations of $\sqrt{\tau} \langle G \rangle_\tau \equiv \frac{1}{\sqrt{\tau}} \sum_{j=0}^{\tau-1} G(S^j x)$, but Eq. (4.5.2) is a much stronger property.

Remarks (1) If the observable G has nonzero SRB-average $\langle G \rangle_\infty \neq 0$ it is convenient to consider, instead, the observable $g = \frac{G}{\langle G \rangle_\infty}$ because it is dimensionless; just as in the case of $\langle G \rangle_\infty = 0$ it is convenient to consider the dimensionless observable $\frac{G}{\sqrt{\langle G^2 \rangle_\infty}}$.

(2) If the dynamics is *reversible*, i.e. there is a smooth, isometric, map I of phase space such that $IS = S^{-1}I$, then any *time reversal odd* observable G, with non zero average and nonzero dispersion $\langle G^2 \rangle_\infty - \langle G \rangle_\infty^2 > 0$, is such that the interval (g_1, g_2) of large deviations for $\frac{G}{\langle G \rangle_\infty}$ is at least $(-1, 1)$ provided there is a dense orbit (which also implies existence of only one attracting set).

(3) The systems in the thermostats model of Sect. (2.2) are all reversible with I being the ordinary time reversal, change in sign of velocity with positions unaltered, and the phase space contraction $-\sigma(x) = \lambda_u(x) + \lambda_s(x)$ is odd under time reversal, see Eq. (4.3.1). Therefore if $\sigma_+ = \langle \sigma \rangle_\infty > 0$ it follows that the observable

$$p' = \frac{1}{\tau} \sum_{j=0}^{\tau-1} \frac{\sigma(S^j x)}{\sigma_+} \tag{4.5.3}$$

has domain of large deviations of the form $(-\overline{g}, \overline{g})$ and contains $(-1, 1)$.

(4) In the thermostats model of Sect. 2.2, see Eq. (2.9.3), σ differs from the entropy production $\varepsilon(x) = \sum_{j>0} \frac{Q_j}{k_B T_j}$ by the time derivative of an observable: it follows that the finite or infinite time averages of σ and of ε have, for large τ, the same distribution. Therefore the same large deviations function $\zeta(p)$ controls the fluctuations of p' in Eq. (4.5.3) and of the entropy production rate:

$$p = \frac{1}{\tau} \sum_{j=0}^{\tau-1} \frac{\varepsilon(S^j x)}{\varepsilon_+}, \qquad \sigma_+ \equiv \langle \sigma \rangle_{SRB} = \langle \varepsilon \rangle_{SRB} \overset{def}{=} \varepsilon_+. \qquad (4.5.4)$$

which is more interesting from the Physics viewpoint.

In the following section an application will be derived providing information about the fluctuations of p, i.e. about the entropy production fluctuations.

4.6 Time Reversal and Fluctuation Theorem

It has been shown, [14, 20] (and interpreted in a mathematical form in [21]), that under the chaotic hypothesis and reversibility of motions on the attracting set, the function $\zeta(p)$ giving the large deviation law, Eq. (4.5.2), for the dimensionless phase space contraction p', Eq. (4.5.3), for SRB states and therefore for the dimensionless entropy production p, Eq. (4.5.4), has *under the chaotic hypothesis*, i.e. strictly speaking for Anosov systems, the *symmetry property*

Theorem (Fluctuation Theorem) *For time reversible Anosov maps there is $\overline{p} \geq 1$ and*

$$\zeta(-p) = \zeta(p) - p\sigma_+, \qquad \text{for all } p \in (-\overline{p}, \overline{p}) \qquad (4.6.1)$$

with $\zeta(p)$ convex and analytic around the segment $(-\overline{p}, \overline{p})$.

The Eq. (4.6.1) expresses the *fluctuation theorem* of [14].[10]

The interest of the theorem is that, as long as chaotic hypothesis and time reversibility hold, it is *universal, model independent* and yields a *parameter free* relation which deals with a quantity which, as mentioned, has the physical meaning of entropy production rate because

$$p\,\varepsilon_+ = \frac{1}{\tau} \sum_{j=0}^{\tau-1} \frac{Q_j}{k_B T_j}, \qquad (4.6.2)$$

and therefore has an independent macroscopic definition, see Sect. 2.8, hence is accessible to experiments.

---

[10] As discussed below, it requires a proof and therefore it should not be confused with several identities to which, for reasons that I fail to understand, the same name has been given, [22] and Appendix L.

The expression for μ_{SRB}, defined via Eq. (4.1.8), can be used to study some statistical properties of p' hence of p. The ratio of the probability of $p' \in [p, p+dp]$ to that of $p' \in [-p, -p+dp]$, using the notations and the approximation under the limit sign in Eqs. (3.8.7), (4.1.8) and setting $a_\tau(x) \overset{def}{=} \frac{1}{2\tau+1} \sum_{j=-\tau}^{\tau} \frac{\sigma(S^j x)}{\sigma_+}$, is

$$\frac{\sum_{\mathbf{q}, a_\tau(x_{\mathbf{q}})=p} e^{-\Lambda_{u,\tau}(x_{\mathbf{q}})}}{\sum_{\mathbf{q}, a_\tau(x_{\mathbf{q}})=-p} e^{-\Lambda_{u,\tau}(x_{\mathbf{q}})}}. \tag{4.6.3}$$

Equation (4.6.3) is studied by establishing a one to one correspondence between addends in the numerator and in the denominator, aiming at showing that corresponding addends have a *constant ratio* which will, therefore, be the value of the ratio in Eq. (4.6.3).

This is made possible by the time reversal symmetry which is the (simple) extra information with respect to [3, 17, 23].

In fact the time reversal symmetry I allows us to suppose, without loss of generality, that the Markovian partition $\mathcal{E}$, hence $\mathcal{E}_\tau$, can be supposed time reversible, see Sect. 4.3: i.e. for each j there is a j' such that $I E_j = E_{j'}$.

The identities $S^{-\tau}(S^\tau x_{\mathbf{q}}) = x_{\mathbf{q}}$, and $S^{-\tau}(I S^{-\tau} x_{\mathbf{q}}) = I x_{\mathbf{q}}$ (time reversal) and $I W^u(x) = W^s(Ix)$, one can deduce, $a_\tau(x_{\mathbf{q}}) = -a_\tau(I x_{\mathbf{q}})$ and $\Lambda_{u,\tau}(I x_{\mathbf{q}}) = -\Lambda_{s,\tau}(x_{\mathbf{q}})$. The ratio Eq. (4.6.3) can therefore be rewritten as:

$$\frac{\sum_{\mathbf{q}, a_\tau(x_{\mathbf{q}})=p} e^{-\Lambda_{u,\tau}(x_{\mathbf{q}})}}{\sum_{\mathbf{q}, a_\tau(x_{\mathbf{q}})=-p} e^{-\Lambda_{u,\tau}(x_{\mathbf{q}})}} \equiv \frac{\sum_{\mathbf{q}, a_\tau(x_{\mathbf{q}})=p} e^{-\Lambda_{u,\tau}(x_{\mathbf{q}})}}{\sum_{\mathbf{q}, a_\tau(x_{\mathbf{q}})=p} e^{\Lambda_{s,\tau}(x_{\mathbf{q}})}} \tag{4.6.4}$$

Then the ratios between corresponding terms in Eq. (4.6.4) are equal to $e^{-\Lambda_{u,\tau}(x_{\mathbf{q}})-\Lambda_{s,\tau}(x_{\mathbf{q}})}$.

This is almost $y = e^{-\sum_{j=-\tau}^{\tau} \sigma(S^{-j} x_{\mathbf{q}})} = e^{-a_\tau(x_{\mathbf{q}})\sigma_+}$. In fact, the latter is the reciprocal of the determinant of the Jacobian matrix of S, i.e. the reciprocal of the total phase space volume variation, while $y' = e^{-\Lambda_{u,\tau}(x_{\mathbf{q}})-\Lambda_{s,\tau}(x_{\mathbf{q}})}$ is only the reciprocal of the product of the variations of two surface elements tangent to the stable and to the unstable manifold in $\mathbf{x}_j$. Hence y and y' differ by a factor related to the sine of the angles between the manifolds at $S^{-\tau/2}\mathbf{x}$ and at $S^{\tau/2}\mathbf{x}$.

But the chaotic hypothesis (i.e. the Anosov property of the motion on the attracting set) implies transversality of their intersections, so that the ratio y/y' is bounded away from 0 and $+\infty$ by $(\mathbf{q}, \tau)$-independent constants.

Therefore the ratio Eq. (4.6.1) is equal to $e^{(2\tau+1) p \sigma_+}$ up to a factor bounded above and below by a (τ, p)-independent constant, i.e. to leading order as $\tau \to \infty$, and the fluctuation theorem for stationary SRB states, Eq. (4.6.1), follows.

Remarks (a) The peculiarity of the result is the linearity in p: we expect that $\zeta(p) - \zeta(-p) = c \langle \sigma \rangle (p + s_3 p^3 + s_5 p^5 + \ldots)$ with $c > 0$ and $s_j \neq 0$, since there is no reason, a priori, to expect a "simple" (i.e. with linear odd part) multifractal distribution.[11] Thus p-linearity (i.e. $s_j \equiv 0$) is a *key test of the theory*, i.e. of the chaotic hypothesis, and a quite unexpected result from the latter viewpoint. Recall, however, that the exponent $(2\tau + 1)\sigma_+ \ p$ is correct up to terms of $O(1)$ in τ (i.e. deviations at small p, r small τ, must be expected).

(b) Equation (4.6.1) requires time reversibility and the chaotic hypothesis and this is a strong assumption: this explains why a few papers have appeared in the literature trying to get rid of the chaotic hypothesis.

(c) Experimental tests can possibly be designed with the aim of checking that the entropy production $\sigma \stackrel{def}{=} \sum_j \frac{Q_j}{k_B T_j}$, defined in experimental situations by the actual measurements of the heat ceded to the thermostats at temperature T_j or, in simulations, by the phase space contraction σ satisfies what will be called the "*fluctuation relation*":

$$\frac{Prob(p \in \Delta)}{Prob(-p \in \Delta)} = e^{p\sigma_+\tau + O(1)} \qquad (4.6.5)$$

where σ_+ is the infinite time average of σ and Δ is an interval small compared to p. A positive result should be interpreted as a confirmation of the chaotic hypothesis, provided time reversibility can be assumed.

(d) In Appendix L a relation often confused with the above fluctuation relation is discussed.

(e) It should be stressed that under the chaotic hypothesis the attracting sets are Anosov systems, but the time reversal symmetry of the motions on the attracting sets is very subtle. As discussed in Sect. 4.3 the fundamental symmetry of time reversal might not hold on the attracting sets $\mathcal{A}$ at strong forcing, when the $\mathcal{A}$'s have dimensionality lower than that of the phase space. Therefore in applying the fluctuation theorem or the fluctuation relation particular care has to be reserved to understanding whether a mechanism of respawning of a time reversal symmetry works: as discussed in Sect. 4.3 this is essentially asking whether the system enjoys the property called there Axiom C.

(f) The fluctuation and the conditional reversibility theorems of the next section can be formulated for maps and flows (i.e. for Anosov maps and Anosov flows). The discrete case is simpler to study than the corresponding Anosov flows because Anosov maps do not have a trivial Lyapunov exponent (the vanishing one associated with the phase space flow direction); the techniques to extend the analysis to Anosov flows, is developed in [25, 26] (where also is achieved the goal of proving the analogue of the fluctuation theorem for such systems).

The conditional reversibility theorem will be presented in the version for flows: the explicit and natural formulation of the fluctuation and the conditional reversibility theorems for maps will be skipped (to avoid repetitions).

---

[11] Actual computation of $\zeta(p)$ is a task possible in the $N = 1$ case considered in [24] but essentially beyond our capabilities in slightly more general systems in the non linear regime.

4.7 Fluctuation Patterns

The fluctuation theorem, Eq. (4.6.1) has several extensions including a remarkable, parameter free relation that concerns the relative probability of *patterns* of evolution of an observable and of their reversed patterns, [27–29], related to the Onsager–Machlup fluctuations theory, which keeps being rediscovered in various forms and variations in the literature.

It is natural to inquire whether there are other physical interpretations of the theorem (hence of the meaning of the chaotic hypothesis) when the external forcing is really different from the value 0.[12] A result in this direction is the *conditional reversibility theorem*, assuming the chaotic hypothesis and $\sigma_+ > 0$, discussed below.

Consider observables F which, for simplicity, have a well-defined time reversal parity: $F(Ix) = \varepsilon_F F(x)$, with $\varepsilon_F = \pm 1$. For simplicity suppose that their time average (i.e. the SRB average) vanishes, $F_+ = 0$. Let $t \to \varphi(t)$ be a smooth function vanishing for $|t|$ large enough; define also $I\varphi$ as the time reversed pattern $I_\tau\varphi(t) \overset{def}{=} \varepsilon_F \varphi(\tau - t)$.

Look at the probability, $P_{\tau;p,\varphi;\eta}$, relative to the SRB distribution (i.e. in the "natural stationary state"), that

$$|F(S_t x) - \varphi(t)| < \eta, \qquad t \in (0, \tau)$$

$$|p - \frac{1}{\tau} \int_0^\tau \frac{\sigma(S_t x)}{\sigma_+} dt| < \eta \qquad\qquad (4.7.1)$$

which will be called the probability that, within tolerance η, F follows the fluctuation pattern $\varphi(t), t \in (0, \tau)$ while there is an average entropy production p. Then the following somewhat unprecise statement (see below), heuristically discussed in [27, 31], can be derived essentially in the same way as the above fluctuation theorem. *Assume the evolution to be a time reversible Anosov flow with phase space contraction rate $\sigma_+ > 0$. Let F and G be observables (time reversal odd for definiteness), let f, g be patterns for F, G respectively and let If, Ig be the time reversed patterns; then:*

$$\frac{P_{\tau,p,f,\eta}}{P_{\tau,p,g,\eta}} = \frac{P_{\tau,-p,If,\eta}}{P_{\tau,-p,Ig,\eta}} \qquad\qquad (4.7.2)$$

for large τ and exactly as $\tau \to \infty$.[13]

No assumption on the fluctuation size (i.e. on the size of φ, see however remark (e) at the end of Sect. (4.6)), nor on the size of the forces keeping the system out of equilibrium, is made.

---

[12] That is, not infinitesimally close to 0 as in the classical theory of nonequilibrium thermodynamics, [30].

[13] Colorfully: *A waterfall will go up, as likely as we see it going down, in a world in which for some reason, or by the deed of a Daemon, the entropy production rate has changed sign during a long enough time* [29, p. 476].

A more mathematical form of the above result, heuristically proved in [27] (however a formal proof is desirable):

Theorem (Fluctuation Patterns) *Under the assumptions of the preceding statement, let $\zeta(p, \varphi)$ the be large deviation function for observing in the time interval $[0, \tau]$ an average contraction of phase space $\frac{1}{\tau} \int_0^\tau \sigma(S_t x)dt = p\sigma_+$ and at the same time $F(S_t x)$ to follow a fluctuation $\varphi(t)$. Then there is $\overline{p} \geq 1$*

$$\zeta(-p, \varepsilon_F I \varphi) - \zeta(p, \varphi) = -p\sigma_+, \quad p \in (-\overline{p}, \overline{p}) \qquad (4.7.3)$$

for all φ, with ζ the joint large deviation rate for p and φ (see below).

Here the rate ζ is defined as the rate that controls the μ_{SRB} probability that the *dimensionless average entropy creation rate p is in an interval $\Delta = (a, b)$ and, at* the same time, $|f(S_t x) - \varphi(t)| < \eta$ by:

$$\sup_{p \in \Delta, |\varphi - \psi| < \eta} e^{-\tau\zeta(p, \psi)} \qquad (4.7.4)$$

to leading order as $\tau \to \infty$ (i.e. the logarithm of the mentioned probability divided by τ converges as $\tau \to \infty$ to $\sup_{p \in \Delta, |\varphi - \psi| < \eta} \zeta(p, \psi)$).

Remarks (1) The result can also be formulated if F is replaced by m observables $\mathbf{F} = (F_1, \ldots, F_m)$, each of well defined parity under time reversal and the pattern φ is correspondingly replaced by m patterns $\varphi = (\varphi_1, \ldots, \varphi_m)$. The r.h.s. of the relation analogous to Eq. (4.7.3) remains unchanged; hence this extension provides in principle arbitrarily many parameter free fluctuation relations. Only few of them can be observed because the difficulty of observing m patterns obviously so rare becomes more and more hard with increasing m.

(2) In other words, in these systems, while it is very difficult to see an "anomalous" average entropy creation rate during a time τ (e.g. $p = -1$), it is also true that *"that is the hardest thing to see"*. Once we see it *all the observables will behave strangely* and the relative probabilities of time reversed patterns will become as likely as those of the corresponding direct patterns under "normal" (e.g. $p = 1$) average entropy creation regime.

(3) It can also be said that the motion in a time symmetric Anosov system is reversible, even in the presence of dissipation, once the dissipation is fixed. Again interesting variations of this property keep being discovered, see for instance [32].

(4) No assumption on the fluctuation size (i.e. on the size of φ), nor on the size of the forces keeping the system out of equilibrium, are made, besides the Anosov property and $\sigma_+ > 0$ (the results hold no matter how small σ_+ is; and they make sense even if $\sigma_+ = 0$, but they become trivial).

(5) The comment (e) in the previous section, about the general case of attracting sets with dimension lower than that of phase space, has to be kept in mind as it might set serious limits to experimental checks (not, of course, of the theorems but of the physical assumption in the chaotic hypothesis which implies the theorems).

There are other remarkable extensions of the fluctuation relation in presence of other symmetries: see [33].

4.8 Onsager Reciprocity, Green-Kubo Formula, Fluctuation Theorem

The fluctuation theorem degenerates in the limit in which σ_+ tends to zero, i.e. when the external forces vanish and dissipation disappears (and the stationary state becomes the equilibrium state).

Since the theorem deals with systems that are time reversible *at and outside* equilibrium, Onsager's hypotheses are certainly satisfied and the system should obey reciprocal response relations at vanishing forcing. This led to the idea[14] that there might be a connection between the fluctuation theorem and Onsager reciprocity and also to the related (stronger) Green-Kubo formula.

This can be checked: switching to continuous time, to simplify the analysis and referring to the finite models of Sects. 2.2, 2.3, define the *microscopic thermodynamic flux* $j(x)$ associated with the *thermodynamic force* E that generates it, i.e. the parameter that measures the strength of the forcing (which makes the system non Hamiltonian), via the relation

$$j(x) = \frac{\partial \sigma(x)}{\partial E} \tag{4.8.1}$$

(not necessarily at $E = 0$) then in [34] a heuristic proof shows that the limit as $E \to 0$ of the fluctuation theorem becomes simply (in the continuous time case) a property of the average, or "macroscopic", *flux* $J = \langle j \rangle_{\mu_E}$:

$$\frac{\partial J}{\partial E}\Big|_{E=0} = \frac{1}{2} \int_{-\infty}^{\infty} \langle j(S_t x) j(x) \rangle_{\mu_E}\Big|_{E=0} dt \tag{4.8.2}$$

where $\langle \cdot \rangle_{\mu_E}$ denotes the average in the stationary state μ_E (i.e. the SRB distribution which, at $E = 0$, is simply the microcanonical ensemble μ_0).

If there are several fields $E_1, E_2, \dots$ acting on the system we can define several thermodynamic fluxes $j_k(x) \overset{def}{=} \partial_{E_k} \sigma(x)$ and their averages $\langle j_k \rangle_{\mu_E}$: in the limit in which all forces E_k vanish a (simple) extension of the fluctuation theorem is shown, [34], to reduce to

$$L_{hk} \overset{def}{=} \frac{\partial J_h}{\partial E_k}\Big|_{E=0} = \frac{1}{2} \int_{-\infty}^{\infty} \langle j_h(S_t x) j_k(x) \rangle_{E=0} dt = L_{kh}, \tag{4.8.3}$$

This extension of the fluctuation theorem was used in [34] and is a particular case of the fluctuation patterns theorem (of Sect. 4.7: the particular case was proved first to

---

[14] Suggested by P. Garrido from the data in the simulation in [12].

derive the 4.8.3 and inspired the later formulation of the general fluctuation patterns theorems).

Therefore we see that the fluctuation theorem can be regarded as *an extension to nonzero forcing* of Onsager reciprocity and, actually, of the Green-Kubo formula.

It is not difficult to see, heuristically, how the fluctuation theorem, in the limit in which the driving forces tend to 0, formally yields the Green-Kubo formula.

Let $I_E(x) \overset{def}{=} \int_0^\tau \sigma_E(S_t x) dt \equiv p \sigma_+ \tau$. We consider time evolution in continuous time and simply note that the fluctuation theorem implies that, for all E (for which the system is satisfies the chaotic hypothesis) $\langle e^{I_E} \rangle = \sum_p \pi_\tau(p) e^{p \tau \sigma_+} = \sum_p \pi_t(-p) e^{O(1)} = e^{O(1)}$ so that:

$$\lim_{\tau \to +\infty} \frac{1}{\tau} \log \langle e^{I_E} \rangle_{\mu_E} = 0 \qquad (4.8.4)$$

where $I_E \overset{def}{=} \int_0^\tau \sigma(S_t x) dt$ with $\sigma(x)$ being the divergence of the equations of motion (i.e. the phase space contraction rate, in the case of continuous time). This remark, [35],[15] can be used to simplify the analysis in [34, 36] as follows.

Differentiating $\frac{1}{\tau} \log \langle e^{I_E} \rangle_{\mu_E} + o(1)$ twice with respect to E, not worrying about interchanging derivatives and limits and the like, one finds that the second derivative with respect to E is a sum of six terms. Supposing that for $E = 0$ it is $\sigma = 0$, hence $I_0 \equiv 0$, the six terms, when evaluated at $E = 0$, are:

$$\frac{1}{\tau} \Big[\langle \partial_E^2 I_E \rangle_{\mu_E} |_{E=0} - \langle (\partial_E I_E)^2 \rangle_{\mu_E} |_{E=0}$$
$$+ \int \partial_E I_E(x) \partial_E \mu_E(x) |_{E=0} - \Big(\langle (\partial_E I_E)^2 \rangle_{\mu_E} \cdot \int 1 \partial_E \mu_E \Big) |_{E=0} +$$
$$+ \int \partial_E I_E(x) \partial_E \mu_E(x) |_{E=0} + \int 1 \cdot \partial_E^2 \mu_E |_{E=0} \Big] \qquad (4.8.5)$$

and we see that the fourth and sixth terms vanish being derivatives of $\int \mu_E(dx) \equiv 1$, and the first vanishes (by integration by parts) because I_E is a divergence and μ_0 is the Liouville distribution (by the assumption that the system is Hamiltonian at $E = 0$ and chaotic). Hence we are left with:

$$\Big(-\frac{1}{\tau} \langle (\partial_E I_E)^2 \rangle_{\mu_E} + \frac{2}{\tau} \int \partial_E I_E(x) \partial_E \mu_E(x) \Big)_{E=0} = 0 \qquad (4.8.6)$$

where the second term is $2 \tau^{-1} \partial_E (\langle \partial_E I_E \rangle_{\mu_E}) |_{E=0} \equiv 2 \partial_E J_E |_{E=0}$, because the SRB distribution μ_E is stationary; and the first term tends to the integral $\int_{-\infty}^{+\infty} \langle j(S_t x) j(x) \rangle_{E=0} dt$ as $\tau \to \infty$. Hence we get the Green-Kubo formula in the case of only one forcing parameter.

---

[15] It says that essentially $\langle e^{I_E} \rangle_{\mu_E} \equiv 1$ or more precisely it is not too far from 1 as $\tau \to \infty$ so that Eq. (4.8.4) holds.

The argument is extended to the case in which the forcing parameter is a vector $\mathbf{E} = (E_1, \ldots, E_n)$ describing the strength of various driving forces acting on the system. One needs a generalization of Eq. (4.8.4) which is a special case of the patterns fluctuation theorem, Eq. (4.7.3), applied to the observable $F_j(x) = E_j \partial_{E_j} \sigma(x)$. The fluctuation patterns theorem, Eq. (4.7.3) can be used instead of Eq. (4.8.4), for the details see [36, Eq. 15–20].

As discussed in Sect. 2.8, it is possible to change the metric (hence the measure of volumes) in phase space $\varXi$ redefining the volume so that $\sigma_\mathbf{E}$ vanishes for $\mathbf{E} = \mathbf{0}$: for the models considered here, see Sects. 2.2, 2.3, the phase space contraction can be transformed (by changing coordinates) into an expression which vanishes for $\mathbf{E} = \mathbf{0}$ and also its derivatives do generate the heat and material currents by differentiation with respect to the external forces or to the temperature differences. At the same time it has the property that it differs by a total derivative from the entropy production rate $\varepsilon(x) = \sum_j \frac{Q_j}{k_B T_j}$.

Hence assumption that $\sigma_\mathbf{E}$ vanishes for $\mathbf{E} = \mathbf{0}$ is less strong than it might seem: $\sigma_\mathbf{E}$ is defined only once a metric on phase space has been introduced.

The above analysis is unsatisfactory because we interchange limits and derivatives quite freely and we even take derivatives of μ_E, which seems to require some imagination as μ_E is concentrated on a set of zero volume.

On the other hand, under the strong hypotheses in which we suppose to be working (that the system is mixing Anosov), we should not need extra assumptions. *Indeed* a non heuristic analysis, [37], is based on the solution of the problem of differentiability with respect to a parameter for SRB distributions, [38].

Certainly assuming reversibility in a system out of equilibrium can be disturbing: one can, thus, inquire if there is a more general connection between the chaotic hypothesis, Onsager reciprocity and the Green-Kubo formula.

This is indeed the case and provides us with a further consequence of the chaotic hypothesis valid, however, only in zero field. It can be shown that the relations Eq. (4.8.3) follow from the sole assumption that at $E = 0$ the system is time reversible and that it satisfies the chaotic hypothesis for E near 0: at $E \neq 0$ it can be, as in Onsager's theory not necessarily reversible, [37].

4.9 Local Fluctuations: An Example

There are cases in which the phase space contraction is an "extensive quantity", because thermostats do not act only across the boundaries but they act also in the midst of the system.

For instance this is the case in the electric conduction models in which dissipation occurs through collisions with the phonons of the underlying lattice. Then heat is generated in the bulk of the system and if a large part of the system is considered the amount of heat generated in the bulk might exceed the amount that exits from the boundaries of the sample thus making necessary a dissipation mechanism that operates also in the system bulk.

In this situation it can be expected that it should be possible to define a local phase space contraction and prove for it a fluctuation relation. This question has been studied in [4, 39] where a model of a chain of N coupled maps has been considered. The results are summarized below.

Consider a collection of N^3 independent identical systems that are imagined located at the sites ξ of a $N \times N \times N$ cubic lattice $\mathcal{L}_N$ centered at the origin: the state of the system located at ξ is determined by a point φ_ξ in a manifold $\mathcal{T}$ which for simplicity will be taken a torus of dimension $2d$. The evolution S_0 of a state $\varphi = (\varphi_\xi)_{\xi \in \Lambda_N}$ is $\varphi \to (S_0 \varphi) = (S_0 \varphi_\xi)_{\xi \in \Lambda_N}$, i.e. the system in each location evolves independently.

Consider a small perturbation of *range $r > 0$* of the evolution

$$(S_\sigma \varphi)_\xi = S_0 \varphi_\xi + \varepsilon \Psi_\xi(\varphi)) \tag{4.9.1}$$

with $\Psi_\xi(\varphi) = \psi((\varphi_\eta)_{|\eta - \xi| < r})$ a smooth "perturbation". It will generate an evolution which generically will not be volume preserving in the sense that

$$\sigma_{\mathcal{L}_N}(\varphi) \overset{def}{=} -\log |\det(\partial_\varphi S(\varphi))| \neq 0) \tag{4.9.2}$$

even when, as it will be assumed here, S_0 is a volume preserving map.

It can be shown that the basic results in [40–42] imply that if the "unperturbed dynamics" S_0 is smooth, hyperbolic, transitive (i.e. if S_0 is an Anosov map) then for ε small enough, *but independently of the system size N*, the system remains an Anosov map.

Therefore it admits a SRB distribution which can also be studied very explicitly by perturbation theory, [43, 44], [4, Sect. 10.4]. For instance the SRB average of the phase space contraction

$$\langle \sigma_{\Lambda_N} \rangle_{SRB} = \sigma_+(\varepsilon, N) = N^3 \overline{\sigma}_+(\varepsilon) + O(N^2) \tag{4.9.3}$$

with $\overline{\sigma}_+(\varepsilon)$ analytic in ε near $\varepsilon = 0$ and generically > 0 there, for small $\varepsilon \neq 0$. It is also *extensive*, i.e. proportional to the volume N^3 of the system up to "boundary corrections" of $O(N^2)$.

By the general theory the fluctuation theorem for $p = \frac{1}{\tau} \sum_{j=0}^{\tau-1} \frac{\sigma(S_\varepsilon^j(\phi))}{\sigma_+(\varepsilon, N)}$ will hold provided the map S_ε is time reversible.

This suggests that given a subvolume $\Lambda \subset \Lambda_N$ and setting

$$\sigma_\Lambda(\varphi) = -\log |\det(\partial_\varphi S(\phi))_\Lambda| \tag{4.9.4}$$

where $(\partial_\varphi S(\varphi))_\Lambda$ denotes the submatrix $(\partial_\varphi S(\varphi))_{\xi\xi'}$ with $\xi, \xi' \in \Lambda$ and

$$p \overset{def}{=} \frac{1}{\tau} \sum_{j=0}^{\tau-1} \frac{\sigma_\Lambda(S^j \varphi)}{\langle \sigma_\Lambda \rangle_{SRB}}. \tag{4.9.5}$$

then the random variable p should obey a large deviation law with respect to the SRB distribution.

It can be shown, [43], that

$$\langle \sigma_\Lambda \rangle_{SRB} = \overline{\sigma}_+ |\Lambda| + O(|\partial \Lambda|) \tag{4.9.6}$$

and the large deviation law of p is *for all* $\Lambda \subseteq \Lambda_N$ a function $\zeta(p)$ defined analytic and convex in an interval (p_1, p_2) which has there the "extensive form":

$$\zeta(p) = |\Lambda| \overline{\zeta}(p) + O(|\partial \Lambda|). \tag{4.9.7}$$

The analogy with the more familiar density fluctuations in a low density gas is manifest: the probability that the number of particles n in a volume Λ subset of the container V, $\Lambda \subseteq V$, in a gas in equilibrium at temperature $T_0 = (k_B \beta_0)^{-1}$ and density $\varrho_0 = \frac{N}{V}$ is such that the random variable $p = \frac{n}{|\Lambda| \varrho_0}$ obeys a large deviations law controlled by an analytic, convex function of p which is extensive, [45]:

$$\text{Prob}_{SRB}(p \in [a, b]) = e^{-|\Lambda| \max_{p \in [a,b]} \beta_0 f_0(\beta_0, p \varrho_0)}, \quad [a, b] \subset \left(0, \frac{\varrho_c}{\varrho_0}\right) \tag{4.9.8}$$

to leading order as $\Lambda \to \infty$, where ϱ_c is the close packing density and

$$f_0(\beta_0, p \varrho_0) \stackrel{def}{=} f(\beta, p \varrho_0) - f(\beta_0, \varrho_0) - \left. \frac{\partial f(\beta_0, \varrho_0)}{\partial \varrho} \right|_{\varrho = \varrho_0} (p \varrho_0 - \varrho_0) \tag{4.9.9}$$

is the difference between the Helmholtz free energy at density ϱ and its linear extrapolation from ρ_0, [46].

The function $f_0(\beta_0, p \varrho_0)$ is *independent* of $\Lambda \subseteq V$. Hence in this example the unobservable density fluctuations in very large volumes V can be measured via density fluctuations in finite regions Λ.

If the map S_ε is *also* time reversible then the validity of the fluctuation theorem for the full system implies that $\overline{\zeta}(-p) = \overline{\zeta}(p) - p \overline{\sigma}_+$ and because of the extensivity of ζ in Eq. (4.9.7), hence the global fluctuation theorem implies (and is implied by) the "local" property of the intensive fluctuation rate $\overline{\zeta}(p)$.

4.10 Local Fluctuations: Generalities

The example in the previous section indicates the direction to follow in discussing large fluctuations in extended systems. A key difference that can be expected in most problems is that in extended systems *in stationary states* dissipation is not a bulk property: due to the conservative nature of the internal forces. For instance in the models in Sect. 2.2 no dissipation occurs in the system proper, C_0, but it occurs "at the boundary" of C_0 where interaction with the thermostats takes place.

Of course we are familiar with the dissipation in gases and fluids modeled by constant friction manifested throughout the system: as, for instance, in the Navier-Stokes equation.

However this is a phenomenologically accounted friction. If the interest is on stationary states of the fluid motion (under stirring forces) then the friction coefficient takes into account phenomenologically that stationarity can be reached because the heat generated by the stirring is transferred across the fluid and dissipated at the boundary, when the latter is in contact with external thermostats.

Therefore in models like the general ones in Sect. 2.2 or in the second in Sect. 2.3 the average dissipation has to be expected to be a boundary effect rather than a bulk effect (as it is in the example in Sect. 4.8 or in the modification of the model in Fig. 2.4, considered in [36]).

Consider an extended system C_0 in contact with thermostats: i.e. a large system enclosed in a volume V large enough so that it males sense to consider subvolumes $\Lambda \subset V$ which still contain many particles. Supposing the system satisfying the chaotic hypothesis and in a stationary state, also the part of the system inside a subvolume Λ will be in a stationary state, i.e. the probability of finding a given microscopic configuration in Λ will be time-independent.

It is natural to try to consider the subvolume $\Lambda \subset V$ as a container in contact with a thermostats in the complementary subvolume V/Λ. However the "wall" of separation between the thermostats and the system, i.e. the boundary $\partial \Lambda$, is only an ideal wall and particles can cross it and do so for two reasons.

First they may cross the boundary because of a macroscopic current established in the system by the action of the stirring forces or by convection; secondly, even in absence of stirring, when the nonequilibrium is only due to differences in temperature at various sectors of the boundary of the global container V, and convection is absent, particles cross back and forth the boundary of $\partial \Lambda$ in their microscopic motion.

It is important to consider also the time scales over which the phenomena occur. The global motion takes often place on a time scale much longer than the microscopic motions: in such case it can be neglected as long as the observations times are short enough. If not so, it may be possible, if the region Λ is small enough, to follow it,[16] because the local "Brownian" motion takes place on a short time scale and it can be neglected only if the free path is much smaller than the size of Λ.

There are a few cases in which the two causes above can be neglected: then, with reference for instance to the models in Sect. 2.2, the region Λ can be considered in contact with reservoirs which at the point $\xi \in \partial \Lambda$ have temperature $T(\xi)$ so that the phase space contraction of the system enclosed in Λ is, see Eq. (2.8.1), up to a total time derivative

$$\varepsilon = \int_{\partial \Lambda} \frac{Q(\xi)}{k_B T(\xi)} ds_\xi, \qquad (4.10.1)$$

---

[16] For a time long enough for being able to consider it as a moving container: for instance while its motion can be considered described by a linear transformation and at the same time long enough to be able to make meaningful observations.

The fluctuations over time intervals *much longer than the time of free flight but much shorter than the time it takes to diffuse over a region of size of the order of the size of Λ* (if such time scales difference is existent) can then be studied by the large deviation laws and we can even expect a fluctuation relation, Eq. (4.6.5), to hold because in Λ there is no friction, provided the time averages are not taken over times too long compared to the above introduced ones.

The situations in which the above idea has chances to work and be observable are dense systems, like fluids, where the free path is short: and an attempt to an application to fluids will be discussed in the next chapter.

The idea and the possibility of local fluctuation theorems has been developed and tested first numerically, [47], and then theoretically, [39], by showing that it indeed works at least in some models (with homogeneous dissipation like the Gaussian Navier-Stokes equations in the OK41 approximation, see Sects. 5.6, 5.7) which are simple enough to allow us to build a formal mathematical theory of the phase space contraction fluctuations.

4.11 Quantum Systems, Thermostats and Nonequilibrium

Recent experiments deal with properties on mesoscopic and atomic scale. In such cases the quantum nature of the systems cannot be always neglected, particularly at low temperature, [28, Chap. 1], and the question is whether a fluctuation analysis parallel to the one just seen in the classical case can be performed in studying quantum phenomena.

Thermostats have a macroscopic phenomenological nature: in a way they could be regarded as classical macroscopic objects in which no quantum phenomena occur. Therefore it seems natural to model them as such and define their temperature as the average kinetic energy of their constituent particles so that the question of how to define it does not arise.

The point of view has been clearly advocated in several papers, for instance in [48] just before the fluctuation theorem and the chaotic hypothesis were developed. Here the analysis is presented with the minor variation that

(a) Gaussian thermostats are used instead of the Nosé-Hoover thermostats and
(b) several different thermostats are allowed to interact with the system,

following [49], *aiming at the application of the chaotic hypothesis to obtain a fluctuation relation for systems with an important quantum component.*

A version of the chaotic hypothesis for quantum systems is already[17] implicit in [48] and in the references preceding it, where the often stated incompatibility of chaotic motions with the discrete spectrum of a confined quantum system is criticized.

---

[17] Writing the paper [49] I was unaware of these works: I thank Dr. M. Campisi for recently pointing this reference out.

Consider the system in Fig. 2.2 when the quantum nature of the particles in the finite container with smooth boundary C_0 cannot be neglected. Suppose for simplicity (see [49]) that the nonconservative force $\mathbf{E}(\mathbf{X}_0)$ acting on C_0 vanishes, i.e. consider the problem of heat flow through C_0. Let H be the operator on $L_2(C_0^{3N_0})$, space of symmetric or antisymmetric wave functions $\Psi(\mathbf{X}_0)$,

$$H = -\frac{\hbar^2}{2m}\Delta_{\mathbf{X}_0} + U_0(\mathbf{X}_0) + \sum_{j>0}\left(U_{0j}(\mathbf{X}_0, \mathbf{X}_j) + U_j(\mathbf{X}_j) + K_j\right) \qquad (4.11.1)$$

where $K_j = \frac{m}{2}\sum_{j>0}\dot{\mathbf{X}}_j^2$ and $\Delta_{\mathbf{X}_0}$ is the Laplacian with 0 boundary conditions (say); and notice that at fixed external configuration $\mathbf{X}_j$ its spectrum consists of eigenvalues $E_n = E_n(\{\mathbf{X}_j\}_{j>0})$ (because the system in C_0 has finite size).

A system–reservoirs model can be the *dynamical system* on the space of the variables $\left(\Psi, (\{\mathbf{X}_j\}, \{\dot{\mathbf{X}}_j\})_{j>0}\right)$ defined by the equations (where $\langle\cdot\rangle_\Psi$ = expectation in the wave function Ψ)

$$-i\hbar\frac{d}{dt}\Psi(\mathbf{X}_0) = (H\Psi)(\mathbf{X}_0), \qquad \text{and for } j > 0$$

$$\ddot{\mathbf{X}}_j = -\left(\partial_j U_j(\mathbf{X}_j) + \langle\partial_j U_j(\cdot, \mathbf{X}_j)\rangle_\Psi\right) - \alpha_j\dot{\mathbf{X}}_j$$

$$\alpha_j \stackrel{def}{=} \frac{\langle W_j\rangle_\Psi - \dot{U}_j}{2K_j}, \qquad W_j \stackrel{def}{=} -\dot{\mathbf{X}}_j \cdot \partial_j U_{0j}(\mathbf{X}_0, \mathbf{X}_j)$$

$$\langle\partial_j U_j(\cdot, \mathbf{X}_j)\rangle_\Psi \stackrel{def}{=} \int_{C_0} d^{N_0}\mathbf{X}_0|\Psi(\mathbf{X}_0)|^2 F(\mathbf{X}_0, \mathbf{X}_j) \qquad (4.11.2)$$

here the first equation is Schrödinger's equation, the second is an equation of motion for the thermostats particles similar to the one in Fig. 2.2, (whose notation for the particles labels is adopted here too). The evolution is time reversible because the map $I(\Psi(\mathbf{X}_0), \{\dot{\mathbf{X}}_j, \mathbf{X}_j\}_{j=1}^n\}) = (\overline{\Psi(\mathbf{X}_0)}, \{-\dot{\mathbf{X}}_j, \mathbf{X}_j\}_{j=1}^n\})$ is a time reversal (isometric in $L_2(C^{3N_0}) \times R^{6\sum_{j>0} N_j}$).

The model, that can be called *Erhenfest dynamics* as it differs from the model in [48, 50] because of the use of a Gaussian rather than a Nosé-Hoover thermostat, has no pretension of providing a physically correct representation of the motions in the thermostats nor of the interaction system-thermostats, see comments at the end of this section.

Evolution maintains the thermostats kinetic energies $K_j \equiv \frac{1}{2}\dot{\mathbf{X}}_j^2$ exactly constant, so that they will be used to define the thermostats temperatures T_j via $K_j = \frac{3}{2}k_B T_j N_j$, as in the classical case.

Let $\mu_0(\{d\Psi\})$ be the *formal* measure on $L_2(C_0^{3N_0})$

$$\left(\prod_{\mathbf{X}_0} d\Psi_r(\mathbf{X}_0)\, d\Psi_i(\mathbf{X}_0)\right)\delta\left(\int_{C_0}|\Psi(\mathbf{Y})|^2\, d\mathbf{Y} - 1\right) \qquad (4.11.3)$$

with Ψ_r, Ψ_i real and imaginary parts of Ψ. The meaning of (4.11.3) can be understood by imagining to introduce an orthonormal basis in the Hilbert's space and to "cut it off" by retaining a large but finite number M of its elements, thus turning the space into a high dimensional space C^M (with $2M$ real dimensions) in which $d\Psi = d\Psi_r(\mathbf{X}_0)\, d\Psi_i(\mathbf{X}_0)$ is simply interpreted as the normalized euclidean volume in C^M.

The formal phase space volume element $\mu_0(\{d\Psi\}) \times \nu(d\mathbf{X}\, d\dot{\mathbf{X}})$ with

$$\nu(d\mathbf{X}\, d\dot{\mathbf{X}}) \overset{def}{=} \prod_{j>0} \left(\delta(\dot{\mathbf{X}}_j^2 - 3N_j k_B T_j)\, d\mathbf{X}_j\, d\dot{\mathbf{X}}_j \right) \tag{4.11.4}$$

is conserved, by the unitary property of the wave functions evolution, just as in the classical case, *up to the volume contraction in the thermostats*, [51].

If $Q_j \overset{def}{=} \langle W_j \rangle_\Psi$, as in Eq. (4.11.2), then the contraction rate σ of the volume element in Eq. (4.11.4) can be computed and is (again):

$$\sigma(\Psi, \dot{\mathbf{X}}, \mathbf{X}) = \varepsilon(\Psi, \dot{\mathbf{X}}, \mathbf{X}) + \dot{R}(\mathbf{X}), \qquad \varepsilon(\Psi, \dot{\mathbf{X}}, \mathbf{X}) = \sum_{j>0} \frac{Q_j}{k_B T_j}, \tag{4.11.5}$$

with R a suitable observable and with ε that will be called *entropy production rate*:

In general solutions of Eq. (4.11.2) *will not be quasi periodic* and the chaotic hypothesis [20, 28, 52], can therefore be assumed (i.e. there is no *a priori* conflict between the quasi periodic motion of an isolated quantum system and the motion of a non isolated system): if so the dynamics should select an invariant distribution μ. The distribution μ will give the statistical properties of the stationary states reached starting the motion in a thermostat configuration $(\mathbf{X}_j, \dot{\mathbf{X}}_j)_{j>0}$, randomly chosen with "uniform distribution" ν on the spheres $m\dot{\mathbf{X}}_j^2 = 3N_j k_B T_j$ and in a random eigenstate of H. The distribution μ, if existing and unique, could be named the *SRB distribution* corresponding to the chaotic motions of Eq. (4.11.2).

In the case of a system *interacting with a single thermostat* at temperature T_1 the latter distribution should attribute expectation value to observables for the particles in $\mathcal{C}_0$, i.e. for the test system hence operators on $L_2(\mathcal{C}_0^{3N_0})$, equivalent to the canonical distribution at temperature T_1, up to boundary terms.

Hence an important *consistency check* for proposing Eq. (4.11.2) as a model of a thermostatted quantum system is that, if the system is in contact with a single thermostat containing configurations $\dot{\mathbf{X}}_1, \mathbf{X}_1$, then there should exist at least one stationary distribution equivalent to the canonical distribution at the appropriate temperature T_1 associated with the (constant) kinetic energy of the thermostat: $K_1 = \frac{3}{2} k_B T_1 N_1$. In the corresponding classical case this is an established result, see comments to Eq. (2.8.7).

A natural candidate for a stationary distribution could be to attribute a probability proportional to $d\Psi\, d\mathbf{X}_1\, d\dot{\mathbf{X}}_1$ times

$$\sum_{n=1}^{\infty} e^{-\beta_1 E_n(\mathbf{X}_1)} \delta(\Psi - \Psi_n(\mathbf{X}_1) e^{i\varphi_n}) \, d\varphi_n \, \delta(\dot{\mathbf{X}}_1^2 - 2K_1) \qquad (4.11.6)$$

where $\beta_1 = 1/k_B T_1$, Ψ are wave functions for the system in $\mathcal{C}_0$, $\dot{\mathbf{X}}_1, \mathbf{X}_1$ are positions and velocities of the thermostat particles and $\varphi_n \in [0, 2\pi]$ is a phase for the eigenfunction $\Psi_n(\mathbf{X}_1)$ of $H(\mathbf{X}_1)$ and $E_n = E_n(\mathbf{X}_1)$ is the corresponding n-th level. The average value of an observable O for the system in $\mathcal{C}_0$ in the distribution μ in (4.11.6) would be

$$\langle O \rangle_\mu = Z^{-1} \int \mathrm{Tr}\, (e^{-\beta H(\mathbf{X}_1)} O) \, \delta(\dot{\mathbf{X}}_1^2 - 2K_1) d\mathbf{X}_1 \, d\dot{\mathbf{X}}_1 \qquad (4.11.7)$$

where Z is the integral in (4.11.7) with 1 replacing O, (normalization factor). Here one recognizes that μ attributes to observables the average values corresponding to a Gibbs state at temperature T_1 with a random boundary condition $\mathbf{X}_1$.

But Eq. (4.11.6) *is not invariant* under the evolution Eq. (4.11.2) and it seems difficult to exhibit explicitly an invariant distribution along the above lines without having recourse to approximations. A simple approximation is possible and is discussed in the next section essentially in the form proposed and used in [48].

Therefore one can say that the SRB distribution[18] for the evolution in (4.11.2) is equivalent to the Gibbs distribution at temperature T_1 with suitable boundary conditions, at least in the limit of infinite thermostats, to the Eq. (4.11.7) in spite of its non stationarity *only as a conjecture*.

Invariant distributions can, however, be constructed following the alternative ideas in [53], as done recently in [50], see remark (5).

4.12 Quantum Adiabatic Approximation and Alternatives

Nevertheless it is interesting to remark that under the *adiabatic approximation* the eigenstates of the Hamiltonian at time 0 evolve by simply following the variations of the Hamiltonian $H(\mathbf{X}(t))$ due to the motion of the thermostats particles, without changing quantum numbers (rather than evolving following the Schrödinger equation and becoming, therefore, *different* from the eigenfunctions of $H(\mathbf{X}(t))$).

In the adiabatic limit in which the classical motion of the thermostat particles takes place on a time scale much slower than the quantum evolution of the system the distribution (4.11.6) *is invariant*, [48].

This can be checked by first order perturbation analysis which shows that, to first order in t, the variation of the energy levels (supposed non degenerate) is compensated by the phase space contraction in the thermostat, [49]. Under time evolution, $\mathbf{X}_1$ changes, at time $t > 0$, into $\mathbf{X}_1 + t\dot{\mathbf{X}}_1 + O(t^2)$ and, assuming non degeneracy, the eigenvalue $E_n(\mathbf{X}_1)$ changes, by perturbation analysis, into $E_n + t\, e_n + O(t^2)$ with

---

[18] Defined, for instance, as the limit of the distributions obtained by evolving in time the Eq. (4.11.6).

$$e_n \overset{def}{=} t\langle \dot{\mathbf{X}}_1 \cdot \partial_{\mathbf{X}_1} U_{01}\rangle_{\psi_n} + t\dot{\mathbf{X}}_1 \cdot \partial_{\mathbf{X}_1} U_1 = -t\,(\langle W_1\rangle_{\psi_n} + \dot{R}_1) = -\frac{1}{\beta_1}\alpha_1 \quad (4.12.1)$$

with α_1 defined in Eq. (4.11.2).

Hence the Gibbs' factor changes by $e^{-\beta t e_n}$ and at the same time phase space contracts by $e^{t\frac{3N_1 e_n}{2K_1}}$, as it follows from the expression of the divergence in Eq. (4.11.5). *Therefore if β is chosen such that $\beta = (k_B T_1)^{-1}$ the state with distribution Eq. (4.11.6) is stationary* in the considered approximation, (recall that for simplicity $O(1/N)$ is neglected, see comment following Eq. (2.8.1)). This shows that, *in the adiabatic approximation*, interaction with only one thermostat at temperature T_1 admits at least one stationary state. The latter is, by construction, a Gibbs state of thermodynamic equilibrium with a special kind (random $\mathbf{X}_1, \dot{\mathbf{X}}_1$) of boundary condition and temperature T_1.

Remarks (1) The interest of the example is to show that even in quantum systems the chaotic hypothesis makes sense and the interpretation of the phase space contraction in terms of entropy production remains unchanged.

(2) In general, under the chaotic hypothesis, the SRB distribution of (4.11.2) (which in presence of forcing, or of more than one thermostat is certainly quite non trivial, as in the classical mechanics cases) will satisfy the fluctuation relation because, besides the chaotic hypothesis, the fluctuation theorem only depends on reversibility: so the model (4.11.2) might be suitable (given its chaoticity) to simulate the steady states of a quantum system in contact with thermostats.

(3) It is certainly unsatisfactory that the simple Eq. (4.11.6) is not a stationary distribution in the single thermostat case (unless the above adiabatic approximation is invoked). However, according to the proposed extension of the chaotic hypothesis, the model does have a stationary distribution which should be equivalent (in the sense of ensembles equivalence) to a Gibbs distribution at the same temperature: the alternative distribution in remark (5) below has the properties of being stationary and at the same time equivalent to the canonical Gibbs distribution for the test system in $\mathcal{C}_0$.

(4) The non quantum nature of the thermostat considered here and the specific choice of the interaction term between system and thermostats should not be important: the very notion of thermostat for a quantum system is not at all well defined and it is natural to think that in the end a thermostat is realized by interaction with a reservoir where quantum effects are not important. Therefore what the analysis really suggests is that, *in experiments in which really microscopic systems are studied, the heat exchanges of the system with the external world should fulfill a fluctuation relation.*

(5) An alternative approach can be based on the quantum mechanics formulation in [53] developed and subsequently implemented in simulations, where it is called *Erhenfest dynamics*, and more recently in [50]. It can be remarked that the equations of motion Eq. (4.11.2) can be derived from the Hamiltonian $\mathcal{H}$ on $L_2(\mathcal{C}_0^{3N_0}) \times \prod_{j=1}^n R^{6N_j}$ imagining a function $\Psi(\mathbf{X}_0) \in L_2(\mathcal{C}_0^{3N_0})$

as $\Psi(\mathbf{X}_0) \stackrel{def}{=} \kappa(\mathbf{X}_0) + i\pi(\mathbf{X}_0)$, with $\pi(\mathbf{X}_0)$, $\kappa(\mathbf{X}_0)$ canonically conjugate and defining $\mathcal{H}$ as:

$$\sum_{j=0}^{n} \frac{\dot{\mathbf{X}}_j^2}{2} + U(\mathbf{X}_j) + \int_{R^{3N_0}} \left(\frac{\partial_{\mathbf{X}_0}\pi(\mathbf{X}_0)^2 + \partial_{\mathbf{X}_0}\kappa(\mathbf{X}_0)^2}{2} \right.$$
$$\left. + \frac{(\pi(\mathbf{X}_0)^2 + \kappa(\mathbf{X}_0)^2)(U(\mathbf{X}_0) + W(\mathbf{X}_0, \mathbf{X}_j))}{2} \right) d\mathbf{X}_0 \stackrel{def}{=} \mathcal{H} \qquad (4.12.2)$$

where $W(\mathbf{X}_0, \mathbf{X}_0) \equiv 0$ and adding to it the constraints $\int |\Psi(\mathbf{X}_0)|^2 d\mathbf{X}_0 = 1$ (which is an integral of motion) and $\frac{1}{2}\dot{\mathbf{X}}_j^2 = 3N_j k_B T_j$, $j = 1, \ldots, n$ by adding to the equations of the thermostats particles $-\alpha_j \dot{\mathbf{X}}_j$ with α_j as in Eq. (4.11.2). In this case, by the same argument leading to the theorem following Eq. (2.8.7), the formal distribution $const\ e^{-\beta\mathcal{H}} d\mathbf{X}_0 d\dot{\mathbf{X}}_0\, d\pi d\kappa$ is stationary and equivalent to the canonical distribution for the test system if the thermostats have all the same temperature. This avoids using the adiabatic approximation. This alternative approach is well suitable for simulations as shown for instance in [50]. The above comment is due to M. Campisi (private communication, see also [54] where a transient fluctuation relation is studied).

(6) It would be interesting to prove for the evolution of the Hamiltonian in Eq. (4.12.2) theorems similar to the corresponding ones for the classical systems in Sect. 5.2, under the same assumptions on the interaction potentials and with Dirichlet boundary conditions for the fields π, κ.

References

1. Ruelle, D.: What are the measures describing turbulence. Prog. Theor. Phys. Suppl. **64**, 339–345 (1978)
2. Ruelle, D.: Measures describing a turbulent flow. Ann. N. Y. Acad. Sci. **357**, 1–9 (1980)
3. Ruelle, D.: A measure associated with axiom a attractors. Am. J. Math. **98**, 619–654 (1976)
4. Gallavotti, G., Bonetto, F., Gentile, G.: Aspects of the ergodic, qualitative and statistical theory of motion. Springer, Berlin (2004)
5. Kernigham, B.W., Ritchie, D.M.: The C Programming Language. Prentice Hall Software Series. Prentice Hall, Engelwood Cliffs (1988)
6. Ornstein, D.: Ergodic theory, randomness and dynamical Systems. Yale Mathematical Monographs, vol. 5. Yale University Press, New Haven (1974)
7. Gallavotti, G.: Ising model and Bernoulli shifts. Commun. Math. Phys. **32**, 183–190 (1973)
8. Ledrappier, F.: Mesure d'equilibre sur un reseau. Commun. Math. Phys. **33**, 119–128 (1973)
9. Bonetto, F., Gallavotti, G.: Reversibility, coarse graining and the chaoticity principle. Commun. Math. Phys. **189**, 263–276 (1997)
10. Gallavotti, G.: Breakdown and regeneration of time reversal symmetry in nonequilibrium statistical mechanics. Physica D **112**, 250–257 (1998)
11. Dettman, C., Morriss, G.: Proof of conjugate pairing for an isokinetic thermostat. Phys. Rev. E **53**, 5545–5549 (1996)
12. Bonetto, F., Gallavotti, G., Garrido, P.: Chaotic principle: an experimental test. Physica D **105**, 226–252 (1997)

13. Eckmann, J.P., Ruelle, D.: Ergodic theory of chaos and strange attractors. Rev. Mod. Phys. **57**, 617–656 (1985)
14. Gallavotti, G., Cohen, E.G.D.: Dynamical ensembles in nonequilibrium statistical mechanics. Phys. Rev. Lett. **74**, 2694–2697 (1995)
15. Gallavotti, G.: Topics in chaotic dynamics. In: Garrido, P., Marro, J. (ed.) Lecture Notes in Physics, vol. 448, pp. 271–311. Springer, Berlin (1995)
16. Sinai, Y.G.: Gibbs measures in ergodic theory. Russ. Math. Surv. **27**, 21–69 (1972)
17. Sinai, Y.G.: Lectures in ergodic theory. Lecture notes in Mathematics. Princeton University Press, Princeton (1977)
18. Sinai, Y.G.: Topics in ergodic theory. Princeton Mathematical Series, vol. 44. Princeton University Press, Princeton (1994)
19. Sinai, Y.G.: Markov partitions and C-diffeomorphisms. Funct. Anal. Appl. **2**(1), 64–89 (1968)
20. Gallavotti, G., Cohen, E.G.D.: Dynamical ensembles in stationary states. J. Stat. Phys. **80**, 931–970 (1995)
21. Gallavotti, G.: Reversible anosov diffeomorphisms and large deviations. Math. Phys. Electron. J. (MPEJ) **1**, 1–12 (1995)
22. Gallavotti, G., Cohen, E.G.D.: Note on nonequilibrium stationary states and entropy. Phys. Rev. E **69**, 035104 (+4) (2004)
23. Bowen, R.: Markov partitions for axiom a diffeomorphisms. Am. J. Math. **92**, 725–747 (1970)
24. Chernov, N.I., Eyink, G.L., Lebowitz, J.L.,Sinai, Y.G.: Derivation of Ohm's law in a deterministic mechanical model. Phys. Rev. Lett. **70**, 2209–2212 (1993)
25. Bowen, R., Ruelle, D.: The ergodic theory of axiom a flows. Inventiones Mathematicae **29**, 181–205 (1975)
26. Gentile, G.: A large deviation theorem for anosov flows. Forum Mathematicum **10**, 89–118 (1998)
27. Gallavotti, G.: Fluctuation patterns and conditional reversibility in nonequilibrium systems. Annales de l' Institut H. Poincaré **70**, 429–443 (1999) (chao-dyn/9703007)
28. Gallavotti, G.: Statistical Mechanics. A Short Treatise. Springer, Berlin (2000)
29. Gallavotti, G.: Foundations of Fluid Dynamics (2nd printing). Springer, Berlin (2005)
30. de Groot, S., Mazur, P.: Nonequilibrium Thermodynamics. Dover, Mineola (1984)
31. Gallavotti, G.: New methods in nonequilibrium gases and fluids. Open Syst. Inf. Dyn. **6**, 101–136 (1999) (preprint chao-dyn/9610018)
32. Gomez-Marin, A., Parondo, J.M.R., Van den Broeck, C.: The footprints of irreversibility. European. Phys. Lett. **82**, 5002(+4) (2008)
33. Hurtado, P., Péres-Espigares, C., Pozo, J., Garrido, P.: Symmetries in fluctuations far from equilibrium. Proc. Nat. Acad. Sci. **108**, 7704G7709 (2011)
34. Gallavotti, G.: Chaotic hypothesis: onsager reciprocity and fluctuation-dissipation theorem . J. Stat. Phys. **84**, 899–926 (1996)
35. Bonetto, F.: Entropy theorem. Private communication, see (9.10.4) in [Ga00] (1997)
36. Gallavotti, G.: Extension of onsager's reciprocity to large fields and the chaotic hypothesis. Phys. Rev. Lett. **77**, 4334–4337 (1996)
37. Gallavotti, G., Ruelle, D.: Srb states and nonequilibrium statistical mechanics close to equilibrium. Commun. Math. Phys. **190**, 279–285 (1997)
38. Ruelle, D.: Differentiation of srb states. Commun. Math. Phys. **187**, 227–241 (1997)
39. Gallavotti, G.: A local fluctuation theorem. Physica A **263**, 39–50 (1999)
40. Pesin, Y.B., Sinai, Y.G.: Space-time chaos in chains of weakly inteacting hyperbolic mappimgs. Adv. Sov. Math. **3**, 165–198 (1991)
41. Bricmont, J., Kupiainen, A.: High temperature expansions and dynamical systems. Commun. Math. Phys. **178**, 703–732 (1996)
42. Jiang, M., Pesin, Y.B.: Equilibrium measures for coupled map lattices: existence, uniqueness and finite-dimensional approximations. Commun. Math. Phys. **193**, 675–711 (1998)
43. Gallavotti, G.: Equivalence of dynamical ensembles and Navier Stokes equations. Phys. Lett. A **223**, 91–95 (1996)

44. Bonetto, F., Gallavotti, G., Gentile, G.: A fluctuation theorem in a random environment. Ergodic Theory Dyn. Syst. **28**, 21–47 (2008)
45. Olla, S.: Large deviations for gibbs random fields. Probab. Theory Relat. Fields **77**, 343–357 (1988)
46. Gallavotti, G., Lebowitz, J.L., Mastropietro, V.: Large deviations in rarefied quantum gases. J. Stat. Phys. **108**, 831–861 (2002)
47. Gallavotti, G., Perroni, F.: An experimental test of the local fluctuation theorem in chains of weakly interacting anosov systems (unpublished, draft). http://ipparcoroma1.infn.it (1999)
48. Mauri, F., Car, R., Tosatti, E.: Canonical statistical averages of coupled quantum-classical systems. Europhys. Lett. **24**, 431–436 (1993)
49. Gallavotti, G.: Heat and fluctuations from order to chaos. Eur. Phys. J. B (EPJB) **61**, 1–24 (2008)
50. Alonso, J.L., Castro, A., Clemente-Gallardo, J., Cuchi, J.C., Echenique, P., Falceto, F.: Statistics and nosé formalism for Ehrenfest dynamics. J. Phys. A **44**, 395004 (2011)
51. Gallavotti, G.: Entropy, thermostats and chaotic hypothesis. Chaos **16**, 043114 (+6) (2006)
52. Gallavotti, G.: The Elements of Mechanics, 2nd edn. http://ipparco.roma1.infn.it, Roma (2008) (I edition was Springer 1984)
53. Strocchi, F.: Complex coordinates and quantum mechanics. Rev. Mod. Phys. **38**, 36–40 (1966)
54. Campisi, M.: Quantum Fluctuation Relations for Ensembles of Wave Functions. arxiv:1306.5557, pp. 1–12 (2013)

Chapter 5
Applications

5.1 Equivalent Thermostats

In Sect. 3.2 two models for the electric conduction have been considered

(1) the classical model of Drude [1, Vol. 2, Sect. 35], [2, p. 139], in which at *every collision* the electron, of charge $e = 1$, velocity is reset to the average velocity at the given temperature, with a random direction.
(2) The Gaussian model in which the total kinetic energy is kept constant by a thermostat force

$$m\ddot{\mathbf{x}}_i = \mathbf{E} - \frac{m\mathbf{E} \cdot \mathbf{J}}{3k_B T}\,\dot{\mathbf{x}}_i + \text{``collisional forces''} \qquad (5.1.1)$$

where $3Nk_B T/2$ is the total kinetic energy (a constant of motion in this model), [3, 4]. A third model could be

(3) a "friction model" in which particles independently experience a constant friction

$$m\ddot{\mathbf{x}}_i = \mathbf{E} - \nu\,\dot{\mathbf{x}}_i + \text{``collisional forces''} \qquad (5.1.2)$$

where ν is a constant tuned so that the *average kinetic energy* is $3Nk_B T/2$, [5, 6].

The first model is a "stochastic model" while the second and third are deterministic: the third is "irreversible" while the second is reversible because the isometry $I(\mathbf{x}_i, \mathbf{v}_i) = (\mathbf{x}_i, -\mathbf{v}_i)$ anticommutes with the time evolution flow S_t defined by the Eq. (5.1.1): $IS_t = S_{-t}I$.

Here the models will be considered in a thermodynamic context in which the number of particles and obstacles are proportional to the system size $L^d = V$ which is large. The chaotic hypothesis will be assumed, hence the systems will admit a SRB distribution which is supposed unique.

Let $\mu_{\delta,T}$ be the SRB distribution for Eq. (5.1.1) for the stationary state that is reached starting from initial data, chosen randomly as in Sect. 2.4, with energy

G. Gallavotti, *Nonequilibrium and Irreversibility*,
Theoretical and Mathematical Physics, DOI: 10.1007/978-3-319-06758-2_5,
© Springer International Publishing Switzerland 2014

$3Nk_BT/2$ and density $\delta = \frac{N}{V}$. The collection of the distributions $\mu_{\delta,T}$ as the kinetic energy T and the density δ vary, define a "statistical ensemble" $\mathcal{E}$ of stationary distributions associated with the Eq. (5.1.1).

Likewise we call $\widetilde{\mu}_{\delta,\nu}$ the class of SRB distributions associated with Eq. (5.1.2) which forms an "ensemble" $\widetilde{\mathcal{E}}$.

A correspondence between distributions of the ensembles $\mathcal{E}$ and $\widetilde{\mathcal{E}}$ can be established by associating $\mu_{\delta,T}$ and $\widetilde{\mu}_{\delta',\nu}$ as "corresponding elements" if

$$\delta = \delta', \qquad \frac{3}{2}k_BT = \int \frac{1}{2}(\sum_j m\dot{\mathbf{x}}_j^2)\,\widetilde{\mu}_{\delta,\nu}(d\mathbf{x}\,d\dot{\mathbf{x}}) \qquad (5.1.3)$$

Then the following conjecture was proposed in [6].

Conjecture 1 (equivalence conjecture) *Let F be a "local observable", i.e. an observable depending solely on the microscopic state of the electrons whose positions is inside some box V_0 fixed as V varies. Then, if $\mathcal{L}$ denotes the local smooth observables, for all $F \in \mathcal{L}$, it is*

$$\lim_{N\to\infty, N/V=\delta} \widetilde{\mu}_{\delta,\nu}(F) = \lim_{N\to\infty, N/V=\delta} \mu_{\delta,T}(F) \qquad (5.1.4)$$

if T and ν are related by Eq. (5.1.3).

This conjecture has been discussed in [7, Sect. 8], [5, Sect. 5], [8, Sects. 2, 5] and [9, Sect. 9.11]: and in [10] arguments in favor of it have been developed. The idea of this kind of ensemble equivalence was present since the beginning as a motivation for the use of thermostats like the Nosé–Hoover's or Gaussian. It is clearly introduced and analyzed in [11], where earlier works are quoted.

The conjecture is very similar to the equivalence, in equilibrium cases, between canonical and microcanonical ensembles: here the friction ν plays the role of the canonical inverse temperature and the kinetic energy that of the microcanonical energy.

It is remarkable that the above equivalence suggests equivalence between a "reversible statistical ensemble", i.e. the collection $\mathcal{E}$ of the SRB distributions associated with Eq. (5.1.1) and a "irreversible statistical ensemble", i.e. the collection $\widetilde{\mathcal{E}}$ of SRB distributions associated with Eq. (5.1.2).

Furthermore it is natural to consider also the collection $\mathcal{E}'$ of stationary distributions for the original stochastic model (1) of Drude, whose elements $\mu'_{\delta,T}$ can be parameterized by the quantities T, temperature (such that $\frac{1}{2}\sum_j m\dot{\mathbf{x}}_j^2 = \frac{3}{2}Nk_BT$), and density ($N/V = \delta$). This is an ensemble $\mathcal{E}'$ whose elements can be put into one to one correspondence with the elements of, say, the ensemble $\mathcal{E}$ associated with model (2), i.e. with Eq. (5.1.1): an element $\mu'_{\delta,T} \in \mathcal{E}'$ corresponds to $\mu_{\delta,T} \in \mathcal{E}$. Then

Conjecture 2 *If $\mu_{\delta,T} \in \mathcal{E}$ and $\mu'_{\delta,T} \in \mathcal{E}'$ are corresponding elements (i.e. Eq. (5.1.3) holds) then*

$$\lim_{N\to\infty,N/V=\delta} \mu_{\delta,T}(F) = \lim_{N\to\infty,N/V=\delta} \mu'_{\delta,T}(F) \qquad (5.1.5)$$

for all local observables $F \in \mathcal{L}$.

Hence we see that there can be statistical equivalence between a viscous irreversible dissipation model and either a stochastic dissipation model or a reversible dissipation model,[1] at least as far as the averages of special observables are concerned.

The argument in [10] in favor of conjecture 1 is that the coefficient α in Fig. 2.3 is essentially the average J of the current over the *whole* box containing the system of particles, $J = N^{-1} \sum_j \dot{x}_i$: hence in the limit $N \to \infty$, $\frac{N}{V} = \delta$ the current J should be constant with probability 1, at least if the stationary SRB distributions can be reasonably supposed to have some property of ergodicity with respect to *space translations*.

In general translation invariance should not be necessary: when a system is large the microscopic evolution time scale becomes much shorter than the macrosopic one and the multiplier α becomes a sum of many quantities rapidly varying (in time as well as in space) and therefore could be considered as essentially constant if only macroscopic time—independent quantities are observed.

5.2 Granular Materials and Friction

The current interest in granular materials properties and the consequent availability of experiments, e.g. [12], suggests trying to apply the ideas on nonequilibrium statistics to derive possible experimental tests of the chaotic hypothesis in the form of a check of whether probabilities of fluctuations agrees with the fluctuation relation, Eq. (4.6.5).

The main problem is that in granular materials collisions are intrinsically *inelastic*. In each collision particles heat up, and the heat is subsequently released through thermal exchange with the walls of the container, sound emission (if the experiment is performed in air), radiation, and so on. If one still wants to deal with a *reversible* system, such as the ones discussed in the previous sections, all these sources of dissipation should be included in the theoretical description. Clearly, this is a *very* hard task, and it seems that it cannot be pursued.

A simplified description, [13], of the system consists in neglecting the internal degrees of freedom of the particles. In this case the inelastic collisions between particles will represent the only source of dissipation in the system. Still the chaotic hypothesis is expected to hold, but in this case the entropy production is strictly positive and there is no hope of observing a fluctuation relation, see e.g. [14], if one looks at the whole system.

Nevertheless, in presence of inelasticity, temperature gradients may be present in the system [12, 15, 16], and heat is transported through different regions of the container. The processes of heat exchange between different regions could be

---

[1] For example a system subject to a Gaussian thermostat.

described assuming that, under suitable conditions, the inelasticity of the collisions can be neglected, and a fluctuation relation for a (suitably defined) entropy production rate might become observable. This could lead to an interesting example of "ensemble equivalence" in nonequilibrium [9], and its possibility will be pursued in detail in the following.

As a concrete model for a granular material experiment let Σ be a container consisting of two flat parallel vertical walls covered at the top and with a piston at the bottom that is kept oscillating by a motor so that its height is

$$z(t) = A \cos \omega t. \tag{5.2.1}$$

The model can be simplified by introducing a sawtooth moving piston as in [16], however the results should not depend too much on the details of the time dependence of $z(t)$.

The container Σ is partially filled with millimeter size balls (a typical size of the faces of Σ is 10 cm and the particle number is of a few hundreds): the vertical walls are so close that the balls almost touch both faces so the problem is effectively two dimensional. The equations of motion of the balls with coordinates (x_i, z_i), $i = 1, \ldots, N$, $z_i \geq z(t)$, are

$$m\ddot{x}_i = f_{x,i}$$
$$m\ddot{z}_i = f_{z,i} - mg + m\delta(z_i - z(t)) \, 2 \, (\dot{z}(t) - \dot{z}_i) \tag{5.2.2}$$

where m=mass, g=gravity acceleration, and the collisions between the balls and the oscillating base of the container are assumed to be elastic [16] (possibly inelasticity of the walls can be included into the model with negligible changes [14]); $\mathbf{f}_i$ is the force describing the particle collisions and the particle-walls or particles-piston collisions.

The force $\mathbf{f}_i = (f_{x,i}, f_{z,i})$ has a part describing the particle collisions: the latter are necessarily inelastic and it will be assumed that their ineslasticity is manifested by a restitution coefficient $\alpha < 1$. A simple model for inelastic collisions with inelasticity α (convenient for numerical implementation) is a model in which collisions take place with the usual elastic collision rule but immediately after the velocities of the particles that have collided are scaled by a factor so that the kinetic energy of the pair is reduced by a factor $1 - \alpha^2$ [14–16].

With in mind the discussion of Sect. 4.9, about the formulation of a local fluctuation relation, the simplest situation that seems accessible to experiments as well as to simulations is to draw ideal horizontal lines at heights $h_1 > h_2$ delimiting a strip Σ_0 in the container and to look at the particles in Σ_0 as a thermostatted system, the thermostats being the regions Σ_1 and Σ_2 at heights larger than h_1 and smaller then h_2, respectively.

After a stationary state has been reached, the average kinetic energy of the particles will depend on the height z, and in particular will decrease on increasing z.

Given the motion of the particles and a time interval t it will be possible to measure the quantity Q_2 of (kinetic) energy that particles entering or exiting the region Σ_0 or colliding with particles inside it from below (the "hotter side") carry out of Σ_0 and the analogous quantity Q_1 carried out by the particles that enter, exit or collide from above (the "colder side").

If T_i, $i = 1, 2$, are the average kinetic energies of the particles in small horizontal corridors above and below Σ_0, a connection between the model of granular material, Eq. (5.2.2), and the general thermostat model in Sect. 2.2 can be established. The connection cannot be exact because of the internal dissipation induced by the inelasticity α and of the fact that the number of particles, and their identity, in Σ_0 depends on time, as particles come and go in the region.

Under suitable assumptions, that can be expected to hold on a specific time scale, the stationary state of Eq. (5.2.2) is effectively described in terms a stationary SRB state of models like the one considered in Sect. 2.2, as discussed below.

Real experiments cannot have an arbitrary duration, [12]: the particles movements are recorded by a digital camera and the number of photograms per second is of the order of a thousand, so that the memory for the data is easily exhausted as each photogram has a size of about 1 Mb in current experiments (<2008). The same holds for numerical simulations where the accessible time scale is limited by the available computational resources.

Hence each experiment lasts up to a few seconds starting after the system has been moving for a while so that a stationary state can be supposed to have been reached. The result of the experiment is the reconstruction of the trajectory, in phase space, of each individual particle inside the observation frame, [12].

In order for the number of particles N_0 in Σ_0 to be approximately constant for the duration of the experiment, the vertical size $(h_1 - h_2)$ of Σ_0 should be chosen large compared to $(Dt)^{1/2}$, where t is the duration of the experiment and D is the diffusion coefficient of the grains. Hence we are assuming, see Sect. 4.10, that the particles motion is diffusive on the scale of Σ_0. Note that at low density the motion could be not diffusive on the scale of Σ_0 (i.e. free path larger than the width of Σ_0): then it would not be possible to divide the degrees of freedom between the subsystem and the rest of the system and moreover the correlation length would be comparable with (or larger than) the size of the subsystem Σ_0. This would completely change the nature of the problem: and violations of the fluctuation relation would occur, [17, 18].

Given the remarks above suppose that in observations of stationary states lasting up to a maximum time θ:

(1) the chaotic hypothesis is accepted,
(2) it is supposed that the result of the observations would be the same if the particles above Σ_0 and below Σ_0 were kept at constant total kinetic energy by reversible thermostats (e.g. Gaussian thermostats), [9–11],
(3) dissipation due to inelastic collisions between particles in Σ_0 is neglected,
(4) fluctuations of the number of particles in Σ_0 is neglected,
(5) dissipation is present in the sense that

$$\sigma_+ \stackrel{def}{=} \frac{1}{\theta}\left(\frac{Q_1(\theta)}{T_1} + \frac{Q_2(\theta)}{T_2}\right) > 0 , \qquad (5.2.3)$$

with $Q_i(t)$ is the total heat ceded to the particles in Σ_i, $i = 1, 2$, in time t.

Chaoticity is expected at least if dissipation is small and evidence for it is provided by the experiment in [12] which indicates that the system evolves to a chaotic stationary state.

For the purpose of checking a fluctuation relation for $\sigma_0 = \frac{1}{\tau}(\frac{Q_1(\tau)}{T_1} + \frac{Q_2(\tau)}{T_2})$, where $Q_i(\tau)$ is the total heat ceded to the particles in Σ_i, $i = 1, 2$, in time τ, the observation time $\tau \leq \theta$ should be not long enough that dissipation due to internal inelastic collisions becomes important. So measurements, starting after the stationary state is reached, can have a duration τ which cannot exceed *a specific time scale* in order that the conditions for a local fluctuation relation can be expected to apply to model Eq. (5.2.2), as discussed below.

Accepting the assumptions above, a fluctuation relation is expected for fluctuations of

$$p = \frac{1}{\tau\,\sigma_+}\left(\frac{Q_1(\tau)}{T_1} + \frac{Q_2(\tau)}{T_2}\right) \tag{5.2.4}$$

in the interval $(-p^*, p^*)$ with p^* equal (at least) to 1, but a discussion of the assumptions is needed, see next section.

The latter is therefore a property that might be accessible to simulations as well as to experimental test. Note however that it is very likely that the hypotheses (2)–(4) above will not be *strictly* verified in real experiments, see the discussion in next section, so that the analysis and interpretation of the experimental results might be non trivial. Nevertheless, a careful test would be rather stringent.

5.3 Neglecting Granular Friction: The Relevant Time Scales

The above analysis assumes, [13], the existence of (at least) two time scales. One is the "equilibrium time scale", θ_e, which is the time scale over which the system evolving at constant energy, equal to the average energy observed, would reach equilibrium in absence of friction and forcing.

An experimental measure of θ_e would be the decorrelation time of self–correlations in the stationary state, and it can be assumed that θ_e is of the order of the mean time between collisions of a selected particle. Note that θ_e also coincides with the time scale over which finite time corrections to the fluctuation relation become irrelevant [19]: this means that in order to be able to measure the large deviations functional for the normalized entropy production rate p in Eq. (5.2.4) one has to choose $t \gg \theta_e$, see also [20] for a detailed discussion of the leading finite time corrections to the large deviation functional.

A second time scale is the "inelasticity time scale" θ_d, which is the scale over which the system reaches a stationary state if the particles are prepared in a random configuration and the piston is switched on at time $t = 0$.

Possibly a third time scale is present: the "diffusion time scale" θ_D which is the scale over which a particle diffuses beyond the width of Σ_0.

The analysis above applies only if the time t in Eq. (5.2.4) verifies $\theta_e \ll t \ll \theta_d, \theta_D$ (note however that the measurement should be started after a time $\gg \theta_d$ since the piston has been switched on in order to have a stationary state); in practice this means that the time for reaching the stationary state has to be quite long compared to θ_e. In this case friction is negligible *for the duration of the measurement* if the measurement is performed starting after the system has reached a stationary state and lasts a time τ between θ_e and $\min(\theta_D, \theta_d)$.

In the setting considered here, the role of friction is "just" that of producing the nonequilibrium stationary state itself and the corresponding gradient of temperature: this is reminiscent of the role played by friction in classical mechanics problems, where periodic orbits (the "stationary states") can be dynamically selected by adding a small friction term to the Hamilton equations. Note that, as discussed below, the temperature gradient produced by friction will be rather small: however smallness of the gradient does not affect the "FR time scale" over which FR is observable [19].

If internal friction were not negligible (that is if $t \geq \theta_d$) the problem would change nature: an explicit model (and theory) should be developed to describe the transport mechanisms (such as radiation, heat exchange between the particles and the container, sound emission, ...) associated with the dissipation of kinetic energy and new thermostats should be correspondingly introduced. The definition of entropy production should be changed, by taking into account the presence of such new thermostats. In this case, even changing the definition of entropy production it is not expected that a fluctuation relation should be satisfied: in fact internal dissipation would not break the chaotic hypothesis, but the necessary time–reversibility assumption would be lost, [13].

The possibility of $\theta_e \ll t \ll \theta_d, \theta_D$ is not obvious, neither in theory nor in experiments. A rough estimate of θ_d can be given as follows: the phase space contraction in a single collision is given by $1 - \alpha$. Thus the average phase space contraction per particle and per unit time is $\sigma_{+,d} = (1 - \alpha)/\theta_e$, where $1/\theta_e$ is the frequency of the collisions for a given particle. It seems natural to assume that θ_d is the time scale at which $\sigma_{+,d}\theta_d$ becomes of order 1: on this time scale inelasticity will become manifest. Thus, we obtain the following estimate:

$$\theta_d \sim \frac{1}{1 - \alpha}\theta_e \tag{5.3.1}$$

In real materials $\alpha \leq 0.95$, so that θ_d can be at most of the order of $20\,\theta_e$. Nevertheless it might be possible that this be already enough to observe a fluctuation relation on intermediate times.

The situation is completely different in numerical simulations where we can play with our freedom in choosing the restitution coefficient α (it can be chosen very close to one [14–16], in order to have $\theta_d \gg \theta_e$) and the size of the container Σ_0 (it can be chosen large, in order to have $\theta_D \gg \theta_e$).

5.4 Simulations for Granular Materials

To check the consistency of the hypotheses in Sect. 5.3, it has to be shown that it is possible to make a choice of parameters so that θ_e and θ_d, θ_D are separated by a large time window. Such choices may be possible, as discussed below, [13]. Let

$$\delta \overset{def}{=} h_1 - h_2 \quad \text{is the width of } \Sigma_0,$$

$$\varepsilon \overset{def}{=} 1 - \alpha,$$

$$\gamma \overset{def}{=} \text{is the temperature gradient in } \Sigma_0, \qquad (5.4.1)$$

$$D \overset{def}{=} \text{is the diffusion coefficient}$$

the following estimates hold:

(a) $\theta_e = O(1)$ as it can be taken of the order of the inverse collision frequency, which is $O(1)$ if density is constant and the forcing on the system is tuned to keep the energy constant as $\varepsilon \to 0$.
(b) $\theta_d = \theta_e O(\varepsilon^{-1})$ as implied by Eq. (5.3.1).
(c) $\theta_D = O(\frac{\delta^2}{D}) = O(\delta^2)$ because D is a constant (if the temperature and the density are kept constant).
(d) $\gamma = O(\sqrt{\varepsilon})$, as long as $\delta \ll \varepsilon^{-1/2}$.

In fact if the density is high enough to allow us to consider the granular material as a fluid, as in Eq. (5) of Brey et al. [16], the temperature profile should be given by the heat equation $\nabla^2 T + c\varepsilon T = 0$ with suitable constant c and suitable boundary conditions on the piston ($T = T_0$) and on the top of the container ($\nabla T = 0$). This equation is solved by a linear combination of $const\, e^{\pm\sqrt{c\varepsilon}z}$, which has gradients of order $O(\sqrt{\varepsilon})$, as long as $\delta \ll 1/\sqrt{\varepsilon}$ and the boundaries of Σ_0 are further than $O(1/\sqrt{\varepsilon})$ from the top.

Choosing $\delta = \varepsilon^{-\beta}$, with $\beta < \frac{1}{2}$, and taking ε small enough, it is $\theta_e \ll \min\{\theta_d, \theta_D\}$ and $\delta \ll O(\varepsilon^{-\frac{1}{2}})$, as required by item (d).

Remark The entropy production rate due to heat transport into Σ_0, in presence of a temperature gradient γ, is given by $\sigma_+ = O(\gamma^2\delta) = O(\varepsilon\delta)$ because the temperature difference is $O(\gamma\delta)$ and the energy flow through the surface is of order $O(\gamma)$ (with $\gamma = O(\sqrt{\varepsilon})$, see item (d)). The order of magnitude of σ_+ is not larger then the average amount σ_d of energy dissipated per unit time in Σ_0 divided by the average kinetic energy T (the latter quantity is of order $O(\theta_e^{-1}\varepsilon\delta)$ because, at constant density, the number of particles in Σ_0 is $O(\delta)$); however the entropy creation due to the dissipative collisions in Σ_0 has fluctuations of order $O(\varepsilon\delta^{\frac{1}{2}})$ because the number of particles in Σ_0 fluctuates by $O(\delta^{\frac{1}{2}})$. This is consistent with neglecting the entropy creation inside the region Σ_0 due to the inelasticity in spite of it being of the same order of the entropy creation due to the heat entering Σ_0 from its upper and lower regions.

The argument supports the proposal that in numerical simulations a fluctuation relation test might be possible by a suitable choice of the parameters. Other choices will be possible: for instance in the high-density limit it is clear that $\theta_D \gg \theta_e$ because the diffusion coefficient will become small. To what extent this can be applied to experiment is a further question.

Remark (1) An explicit computation of the large deviation function of the dissipated power, in the regime $t \gg \theta_d$ (i.e. when the dissipation is mainly due to inelastic collisions) recently appeared in [21]. However in the model only the dissipation due to the collisions was taken into account, [13]: so that it is not clear how the heat produced in the collisions is removed from the system, see the discussion above. It turned out that in this regime no negative values of p are observed so that the fluctuation relation, Eq. (4.6.5), p. 81, cannot hold. This is interesting and expected on the basis of the considerations above. It is not clear if, including the additional thermostats required to remove heat from the particles and prevent them to warm up indefinitely, the fluctuation relation, Eq. (4.6.5), is recovered.

(2) There has also been some debate on the interpretation of the experimental results of [12]. In [14] a simplified model, very similar to the one discussed above, was proposed and showed to reproduce the experimental data of [12]. The prediction of the model is that the fluctuation relation is not satisfied. Note however that the geometry considered in [12, 14] is different from the one considered here: the whole box is vibrated, so that the the temperature profile is symmetric, and a region Σ_0 in the center of the box is considered. Heat exchange is due to "hot" particles entering Σ_0 (i.e. Q_+) and "cold" particles exiting Σ_0 (i.e. Q_-). One has $Q = Q_+ + Q_- \neq 0$ because of the dissipation in Σ_0.

In this regime, again, the fluctuation relation is not expected to hold if the thermostat dissipating the heat produced in the collisions is not included in the model: it is an interesting remark of [14] that partially motivated the present discussion. Different experiments can be designed in which the dissipation is mainly due to heat exchanges and the inelasticity is negligible, as the one proposed above as an example.

(3) Even in situations in which the dissipation is entirely due to irreversible inelastic collisions between particles, such as the ones considered in [14, 21], the chaotic hypothesis is expected to hold, and the stationary state to be described by a SRB distribution. But in these cases failure of the fluctuation relation is not contradictory, due to the irreversibility of the equations of motion.

(4) In cases like the Gaussian isoenergetic models, or in other models in which the kinetic energy fluctuates (e.g. in the proposal above) care has to be paid to measure the fluctuations of the ratios $\frac{Q}{K}$ rather than those of Q and K separately because there might not be an "effective temperature" of the thermostats (i.e. fluctuations of K may be important).

(6) Finally it is important to keep track of the errors due to the size of Δ in the fluctuation relation, Eq. (4.6.5) p. 81: the condition that $\Delta \ll p$ make it very difficult to test the fluctuation relation near $p = 0$: this may lead to interpretation problems and, in fact, in many experimental works or simulations the fluctuation

relation is written for the non normalized entropy production or phase space contraction

$$\frac{Prob(A \in \Delta)}{Prob(-A \in \Delta)} \simeq e^A \qquad (5.4.2)$$

where A is the total phase space contraction in the observation time ([22], and see Appendix L). This relation may lead to illusory agreement with data, unless a detailed error analysis is done, as it can be confused (and it has been often confused) with the linearity at small A due to the extrapolation at $p = 0$ of the central limit theorem[2] or just to the linearity of the large deviation function near $p = 0$. Furthermore the $A = p\tau\sigma_+$ depends on the observation time τ and on the dissipation rate σ_+ with $p = O(1)$: all this is hidden in Eq. (5.4.2).

5.5 Fluids

The ideas in Sect. 5.1 show that the negation of the notion of reversibility is not "irreversibility": *it is instead the property that the natural time reversal map I does not verify $IS = S^{-1}I$*, i.e. does not anticommute with the evolution map S. This is likely to generate misunderstandings as the word irreversibility usually refers to lack of velocity reversal symmetry in systems whose microscopic description is or should be velocity reversal symmetric.

The typical phenomenon of reversibility (i.e. the indefinite repetition, or "*recurrence*", of "*impossible*" states) in isolated systems should indeed manifest itself, but on time scales much longer and/or on scales of space much smaller than those interesting for the class of motions considered here: where motions of the system could be considered as a continuous fluid.

The transport coefficients (such as viscosity or conductivity or other) *do not have a fundamental nature*: rather they must be thought of as macroscopic parameters related to the disorder at molecular level.

Therefore it should be possible to describe in different ways the same systems, simply by replacing the macroscopic coefficients with quantities that vary in time or in space but rapidly enough to make it possible identifying them with some average values (at least on suitable scales of time and space). *The equations thus obtained would then be physically equivalent to the previous.*

Obviously we can *neither expect nor hope* that, by modifying the equations and replacing various constant with variable quantities, simpler or easier equations will result (on the contrary!). However imposing that equations that should describe the same phenomena do give, actually, the same results can be expected to lead to *nontrivial relations* between properties of the solutions (of both equations).

---

[2] Which instead can be applied only to $|p - 1| \lesssim \frac{1}{\sqrt{\tau\sigma_+}}$.

And providing different descriptions of the same system is not only possible but it can even lead to laws and deductions that would be impossible (or at least difficult) to derive if one did confine himself to consider just a single description of the system (here I think for instance to the description of equilibrium by the microcanonical or the canonical ensembles).

What just said *has not been systematically applied to the mechanics of fluids*, although by now there are several deductions of macroscopic irreversible equations starting from microscopic velocity reversible dynamics, for instance Lanford's derivation of the Boltzmann equation, [23].

Therefore keeping in mind the above considerations we shall imagine other equations that should be "equivalent" to the Navier–Stokes incompressible equation (in a container Ω with some boundary conditions).

Viscosity will be regarded as a phenomenological quantity whose role is to forbid to a fluid to increase indefinitely its energy if subject to non conservative external forces. Hence we regard the incompressible Navier Stokes equations as obtained from the incompressible Euler equations by requiring that the dissipation per unit time is constant and we do that by imposing the constraint via Gauss' least effort principle.

The equivalence viewpoint between irreversible and reversible equations in fluid mechanics is first suggested by the corresponding equivalence, Sect. 5.1, for the thermostatted systems and by the work [24] where it is checked in a special case in which the Navier Stokes equations in 3 dimensions, in a periodic container with side size 1, are simulated (with 128^3 modes) and the results are compared with corresponding ones on similar equations with a constraint forcing the energy content of the velocity field in a momentum shell $2^{n-1} < |\mathbf{k}| < 2^n$ to follow the Kolmogorov-Obukov $\frac{5}{3}$-law: remarkably showing remarkable agreement.

It has to be stressed that, aside from the reversibility question, the idea that the Navier Stokes equations can be profitably replaced by equations that should be equivalent is widely, [25], and successfully used in computational approaches (even in engineering applications)[3]; a prominent example are the "large eddy simulations" where effective viscosities may be introduced (usually not reversible, [26]) to take into account terms that are neglected in the process of cut-off to eliminate the short wavelength modes, although a fundamental approach does not seem to have been developed.[4]

The basic difference between the large eddies simulations approach and the equivalence idea discussed in this and the following sections is that it is not meant as a method to reduce the number of equations and to correct the reduction by adding

---

[3] "The action of the subgrid scales on the resolved scales is essentially an energetic action, so that the balance of the energy transfers alone between the two scale ranges is sufficient to describe the action of the subgrid scales", [25, p. 104]. And about the large eddy simulations: "Explicit modeling of the desired effects, i.e. including them by adding additional terms to the equations: the actual subgrid models", [25, p. 105].

[4] About one of the many important methods: "There is no particular justification for this local use of relations that are on average true for the whole, since they only ensure that the energy ransfers through the cutoff are expressed correctly on the average, and not locally, [25, p. 124].

extra terms in the simplified equations; it deals with the full equation and tries to establish an equivalence with corresponding reversible equations. In particular the new equations are not computationally easier.

For simplicity consider the fluid in a container $\mathcal{C}_0$ with size L and smooth boundary or in a cubic container with periodic boundaries subject to a non conservative volume force $\mathbf{g}$: the fluid can be described by a velocity field $\mathbf{u}(x)$, $x \in \mathcal{T}^3$ with zero divergence $\partial \cdot \mathbf{u} = 0$, "Eulerian description". It can also be represented by twice as many variables, i.e. by two fields $\boldsymbol{\delta}(x)$, $\dot{\boldsymbol{\delta}}(x)$, with 0 divergence, which represent the displacement $\boldsymbol{\delta}(x)$ from a reference position, fixed once and for all, of a fluid element and its velocity $\dot{\boldsymbol{\delta}}(x)$, "Lagrangian description", for details see Appendix J.

The relation between the two representations is $\dot{\boldsymbol{\delta}}(x) = \mathbf{u}(\boldsymbol{\delta}(x))$. In the Lagrangian representation the fluid is thought of as a system of moving points: and remarkably it is a Hamiltonian system. So that a non holonomic constraint, like to keep constant dissipation per unit time or constant $\mathcal{D}$:

$$\mathcal{D} \stackrel{def}{=} \int_{\mathcal{T}^3} (\partial \mathbf{u}(x))^2 d^3x = const \tag{5.5.1}$$

can be imposed, naturally, via Gauss' least constraint principle, Appendix E.

In Appendix J it is shown (in the periodic boundary conditions case) that imposing the constraint $\mathcal{D} = const$ on a perfect fluid leads to the "Gaussian Navier-Stokes" equation:

$$\dot{\mathbf{u}} + \underset{\sim}{\mathbf{u}} \cdot \underset{\sim}{\partial}\mathbf{u} = \alpha(\mathbf{u})\Delta\mathbf{u} - \partial p + \mathbf{g},$$

$$\alpha(\mathbf{u}) \stackrel{def}{=} -\frac{\int \widehat{\partial}\mathbf{u} \cdot (\widehat{\partial}((\underset{\sim}{\mathbf{u}} \cdot \underset{\sim}{\partial})\mathbf{u}))\, dx + \int \Delta\mathbf{u} \cdot \mathbf{g}\, dx}{\int (\Delta\mathbf{u}(x))^2\, dx} \tag{5.5.2}$$

The above equation is *reversible* and time reversal is simply $I\mathbf{u}(x) = -\mathbf{u}(x)$ which means that "fluid elements" retrace their paths with opposite velocity, unlike the classical Navier-Stokes equation:

$$\dot{\mathbf{u}} + \underset{\sim}{\mathbf{u}} \cdot \underset{\sim}{\partial}\mathbf{u} = \nu\Delta\mathbf{u} - \partial p + \mathbf{g}, \qquad \partial \cdot \mathbf{u} = 0 \tag{5.5.3}$$

in which ν is a viscosity constant.

A further equation is obtained by requiring that $\mathcal{E} = \int \mathbf{u}^2 dx = const$ and

$$\dot{\mathbf{u}} + \underset{\sim}{\mathbf{u}} \cdot \underset{\sim}{\partial}\mathbf{u} = \alpha(\mathbf{u})\Delta\mathbf{u} - \partial p + \mathbf{g},$$

$$\alpha(\mathbf{u}) \stackrel{def}{=} \frac{\int \mathbf{u} \cdot \mathbf{g}\, dx}{\int (\partial \mathbf{u}(x))^2\, dx} \tag{5.5.4}$$

which is interesting although it does not follow from Gauss' principle as the previous Eq. (5.5.1).

The Eqs. (5.5.2), (5.5.3) and (5.5.4) fit quite naturally in the frame of the theory of non equilibrium statistical mechanics even though the model is not based on particles systems.

Showing this will be attempted in the following section.

5.6 Developed Turbulence

Introduce the "local observables" $F(\mathbf{u})$ as functions depending only upon finitely many Fourier components of $\mathbf{u}$, *i.e.* depending on the "large scale" properties of the velocity field $\mathbf{u}$.[5] In periodic boundary conditions it will also be supposed that $\int \mathbf{u}\, dx = 0$, to eliminate an uninteresting conserved quantity.

Then, *conjecture*, [27], in periodic boundary conditions (for simplicity) the two Eqs. (5.5.2), (5.5.3) should have "same large scale statistics" in the limit in which $\nu \to 0$ or, more physically and defining the *Reynolds number* $R = \frac{\sqrt{|\mathbf{g}|L^3}}{\nu}$, with $|\mathbf{g}| = \max |\mathbf{g}(x)|$ and $L = $ container size, as $R \to \infty$.

Assuming that the statistics μ_ν and $\widetilde{\mu}_\mathcal{D}$ for the Eqs. (5.5.2), (5.5.3) exist, by *"same statistics"* as $R \to \infty$ it is meant that

(1) if the dissipation $\mathcal{D}$ of the initial datum $\mathbf{u}(0)$ for the first equation is chosen equal
to the average $\langle \int (\partial \mathbf{u})^2\, d\mathbf{x} \rangle_{\mu_\nu}$ for the SRB distribution μ_ν of the second equation,
(1') or if the average $\langle \alpha \rangle_{\widetilde{\mu}_\mathcal{D}} = \nu$ then,
(2) in the limit $R \to \infty$ the difference $\langle F \rangle_{\mu_\nu} - \langle F \rangle_{\widetilde{\mu}_\mathcal{D}} \xrightarrow[R \to +\infty]{} 0$.

If the chaotic hypothesis is supposed to hold it is possible to use the fluctuation theorem, which is a consequence of reversibility, to estimate the probability that, say, the value of α is very different from ν. For this purpose the attracting set has to be determined.

Of course a first problem is that the equations are infinite dimensional and it is even unknown whether they admit smooth solutions so that extending the theory to cover their stationary statistics is simply out of question. This can be partly bypassed adopting a phenomenological approach.

Assuming the "OK41 theory of turbulence" [28, Chap. 7], the attracting set will be taken to be the set of fields with Fourier components $\mathbf{u_k} = 0$ unless $|\mathbf{k}| \leq R^{\frac{3}{4}}$, fulfilling the equations $\partial \cdot \mathbf{u} = 0$ and if $\mathcal{P}_\mathbf{k}$ is the orthogonal projection on the plane orthogonal to $\mathbf{k}$ (see Eqs. (5.2.2), (5.5.3)):

$$\dot{\mathbf{u}}_\mathbf{k} = -i\mathcal{P}_\mathbf{k}\left(\sum_{0<|\mathbf{k}'|<R^{\frac{3}{4}}} \mathbf{k}' \cdot \underline{\mathbf{u}}_{\mathbf{k}'}\mathbf{u}_{\mathbf{k}-\mathbf{k}'} \right) - \alpha(\mathbf{u})\mathbf{k}^2\mathbf{u}_\mathbf{k} + \mathbf{g}_\mathbf{k}$$

$$\alpha(\mathbf{u}) = \nu \quad \text{or} \quad \alpha(\mathbf{u}) = \frac{\sum_k k^2 \bar{\mathbf{g}}_k \cdot \mathbf{u}_k + i \widehat{\mathbf{k}} \bar{\mathbf{u}}_k \cdot \sum_h \widehat{\mathbf{k}}(\mathbf{u}_{k-h}) \cdot \mathbf{h}\, \mathbf{u}_h)}{\sum_{|\mathbf{k}|<R^{\frac{3}{4}}} |\mathbf{k}|^2 |\mathbf{u}_\mathbf{k}|^2}$$

$$(5.6.1)$$

---

[5] Here local refers to locality in momentum space.

where $\bar{\mathbf{u}}_k$ is the complex conjugate of $\mathbf{u}_k$, $0 < |\mathbf{k}|, |\mathbf{k}'|, |\mathbf{k} - \mathbf{k}'|, |\mathbf{h}| < R^{\frac{3}{4}}$ and $\alpha(\mathbf{u}) = \nu$ in the case of the Navier-Stokes equations while the second possibility for $\alpha(\mathbf{u})$ is for the Gaussian Navier-Stokes case.

Then the expected identity $\langle \alpha \rangle = \nu$, between the average friction $\langle \alpha \rangle$ in the second of Eq. (5.5.1) and the viscosity ν in the first, implies that the divergence of the evolution equation in the second of Eq. (5.6.1) is in average

$$\sigma \sim \nu \sum_{|\mathbf{k}| \le R^{3/4}} 2|\mathbf{k}|^2 \sim \nu \left(\frac{2\pi}{L}\right)^2 \frac{4\pi}{5} R^{15/4} \tag{5.6.2}$$

because the momenta $\mathbf{k}$ are integer multiples of $\frac{2\pi}{L}$.

The equations are now a finite dimensional system, reversible in the second case, then assuming the chaotic hypothesis the fluctuation theorem can be applied to the SRB distribution.

Therefore the SRB-probability to see, in motions following the second equation in Eq. (5.6.1), a *"wrong"* average friction $-\nu$ (instead of the "right" ν) for a time τ is

$$\text{Prob}_{srb} \sim \exp\left(-\tau\nu \frac{16\pi^3}{5L^2} R^{\frac{15}{4}}\right) \overset{def}{=} e^{-\gamma\tau} \tag{5.6.3}$$

It can be estimated in the situation considered below for a flow in air:

$$\begin{cases} \nu = 1.5 \, 10^{-2} \, \frac{cm^2}{s}, & v = 10 \, \frac{cm}{s} \quad L = 100 \, cm \\ R = 6.67 \, 10^4, & g = 3.66 \, 10^{14} \, sec^{-1} \\ P \overset{def}{=} \text{Prob}_{srb} = e^{-g\tau} = e^{-1.8 \, 10^8}, & \text{if} \quad \tau = 10^{-6}s \end{cases} \tag{5.6.4}$$

where the first line are data of an example of fluid motion and the other two lines follow from Eq. (5.6.3). They show that, by the fluctuation relation, viscosity can be $-\nu$ during 10^{-6} s (*say*) with probability P as in Eq. (5.6.4): the unlikelihood is similar, in spirit, to the estimates about Poincaré's recurrences, [28].

In the next section we discuss the possibility of drawing some observable conclusions from the Gaussian Navier-Stokes equations hence, by the above equivalence conjecture, on the Navier-Stokes equations.

5.7 Intermittency

Imagine that the fluid consists of particles in a container $\mathcal{C}_0$ with smooth boundaries in contact with external thermostats like in the models in Sect. 2.2. The particles are so many that the system can be well described by a macroscopic equation, like for instance:

$$\begin{aligned}
&\text{(1)} \quad \partial_t \varrho + \partial \cdot (\varrho \mathbf{u}) = 0 \\
&\text{(2)} \quad \partial_t \mathbf{u} + \underaccent{\tilde}{u} \cdot \partial \underaccent{\tilde}{u} = -\tfrac{1}{\varrho} \partial p + \tfrac{\eta}{\varrho} \Delta \mathbf{u} + \mathbf{g} \\
&\text{(3)} \quad \partial_t U + \partial \cdot (\mathbf{u}U) = \eta \underaccent{\tilde}{\tau}' \partial \underaccent{\tilde}{u} + \kappa \Delta T - p \partial \cdot \mathbf{u} \\
&\text{(4)} \quad T \left(\partial_t s + \partial \cdot (\mathbf{u}s) \right) = \eta \underaccent{\tilde}{\tau}' \partial \underaccent{\tilde}{u} + \kappa \Delta T
\end{aligned} \qquad (5.7.1)$$

here $\varrho(x)$ is the density field, $\mathbf{u}(x)$ is the velocity field, $s(x)$ the entropy density, $\mathbf{g}(x)$ is a (nonconservative) external force generating the fluid motion and $p(x)$ is the physical pressure, the Navier-Stokes stress tensor $\underaccent{\tilde}{\tau}'$ is $\tau'_{ij} = (\partial_i u_j + \partial_j u_i)$ and $T(x)$ is the temperature and η, κ are transport coefficients (dynamical viscosity and conductivity). The conditions at the boundary of the fluid container $\mathcal{C}_0$ will be time independent, $T = T(\underaccent{\tilde}{\xi})$ and $\mathbf{u} = 0$ (no slip boundary).

Equation (5.7.1) are macroscopic equations that can be valid only in some limiting regime, [29]. Given a system of unit mass particles with short range pair interactions let δ be a dimensionless scaling parameter. Then a typical conjecture is: for suitably restricted and close to local equilibrium initial data (see [28, p. 21] for examples) *on time scales of $O(\delta^{-2})$ and space scales $O(\delta^{-1})$ the evolution of $\varrho, \mathbf{u}, T, U, s$ follows the incompressible NS equation*, [28, p. 30], given by (1) and (4) above with $\mathbf{u}(x) \overset{def}{=} \sum_{\xi \in \delta^{-1}\Delta(x)} \frac{\dot{\xi}}{\delta^{-1}|\Delta(x)|}$ and $T(x) = \sum_{\xi \in \delta^{-1}\Delta(x)} \frac{\dot{\xi}^2}{\delta^{-1}|\Delta(x)|}$, with $\Delta(x)$ a unit cube centered at x.

Then there will be two ways of computing the entropy creation rate. The first would be the classic one described for instance in [30], and the second would simply be the divergence of the microscopic equations of motion in the model of Fig. 2.1, under the assumption that the motion is closely described by macroscopic equations for a fluid in local thermodynamic equilibrium, like the Navier-Stokes equations, Eq. (5.5.3).

The classical entropy production rate in nonequilibrium thermodynamics of an *incompressible thermoconducting fluid* is [28, p. 6],

$$k_B \varepsilon_{classic} = \int_{\mathcal{C}_0} \left(\kappa \left(\frac{\partial T}{T} \right)^2 + \eta \frac{1}{T} \underaccent{\tilde}{\tau}' \partial \underaccent{\tilde}{u} \right) dx. \qquad (5.7.2)$$

By integration by parts and use of the first and fourth of Eq. (5.7.1), $k_B \varepsilon_{classic}$ becomes, if $S \overset{def}{=} \int_{\mathcal{C}_0} s\, d\mathbf{x}$ is the total thermodynamic entropy of the fluid,

$$\begin{aligned}
\int_{\mathcal{C}_0} &\left(-\kappa \partial T \cdot \partial T^{-1} + \eta \frac{1}{T} \underaccent{\tilde}{\tau}' \partial \underaccent{\tilde}{u} \right) d\mathbf{x} \\
&= -\int_{\partial \mathcal{C}_0} \kappa \frac{\mathbf{n} \cdot \partial T}{T}\, ds_\xi + \int_{\mathcal{C}_0} \frac{(\kappa \Delta T + \eta \underaccent{\tilde}{\tau}' \partial \underaccent{\tilde}{u})}{T}\, d\mathbf{x} \\
&= -\int_{\partial \mathcal{C}_0} \kappa \frac{\mathbf{n} \cdot \partial T}{T}\, ds_\xi + \dot{S} + \int_{\mathcal{C}_0} \mathbf{u} \cdot \partial s\, d\mathbf{x} \\
&= -\int_{\partial \mathcal{C}_0} \kappa \frac{\partial T \cdot \mathbf{n}}{T}\, ds_\xi + \dot{S}
\end{aligned} \qquad (5.7.3)$$

where $S = \int s(\mathbf{x}) \, d\mathbf{x}$ is the total entropy, $\mathbf{n}$ is the outer normal to $\partial \mathcal{C}_0$.

This can be naturally compared with the general expression in Eq.(2.8.3): in the limit $\delta \to 0$ each volume element will contain an infinite number of particles and fluctuations will be suppressed; the *average* entropy production will be defined and, up to a time derivative of a suitable quantity, see Sect. 2.8, it will be

$$\langle \varepsilon \rangle_\mu = - \int_{\partial \mathcal{C}_0} \kappa \, \frac{\mathbf{n}(\boldsymbol{\xi}) \cdot \partial \, T(\boldsymbol{\xi})}{T(\boldsymbol{\xi})} ds_{\boldsymbol{\xi}} = \varepsilon_{classic} - \dot{S} \qquad (5.7.4)$$

the average is intended over a time scale long compared to the microscopic time evolution but macroscopically short.

That is this *leads to* the expression Eq.(5.7.3), "local on the boundary" or "localized at the contact between system and thermostats", since $\mathbf{u} \cdot \mathbf{n} \equiv 0$ by the boundary conditions, *plus the time derivative of the total "thermodynamic entropy" S of the fluid*.

Returning to the observability question suppose the validity of the chaotic hypothesis for the reversible equations on their attracting sets where they can be modeled by finite dimensional motions, like the ones considered in Sect. 5.6 via the OK41 theory. Then there will be a function $\zeta(p)$ controlling the fluctuations of entropy production and satisfying the fluctuation theorem symmetry, Eq.(4.6.1). The nontrivial dependence of ζ on p is sometimes referred to as an "intermittency phenomenon".

By *intermittency* it is meant here, [31], an event that is realized rarely and randomly: rarity can be in time, time intermittency, in the sense that the interval of time in which the event is realized has a small frequency among time intervals δt of equal length into which we divide the time axis; it can be in space, spatial intermittency, if the event is rarely and randomly verified inside cubes δx of a lattice into which we imagine to divide the space R^3. Rarity can be also in space time, space time intermittency, if the event is rarely and randomly verified inside regions $\delta t \times \delta x$ forming a lattice into which we imagine partitioned the space time $R \times R^3$ or, in the case of discrete evolutions, $Z \times R^3$.

We now address the question: "is this intermittency observable"? is its rate function $\zeta(p)$ measurable? [32].

This can be discussed in the case of an incompressible fluid satisfying (2), (4) in Eq.(5.7.1) with $\varrho = 1$ and $\partial \cdot \mathbf{u} = 0$ and describing the attractor via the OK41 theory as in the previous section: hence (2) becomes Eq.(5.6.1).

Clearly σ_+ and $\zeta(p)$ will grow with the size of the system i.e. with the number of degrees of freedom, at least, which approaches ∞ as $R \to \infty$ so that there should be serious doubts about the observability of so rare fluctuations.

However if we look at a small subsystem in a little volume V_0 of linear size L_0 we can regard it, under suitable conditions, again as a fluid enclosed in a box V_0 described by the same reversible Gaussian Navier-Stokes equations. The above analysis leading to the expression of the entropy production in a region $\mathcal{C}_0$ in Eqs.(5.7.2), (5.7.3), (5.7.4) showing that it depends on the temperature and its gradient *at the boundary* of $\mathcal{C}_0$, only plays of course an essential role in clarifying the conditions under which a small volume in a fluid can be considered as a system thermostatted by the neighboring fluid. For a discussion of the physical conditions see Sect. 4.10.

We imagine, therefore, that this small system also verifies a fluctuation relation in the sense that if the interpretation of the OK41 theory, [28, Sect. 6.2], as determining the attracting set is accepted. Then the fluctuating viscosity term contributes to the phase space contraction $K_{L_0}(R)\alpha_{L_0}(\mathbf{u})$ with $K_{L_0}(R) \overset{def}{=} \sum_{|\mathbf{k}| \le R_{L_0}^{3/4}} |\mathbf{k}|^2$ and $\mathbf{k} = 2\pi \frac{\mathbf{n}}{L_0}$, where R_{L_0} is the Reynolds number on scale L_0 which, from the OK41 theory, is $R_{L_0} = (L_0/L)^{4/3} R$, i.e.

$$
\begin{aligned}
\sigma_{V_0}(\mathbf{u}) &= K_{L_0}(R)\, \alpha_{L_0}(\mathbf{u}) \\
\alpha_{L_0}(\mathbf{u}) &= \frac{\int_{V_0} \left(-\mathbf{g}\cdot\Delta\mathbf{u} - \Delta\mathbf{u}\cdot(\mathbf{u}\cdot\partial\mathbf{u}) \right) dx}{\int_{V_0} (\Delta\mathbf{u})^2\, dx}
\end{aligned}
\tag{5.7.5}
$$

see Eqs. (5.5.2), (5.6.1), then it should be that the fluctuations of σ averaged over a time span τ are controlled by rate functions $\zeta_V(p)$ and $\zeta_{V_0}(p)$ that we can expect to be, for R large

$$
\begin{aligned}
\zeta_V(p) &= \overline{\zeta}(p)\, K_L(R), \qquad \zeta_{V_0}(p) = \overline{\zeta}(p)\, K_{L_0}(R), \\
\langle \sigma_{V_0} \rangle_+ &= \overline{\sigma}_+ \, K_{L_0}(R)
\end{aligned}
\tag{5.7.6}
$$

Hence, if we consider observables dependent on what happens inside V_0 and if L_0 is small so that $K_{L_0}(R)$ is not too large and we observe them in time intervals of size τ, then the time frequency during which we can observe a deviation "of size" $1 - p$ from irreversibility will be small of the order of

$$
e^{(\overline{\zeta}(p) - \overline{\zeta}(1))\, \tau\, K_{L_0}(R)}
\tag{5.7.7}
$$

for τ large, where the local fluctuation rate $\overline{\zeta}(p)$ verifies, assuming the chaotic hypothesis for the Eq. (5.7.1):

$$
\overline{\zeta}(-p) = \overline{\zeta}(p) - \overline{\sigma}_+ \, p
\tag{5.7.8}
$$

Therefore by observing the frequency of intermittency one can gain some access to the function $\overline{\zeta}(p)$.

Note that one *will necessarily observe a given fluctuation somewhere* in the fluid if L_0 is taken small enough and the size L of the container large enough: in fact the entropy driven intermittency takes place not only in time but also in space. Thus we shall observe, inside a box of size L_0 "somewhere" in the total volume V of the system and in a time interval τ "sometime" in a large time interval T, a fluctuation of size $1 - p$ with high probability if

$$
(T/\tau)(L/L_0)^3 e^{(\overline{\zeta}(p) - \overline{\zeta}(1))\, \tau\, K_{L_0}(R)} \simeq 1
\tag{5.7.9}
$$

and the special event $p = -1$ will occur with high probability if

$$(T/\tau)(L/L_0)^3 e^{-\bar{\sigma}+\tau K_{L_0}(R)} \simeq 1 \qquad (5.7.10)$$

by Eq. (5.7.8). Once this event is realized the fluctuation patterns will have relative probabilities as described in the fluctuations pattern theorem, Eq. (4.7.2), [31].

Hence the intermittency described here is an example of *space-time intermittency*.

5.8 Stochastic Evolutions

Time reversal symmetry plays an essential role in the fluctuations theory. Therefore the question whether a kind of fluctuation theorem could hold even in cases in which the symmetry is absent has been studied in several works with particular attention devoted to problems in which stochastic forces act.

The first ideas and results appear in [33], followed by [34], [35]. The natural approach would be to consider stochastic models as special cases of the deterministic ones: taking the viewpoint that noise can (and *should*) be thought as generated by a chaotic evolution. In any case this is precisely what is done in simulations where it is generated by a random number generator which is a program, e.g. see [36], that simulates a chaotic evolution.

The latter approach is more recent and has also given results, [37–39], showing that extensions of the fluctuation theorem can be derived in special examples which although stochastic nevertheless can be mapped into a reversible deterministic dynamical system which includes among the phase space variables the coordinates describing the noise generator system.

However the path followed in the literature has mostly been along different lines although, of course, it has provided important insights even allowing the treatment of problems in which a phenomenological, constant, friction unavoidably destroys time reversal symmetry.

The paradigmatic case is the equation $(i = 1, 2, \ldots, N)$

$$m\ddot{x}_i + \gamma\dot{x}_i + \partial_{x_i} U(\mathbf{x}) - f_i = \xi_{1,i}$$
$$\ddot{\xi}_i = F(\xi_i) \qquad (5.8.1)$$

where ξ_i are independent chaotic motions, Hamiltonian for simplicity, on a d dimensional manifold under the action of a force F_i and $\xi_{1,i}$ is one component of ξ_i or a function of it.[6]

The systems describing ξ_i can be considered a model for a family of random number generators. In this section the variables x_i are angles, so that the above

---

[6] For instance the ξ_i, $\omega_i \overset{def}{=} \dot{\xi}_i$ could be the coordinates of a geodesic flow on a manifold of constant negative curvature, [40].

system is a family of coupled pendulums subject to friction, with a *phenomenological friction coefficient* γ, and stirring by the torques f_i.

There is no way to consider Eq. (5.8.1) as time reversible: however the equivalence conjectures of Sect. 5.1 suggest to consider the model

$$m\ddot{x}_i + \alpha(\mathbf{x}, \dot{\mathbf{x}}, \mathbf{w})\dot{x}_i + \partial_{x_i} U(x) - f_i = \xi_{1,i}$$

$$\alpha(\mathbf{x}, \dot{\mathbf{x}}, \mathbf{w}) = \frac{\mathbf{f} \cdot \dot{\mathbf{x}} + \sum_i \xi_{1,i}\dot{x}_i - \dot{U}}{NT/m} \tag{5.8.2}$$

$$\dot{\xi}_i = \omega_i, \qquad \dot{\omega}_i = F(\xi_i)$$

which has $\frac{1}{2}m\dot{\mathbf{x}}^2 = \frac{1}{2}T$ as an exact constant of motion, if the initial data are on the surface $\frac{1}{2}m\dot{\mathbf{x}}^2 = \frac{1}{2}T$.

The equivalence conjectures considered in several cases in the previous sections indicate that in this case for small γ it should be $\langle\alpha\rangle_{SRB} = \gamma$, if the initial T for Eq. (5.8.2) is the average value of the kinetic energy for Eq. (5.8.1), and the corresponding stationary states should be equivalent.

The model in Eq. (5.8.2) is reversible and the map $I(\mathbf{x}, \mathbf{v}, \boldsymbol{\xi}, \boldsymbol{\omega}) = (\mathbf{x}, -\mathbf{v}, \boldsymbol{\xi}, -\boldsymbol{\omega})$, where $\dot{\mathbf{x}} = \mathbf{v}$ and $\dot{\boldsymbol{\xi}} = \boldsymbol{\omega}$ is a time reversal symmetry

$$S_t I = I S_{-t} \tag{5.8.3}$$

The SRB distribution can be defined as the statistics of the data chosen with distribution

$$\mu(d\mathbf{x}\,d\dot{\mathbf{x}}\,d\boldsymbol{\xi}\,d\boldsymbol{\omega}) = \varrho(\mathbf{x}, \dot{\mathbf{x}}, \boldsymbol{\xi}, \boldsymbol{\omega})d\mathbf{x}\,d\dot{\mathbf{x}}\,\delta(m\dot{\mathbf{x}}^2 - T)\,d\boldsymbol{\xi}\,d\boldsymbol{\omega} \tag{5.8.4}$$

where ϱ is an arbitrary (regular) function.

The chaotic hypothesis is naturally extended to such systems and we expect that an invariant distribution μ_{SRB} exists and describes the statistics of almost all initial data chosen with the distribution in Eq. (5.8.4). And since reversibility holds we can expect that the phase space contraction[7]

$$\sigma(\mathbf{x}, \dot{\mathbf{x}}, \mathbf{w}) = N\alpha(\mathbf{x}, \dot{\mathbf{x}}, \mathbf{w}) = \frac{\mathbf{f} \cdot \dot{\mathbf{x}} + \sum_i \xi_{1,i} \cdot x_i - \dot{U}}{T/m} \tag{5.8.5}$$

has a positive time average $\sigma_+ = N\gamma$ and its finite time averages

$$p = \frac{1}{\tau} \int_0^\tau dt \frac{\sigma(S_t(\mathbf{x}, \dot{\mathbf{x}}, \mathbf{w}))}{\sigma_+} \tag{5.8.6}$$

obeys a fluctuation relation with large deviations rate $\zeta(p)$:

---

[7] Notice that in this case no contraction occurs in the $\boldsymbol{\xi}, \boldsymbol{\omega}$ space because the evolution there is Hamiltonian.

$$\zeta(-p) = \zeta(p) - p\sigma_+ \tag{5.8.7}$$

for p in a suitable interval $(-p^*, p^*)$.

An example in which all the above argument can be followed with mathematical rigor is in [38].

But reversibility is also *not always necessary*. The cases of Eq. (5.8.1) *under the assumption that the potential energy $U(\mathbf{x})$ is bounded, x_i are not angles but real variables, and $\xi_{1,i}$ is a white noise* are simpler and have been treated in [33, 34]. They provide remarkable examples in which time reversal does not hold and nevertheless the fluctuation relation is obeyed and can be proved. And in [34, 35] a general theory of the fluctuation relation is developed (far beyond the idea of the "Ising model analogy", [37, Sect. 3]).

The above discussion makes clear that there is little difference between the stochastic cases and the deterministic ones. It can be said that the theory of Markov partitions and coarse graining turns deterministic systems into stochastic ones and, viceversa, the equivalence conjectures of Sects. 5.1, 5.6 do the converse. The Langevin equation is a paradigmatic case in which a stochastic system appears to be equivalent (as far as the entropy production fluctuations are concerned) to a deterministic reversible one.

A very interesting case is Eq. (5.8.1) in which $i = 1$ and x is an angle and $U(x) = -2gV\cos x$ and the noise is a white noise: i.e. a forced pendulum subject to white noise and torque, see Appendix M. That this is *surprisingly* a nontrivial case shows that even the simplest non equilibrium cases can be quite difficult and interesting.

5.9 Very Large Fluctuations

The importance of the boundedness assumption on the potential energy U in the theory of the fluctuation relation has been stressed in [41]. In this section an interesting example in which an unbounded potential acts and a kind of fluctuation relation holds is analyzed, [42, 43], to exhibit the problems that may arise and gave rise to [41].

The system is a particle trapped in a "harmonic potential" and subject to random forcing due to a Brownian interaction with a background modeled by a white noise $\zeta(t)$ and "overdamped", i.e. described by a Langevin equation

$$\dot{x} = -(x - vt) + \zeta(t), \qquad x \in R \tag{5.9.1}$$

in dimensionless units, with the average $\langle \zeta(t)\zeta(t')\rangle = \delta(t - t')$. Here v is a constant "drag velocity" and the model represents a particle in a harmonic well dragged by an external force at constant speed v: a situation that can be experimentally realized as described in [44].

The driving force that is exercised by external forces balancing the reaction due to the climbing of the harmonic well is $x - vt$ and the energy U is $\frac{1}{2}(x - vt)^2$. The work done by the external force and the harmonic energy variation during the time interval $(0, t)$ are therefore

$$W = \int_0^t v \cdot (x(\tau') - vt')dt', \qquad \Delta U = \frac{1}{2}((x(t) - vt)^2 - \frac{1}{2}x(0)^2, \qquad (5.9.2)$$

and the quantity $Q = W - \Delta U$ is the work that the system performs on the "outside", [42].

Also in this case there is no time reversal symmetry but a fluctuation relation could be expected for the dimensionless entropy production rate $p = \frac{1}{\tau}\int_0^\tau \frac{Q(t)}{\langle Q \rangle}dt$ on the basis of the equivalence discussed in Sect. 5.5. This model is extremely simple due to the linearity of the force and to the Gaussian noise and the fluctuations of p have a rate that can be quite explicitly evaluated. By the choice of the units it is $\langle W \rangle = \langle Q \rangle = 1$.

However this is a case with U unbounded: and the finite time average $\frac{1}{\tau}\int_0^\tau Q(t)dt$ differs from that of W, given by $\frac{1}{\tau}\int_0^\tau W(t)dt$, by a "boundary term", namely $\frac{1}{\tau}\Delta U$. If U were bounded this would have no effect on the fluctuation rate $\zeta(p)$: but since U is not bounded care is needed. An accurate analysis, possible because the model can be exactly solved, [42], shows that $\zeta(p)$ only satisfies the fluctuation relation $\zeta(-p) = \zeta(p) - p$ for $|p| \leq 1$.

This important remark has consequences in a much more general context including the thermostatted models of Sect. 2.2 as it appeared also in [45]. The reason for the problem was ascribed correctly to the large values that U can assume even in the stationary distribution.

Since the fluctuation theorem proof requires that phase space be bounded, and the system equations be smooth, hence p be bounded, the proof cannot be directly applied. A detailed analysis of the problem in the more general context of Sect. 2.2 has been discussed in [41]: there a simple solution is given to the "problem" of apparent violation of the fluctuation relation in important cases in which the forces can be unbounded (e.g. when the interparticle potentials are of Lennard-Jones type).

Just study the motion rather than in continuous time via timed observations timed to events x in which $U(x)$ is below a prefixed bound. This means studying the motion via a Poincaré's section which is not too close to configurations x where $U(x)$ is too large. In this case the contribution to the phase space contraction due to "boundary terms" like $\frac{1}{\tau}(U(x(\tau)) - U(x(0))$ vanishes as $\tau \to \infty$ (and in a controlled way) and a fluctuation relation can be expected if the other conditions are met (i.e. chaoticity and reversibility).

More detailed analysis of the problem of the fluctuation relation in cases in which unbounded forces can act is in [46]. There a general theory of the influence of the singularities in the equations of motion is presented: the most remarkable phenomena are that the fluctuation relation should be expected to hold but only for a limited range $p \in (-p^*, p^*)$, less than the maximal observable with positive probability, and beyond it observable deviations occur (unlike the case in which the fluctuation relation holds as a theorem, i.e. for Anosov systems) and it becomes even possible that the function $\zeta(p)$ can become non convex: this is a property that appears in almost all attempts at testing fluctuation relations.

Finally it can be remarked that Lennard-Jones forces are an idealization and in nature the true singularities (if at all present) can be very difficult to see. This is also true in simulations: no matter which precision is chosen in the program (usually not very high) there will be a cut off to the values of any observable, in particular of the maximum value of the potential energy. This means that, if really long observation times could be accessed, eventually the boundary terms become negligible (in experiments because Nature forbids singularities or, in simulations, because computers have not enough digits). This means that eventually the problem is not a real one: *but* time scales far beyond interest could be needed to realize this. Therefore the theory based on timed observations is not only more satisfactory but it also deals with properties that are closer to possible observations, [46].

5.10 Thermometry

The proposal that the model in Sects. 4.11, 4.12 can represent correctly a thermostatted quantum system is based on the image that I have of a thermostat as a classical object in which details like the internal interaction are not relevant. And the definition of temperature is not really obvious particularly at low temperatures or on nano-scale systems when quantum phenomena become relevant: furthermore in quantum systems the identity between average kinetic energy and absolute temperature ceases to hold, [9, Chap. 2].

Also a basic question is the very notion of temperature in non equilibrium systems. An idea is inspired by an earlier proposal for using fluctuation measurements to define temperature in spin glasses, [47], [48, p. 216].

If the models can be considered valid at least until it makes sense to measure temperatures via gas thermometers, i.e. optimistically down to the $\sim 3\,^{\circ}K$ scale but certainly at room temperature or somewhat higher, then the chaotic hypothesis can probably be tested with present day technology with suitable thermometric devices.

If verified it could be used to develop a "fluctuation thermometer" to perform temperature measurements below $3\,^{\circ}K$ which are *device independent* in the same sense in which the gas thermometers are device independent (*i.e.* do not require, in principle, "calibration" of a scale and "comparison" procedures).

To fix ideas a recent device, "active scanning thermal microscopy" [49], to measure temperature of a test system can be used for illustration purposes. The device was developed to measure temperature in a region of 100 nm, linear size, of the surface of the test system supposed in a stationary state (on a time scale > 10 ms).

Consider a sample in a stationary equilibrium state, and put it in contact with a bowing arm ("cantilever", see Fig. 5.1): monitor, via a differential thermocouple, the temperature at the arm extremes and signal the differences to a "square root device", [50], which drives another device that can inject or take out heat from the arm and keep, by feedback, the temperature differences in the arm $\Delta T = 0$. The arm temperature is then measured by conventional methods (again through a thermocouple):

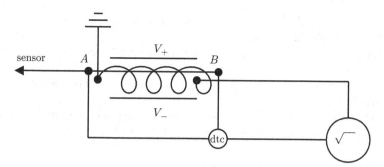

Fig. 5.1 The (microscopic) sensor is attached to the arm AB: a differential termocouple (dtc) is at the extremes of AB, and through AB is also maintained a current at small constant voltage $V_+ - V_-$; the thermocouple sends signals to a "square root circuit" which controls a "coil" (a device that can heat or cool the arm). The circuit that fixes the voltage is not represented, and also not represented are the amplifiers needed (there has to be one at the exit of the differential thermocouple and one after the square root circuit); furthermore there has to be also a device that records the output of the square root circuit hence the power fluctuations

the method is called "active scanning thermal microscopy", [49, p. 729].[8] The test system is supposed in a stationary state.

A concrete example of a nanoscale device to measure temperature (above room temperature) on a scale of $100\,\text{nm} = 10^3\,A^o$ can be found in [49].

With other earlier methods it is possible to measure temperatures, on a scale of $30\,\text{nm}$ on a time scale $>1\,\text{ms}$, closer to the quantum regime: but the technology ("passive scanning thermal microscopy") seems more delicate, [49].

This device suggests a similar one, Fig. 5.1, for a different use: a very schematic description follows. A small electric current could be kept flowing through the arm AB, by an applied constant voltage difference $V_+ - V_-$, to keep the arm in a non-equilibrium steady state; contact between the arm and the sample is maintained via the sensor (without allowing heat exchanges between the two, after their equilibrium is reached); and the heat flow Q (from the "heater", which should actually be a pair of devices, heater + cooler) to keep the arm temperature constant, at the value of the sample temperature is controlled, via a feedback mechanism driven by the square root circuit. The heat fluctuations could be revealed through measurements of the electric current flowing out of the square root circuit or by monitoring the heater output.[9]

The steady heat output Q_+ can be compared to the instantaneous heat output Q and the statistics $P_\tau(p)$ of the ratio $p = \frac{Q}{Q_+}$ over a time span τ might be measured. The temperature (of the bowing arm, hence of the system) can be read from the slope

---

[8] The method consists in "*detecting the heat flow along the cantilever and feeding power proportional to it to the cantilever. Feedback with sufficient gain that keeps the arm at the same temperature as the sample contact point, then cantilever temperature is measured by another thermocouple on the middle of the cantilever*", [49, p. 729].

[9] This is a device turning the arm into a test system and the attached circuits into a thermostat.

of the function $\frac{1}{\tau} \log \frac{P_\tau(p)}{P_\tau(-p)} = p \frac{Q_+}{k_B T}$ from the fluctuation relation: alternatively this could be a test of the fluctuation relation.

The arm and the sensor should be as small as possible: hence 100 nm linear size and 10 ms for response time [49] are too large for observing important fluctuations: hopefully the delicate technology has now improved and it might be possible to build a working device.

The idea is inspired by a similar earlier proposal for using fluctuation measurements to define temperature in spin glasses, [47], [48, p. 216].

5.11 Processes Time Scale and Irreversibility

A *process*, denoted Γ, transforming an initial stationary state $\mu_{ini} \equiv \mu_0$ for an evolution like the one in Fig. 2.2 $\dot{x} = F(x) + \Phi(x, t)$ under initial forcing $\Phi_{ini} \equiv \Phi(x, 0)$ into a final stationary state $\mu_{fin} \equiv \mu_\infty$ under final forcing $\Phi_{fin} \equiv \Phi(x, \infty)$ will be defined by a piecewise smooth function $t \to \Phi(t)$, $t \in [0, +\infty)$, varying between $\Phi(x, 0) = \Phi_0(x)$ to $\Phi(x, +\infty) = \Phi_\infty(x)$.

For intermediate times $0 < t < \infty$ the time evolution $x = (\dot{\mathbf{X}}, \mathbf{X}) \to x(t) = S_{0,t} x$ is generated by the equations $\dot{x} = F(x) + \Phi(x, t)$ with initial state in phase space $\mathcal{F}$: it is a non autonomous equation.

The time dependence of $\Phi(t)$ could for instance be due to a motion of the container walls which changes the volume from the initial $C_0 = V_0$ to V_t to $C_0' = V_\infty$: hence the points $x = (\dot{\mathbf{X}}, \mathbf{X})$ evolve at time t in a space $\mathcal{F}(t)$ which also may depend on t.

During the process the initial state evolves into a state μ_t attributing to an observable $F_t(x)$ defined on $\mathcal{F}(t)$ an average value given by

$$\langle F_t \rangle = \int_{\mathcal{F}(t)} \mu_t(dx) F_t(x) \stackrel{def}{=} \int_{\mathcal{F}(0)} \mu_0(dx) F_t(S_{0,t} x) \qquad (5.11.1)$$

We shall also consider the probability distribution $\mu_{SRB,t}$ which is defined as the SRB distribution of the dynamical system obtained by "freezing" $\Phi(t)$ at the value that is taken at time t and imagining the time to evolve further until the stationary state $\mu_{SRB,t}$ is reached: *in general* $\mu_t \neq \mu_{SRB,t}$.

Forces and potentials will be supposed smooth, i.e. analytic, in their variables aside from impulsive elastic forces describing shocks, allowed here to model shocks with the containers walls and possible shocks between hard core particles.

Chaotic hypothesis will be assumed: this means that in the physical problems just posed on equations of motion written symbolically $\dot{x} = F(x) + \Phi(x, t)$ with Φ time dependent, the motions are so chaotic that the attracting sets on which their long time motion would take place, if $\Phi(x, t)$ was fixed at the value taken at time t, can be regarded as smooth surfaces on which motion is highly unstable.

It is one of the basic tenets in Thermodynamics that all (nontrivial) processes between equilibrium states are "irreversible": only idealized (strictly speaking nonexistent) "quasi static" processes through equilibrium states can be reversible.

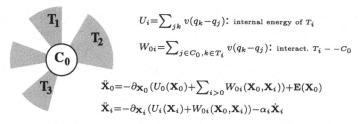

$$U_i = \sum_{jk} v(q_k - q_j): \text{ internal energy of } T_i$$

$$W_{0i} = \sum_{j \in C_0, k \in T_i} v(q_k - q_j): \text{ interact. } T_i - -C_0$$

$$\ddot{\mathbf{X}}_0 = -\partial_{\mathbf{X}_0}\left(U_0(\mathbf{X}_0) + \sum_{i>0} W_{0i}(\mathbf{X}_0, \mathbf{X}_i)\right) + \mathbf{E}(\mathbf{X}_0)$$

$$\ddot{\mathbf{X}}_i = -\partial_{\mathbf{X}_i}(U_i(\mathbf{X}_i) + W_{0i}(\mathbf{X}_0, \mathbf{X}_i)) - \alpha_i \dot{\mathbf{X}}_i$$

Fig. 5.2 α_i s.t. $\frac{m}{2}\sum_{i>0}\dot{\mathbf{X}}_i^2 = \frac{1}{2}N_i k_B T_i(t)$: $\alpha_i = \frac{Q_i - \dot{U}_i}{N_i k_B T_i(t)}$, see Eq. (2.2.1)

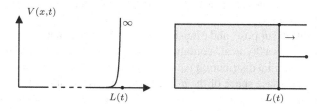

Fig. 5.3 Piston extension is $L(t)$ and $V(x,t)$ is a potential modeling its wall. A sudden doubling of $L(t)$ would correspond to a Joule-Thomson expansion

The question addressed here is whether irreversibility can be made a quantitative notion at least in models based on microscopic evolution, like the model in Fig. 2.2 and in processes between equilibrium states.

Some examples:

(1) Gas in contact with reservoirs with varying temperature, see Sect. 2.2 (Fig. 5.2):
(2) Gas in a container with moving wall (Fig. 5.3).
(3) Paddle wheels stirring a liquid (Fig. 5.4).

In the examples the t dependence of $\Phi(x,t)$ vanishes as t becomes large. Example 1 is a process transforming a stationary state into a new stationary state; while examples 2,3 are processes transforming an equilibrium state into an equilibrium state.

The work $Q_a \overset{def}{=} \sum_{j=1}^{N_a} -\dot{\mathbf{x}}_{a,j} \cdot \partial_{x_{a,j}} U_{0,a}$ in example 1 will be interpreted as *heat* Q_a ceded, per unit time, by the particles in C_0 to the a-th thermostat (because the "temperature" of C_a, $a > 0$ remains constant). The phase space contraction rate due

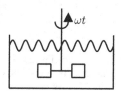

Fig. 5.4 The wavy lines symbolizes the surface of the water. Slow rotation for a long time would correspond to the Joule paddle wheels measurement of the heat-work conversion factor

to heat exchanges between the system and the thermostats can, therefore, be naturally defined as in Eq. (2.8.1):

$$\sigma^\Gamma(\dot{\mathbf{X}}, \mathbf{X}) \overset{def}{=} \sum_{a=1}^{N_a} \frac{Q_a}{k_B T_a} + \dot{R} \qquad (5.11.2)$$

where $R = \sum_{a>0} \frac{U_a}{k_B T_a}$.

Phase space volume can also change because new regions become accessible (or inaccessible)[10] so that the total phase space contraction rate, denoted $\sigma_{tot,t}$, in general will be different from σ_t^Γ.

It is reasonable to suppose, and often it can even be proved, that at every time t the configuration $S_{0,t}x$ is a "typical" configuration of the "frozen" system if the initial x was typical for the initial distribution μ_0: i.e. it will be a point in $\mathcal{F}(t) = V_t^N \times R^{3N} \times \prod \mathcal{F}_a$, if $\mathcal{F}_a$ is the phase space of the a-th thermostat, whose statistics under forces imagined frozen at their values at time t will be $\mu_{SRB,t}$, see comments following Eq. (5.11.1).[11] Since we must consider as accessible the phase space occupied by the attractor of a typical phase space point, the volume variation contributes an extra $\sigma_t^v(x)$ to the phase space variation, via the rate at which the phase space volume $|\mathcal{F}_t|$ contracts, namely:

$$\sigma_t^v(x) = -\frac{1}{|\mathcal{F}_t|} \frac{d\,|\mathcal{F}_t|}{dt} = -N \frac{\dot{V}_t}{V_t} \qquad (5.11.3)$$

which does not depend on x as it is a property of the phase space available to (any typical) x.

Therefore the total phase space contraction per unit time can be expressed as, see Eqs. (5.11.3), (5.11.2),

$$\sigma_{tot}(\dot{\mathbf{X}}, \mathbf{X}) = \sum_a \frac{Q_a}{k_B T_a} - N \frac{\dot{V}_t}{V_t} + \dot{R}(\dot{\mathbf{X}}, \mathbf{X}) \qquad (5.11.4)$$

i.e. there is a simple and direct relation between phase space contraction and entropy production rate, [51]. Eq. (5.11.4) shows that their difference is a "total time derivative".

In studying stationary states with a fixed forcing $F(x) + \Phi(x, t)$ frozen at the value it has at time t it is $N\dot{V}_t/V_t = 0$, the interpretation of $\dot{R}$ is of "reversible" heat exchange between system and thermostats. In this case, in some respects, the difference $\dot{R}$ can be ignored. For instance in the study of the fluctuations of the average over time of entropy production rate in a *stationary state* the term $\dot{R}$ gives no

---

[10] For instance this typically means that the external potential acting on the particles undergoes a change, e.g. a moving container wall means that the external potential due to the wall changes from 0 to $+\infty$ or from $+\infty$ to 0. This is example 2 or, since the total energy varies, also example 3.

[11] If for all t the "frozen" system is Anosov, then any initial distribution of data which admits a density on phase space will remain such, and therefore with full probability its configurations will be typical for the corresponding SRB distributions. Hence if $\Phi(t)$ is piecewise constant the claim holds.

contribution, or it affects only the very large fluctuations, [41, 42] if the containers C_a are very large (or if the forces between their particles can be unbounded, which we are excluding here for simplicity, [41], see also Sect. 5.9).

Even in the case of processes the quantity $\dot{R}$ has to oscillate in time with 0 average on any interval of time (t, ∞) if the system starts and ends in a stationary state.

For the above reasons we define the *entropy production rate* in a process to be Eq. (5.11.4) *without* the $\dot{R}$ term (in Sect. 2.8 it was remarked that it depends on the coordinate system used):

$$\varepsilon(\dot{\mathbf{X}}, \mathbf{X}) = \sum_a \frac{Q_a}{k_B T_a} - N\frac{\dot{V}_t}{V_t} \overset{def}{=} \varepsilon_t^{srb}(\dot{\mathbf{X}}, \mathbf{X}) - N\frac{\dot{V}_t}{V_t} \qquad (5.11.5)$$

where ε_t^{srb} is defined by the last equality and the name is chosen to remind that if there was no volume change ($V_t = const$) and the external forces were constant (at the value taken at time t) then ε_t^{srb} would be the phase space contraction natural in the theory for the SRB distributions when the external parameters are frozen at the value that they have at time t.

It is interesting, and necessary, to remark that in a stationary state the time averages of ε, denoted ε_+, and of $\sum_a \frac{Q_a}{k_B T_a}$, denoted $\sum_a \frac{\langle Q_a \rangle_+}{k_B T_a}$, coincide because $N\dot{V}_t/V_t = 0$, as $V_t = const$, and $\dot{R}$ has zero time average being a total derivative. On the other hand under very general assumptions, much weaker than the chaotic hypothesis, the time average of the phase space contraction rate is ≥ 0, [52, 53], so that in a stationary state: $\sum_a \frac{\langle Q_a \rangle_+}{k_B T_a} \geq 0$. which is a consistency property that has to be required for any proposal of definition of entropy production rate.

Remark (1) In stationary states the above models are a realization of *Carnot's machines* : the machine being the system in C_0 on which external forces Φ work leaving the system in the same stationary state (a special "cycle") but achieving a transfer of heat between the various thermostats (in agreement with the second law only if $\varepsilon_+ \geq 0$).

(2) The fluctuation relation becomes observable for the entropy production Eq. (5.11.5), over a time scale independent of the size of the thermostats, because the heat exchanged is a boundary effect (due to the interaction of the test system particles and those of the thermostats in contact with it, see Sect. 2.8).

Coming back to the question of defining an irreversibility degree of a process Γ we distinguish between the (non stationary) state μ_t into which the initial state μ_0 evolves in time t, under varying forces and volume, and the state $\mu_{SRB,t}$ obtained by "freezing" forces and volume at time t and letting the system settle to become stationary, see comments following Eq. (5.11.1). We call ε_t the entropy production rate Eq. (5.11.5) and ε_t^{srb} the entropy production rate in the "frozen" state $\mu_{SRB,t}$, as in Eq. (5.11.5).

The proposal is to define the *process time scale* and, in processes leading from an equilibrium state to an equilibrium state, the *irreversibility time scale* $\mathcal{I}(\Gamma)^{-1}$ of a process Γ by setting:

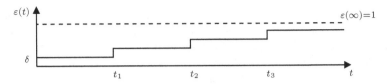

Fig. 5.5 An example of a process proceeding at jumps of size δ at times $t_0 = 0, t_1, \ldots$: the final value $\varepsilon = 1$ can be reached in a finite time or in an infinite time

$$\mathcal{I}(\Gamma) = \int_0^\infty \left(\langle \varepsilon_t \rangle_{\mu_t} - \langle \varepsilon_t^{srb} \rangle_{SRB,t} \right)^2 dt \qquad (5.11.6)$$

If the chaotic hypothesis is assumed then the state μ_t will evolve exponentially fast under the "frozen evolution" to $\mu_{SRB,t}$. Therefore the integral in Eq. (5.11.6) will converge for reasonable t dependences of $\boldsymbol{\Phi}, V$.

A physical definition of "quasi static" transformation is a transformation that is "very slow". This can be translated mathematically, for instance, into an evolution in which $\boldsymbol{\Phi}_t$ evolves like, if not exactly, as

$$\boldsymbol{\Phi}_t = \boldsymbol{\Phi}_0 + (1 - e^{-\gamma t})(\boldsymbol{\Phi}_\infty - \boldsymbol{\Phi}_0) \equiv \boldsymbol{\Phi}_0 + (1 - e^{-\gamma t})\Delta. \qquad (5.11.7)$$

An evolution Γ close to quasi static, but simpler for computing $\mathcal{I}(\Gamma)$, would proceed changing $\boldsymbol{\Phi}_0$ into $\boldsymbol{\Phi}_\infty = \boldsymbol{\Phi}_0 + \varepsilon(t)\Delta$ by $1/\delta$ steps of size δ, each of which has a time duration t_δ long enough so that, at the k-th step, the evolving system settles onto its stationary state at forcing $\boldsymbol{\Phi}_0 + k\delta\Delta$ (Fig. 5.5).

If the corresponding time scale can be taken $= \kappa^{-1}$, independent of the value of the forces so that t_δ can be defined by $\delta e^{-\kappa t_\delta} \ll 1$, then $\mathcal{I}(\Gamma) = const\, \delta^{-1}\delta^2 \log \delta^{-1}$ because the variation of $\sigma_{(k+1)\delta,+} - \sigma_{k\delta,+}$ is, in general, of order δ as a consequence of the differentiability of the SRB states with respect to the parameters, [54].

Remark (1) A drawback of the definition proposed in Sect. (4) is that although $\langle \varepsilon_t^{srb} \rangle_{SRB,t}$ is *independent* on the metric that is used to measure volumes in phase space the quantity $\langle \varepsilon_t \rangle_{\mu_t}$ *depends* on it. Hence the irreversibility degree considered here reflects also properties of our ability or method to measure (or to imagine to measure) distances in phase space. One can keep in mind that a metric independent definition can be simply obtained by minimizing over all possible choices of the metric: but the above simpler definition seems, nevertheless, preferable.

(2) Suppose that a process takes place because of the variation of an acting conservative force, for instance because a gravitational force changes as a large mass is brought close to the system, while no change in volume occurs and the thermostats have all the same temperature. Then the "frozen" SRB distribution, for all t, is such that $\langle \varepsilon^{srb} \rangle_{SRB,t} = 0$ (because the "frozen equations", being Hamiltonian equations, admit a SRB distribution which has a density in phase space). The isothermal process thus defined has *therefore* (and *nevertheless*) $\mathcal{I}(\Gamma) > 0$.

(3) Consider a typical irreversible process. Imagine a gas in an adiabatic cylinder covered by an adiabatic piston and imagine to move the piston. The simplest situation arises if gravity is neglected and the piston is suddenly moved at speed w.

Unlike the cases considered so far, the absence of thermostats (adiabaticity of the cylinder) imposes an extension of the analysis. The simplest situation arises when the piston is moved at speed so large, or it is so heavy, that no energy is gained or lost by the particles because of the collisions with the moving wall (this is, in fact, a case in which there are no such collisions). This is an extreme idealization of the classic Joule-Thomson experiment.

Let S be the section of the cylinder and $H_t = H_0 + w\,t$ be the distance between the moving lid and the opposite base. Let $\Omega = S\,H_t$ be the cylinder volume. In this case, if the speed $w \gg \sqrt{k_B T}$ the volume of phase space changes because the boundary moves and it increases by $N\,w\,S\,\Omega^{N-1}$ per unit time, i.e. its rate of increase is (essentially, see remark 5 below) $N\frac{w}{H_t}$.

Hence $\langle \varepsilon_t \rangle_t$ is $-N\frac{w}{H_t}$, while $\varepsilon_t^{srb} \equiv 0$. If $T = \frac{L}{w}$ is the duration of the transformation ("Joule-Thomson" process) increasing the cylinder length by L at speed w, then

$$\mathcal{I}(\Gamma) = \int_0^T N^2 \left(\frac{w}{H_t}\right)^2 dt \xrightarrow[T\to\infty]{} w\frac{L}{H_0(H_0+L)}N^2 \qquad (5.11.8)$$

and the transformation is irreversible. The irreversibility time scale approaches 0 as $w \to \infty$, as possibly expected. If $H_0 = L$, i.e. if the volume of the container is doubled, then $I(\Gamma) = \frac{w}{2L}$ and the irreversibility time scale of the process coincides with its "duration".

(4) If in the context of (3) the piston is replaced by a sliding lid which divides the cylinder in two halves of height L each: one empty at time zero and the other containing the gas in equilibrium. At time 0 the lid is lifted and a process Γ' takes place. In this case $\frac{dV_t}{dt} = V\delta(t)$ because the volume $V = S\,L$ becomes suddenly double (this amounts at a lid receding at infinite speed). Therefore the evaluation of the irreversibility scale yields

$$\mathcal{I}(\Gamma') = \int_0^\infty N^2 \delta(t)^2 dt \to +\infty \qquad (5.11.9)$$

so that the irreversibility becomes immediately manifest, $I(\Gamma') = +\infty$, $\mathcal{I}(\Gamma')^{-1} = 0$. This idealized experiment is rather close to the actual Joule-Thomson experiment.

In the latter example it is customary to estimate the degree of irreversibility at the lift of the lid by the *thermodynamic equilibrium entropy* variation between initial and final states. It would of course be interesting to have a general definition of entropy of a non stationary state (like the states μ_t at times ($t \in (0, \infty)$ in the example just discussed) that would allow connecting the degree of irreversibility to the thermodynamic entropy variation in processes leading from an initial equilibrium state to a final equilibrium state, see [55].

(5) The case with $w \ll \sqrt{k_B T}$ can also be considered. The lid mass being supposed infinite, a particle hits it if its perpendicular speed $v_1 > w$ and rebounds with a kinetic energy decreased by $2v_1 w$: if ϱ is the density of the gas and N the particles number the phase space contraction receives an extra contribution $\simeq$

$$N \varrho S \int_w^\infty d^3 v\, v_1 \frac{2v_1 w}{\frac{3}{2}k_B T} e^{-\frac{\beta}{2}v^2} \left(\frac{\sqrt{\beta}}{2\pi}\right)^3 = N w \varrho S \text{ const.}$$

(6) The Joule experiment for the measurement of the conversion factor of calories into ergs can be treated in a similar way: but there is no volume change and the phase space contraction is similar to the "extra" contribution in remark (5).
(7) In processes leading to or starting from a stationary nonequilibrium state the process time scale can become as long as wished but it would not be appropriate to call it the "irreversibility time scale": the process time scale is, in all cases, a measurement of how far a process is from a quasi static one with the same initial and final states.

It might be interesting, and possible, to study a geodesic flow on a surface of constant negative curvature under the action of a slowly varying electric field and subject to a isokinetic thermostat: the case of constant field is studied in [56], but more work will be necessary to obtain results (even in weak coupling) on the process in which the electric field $E(t)$ varies.

References

1. Becker, R.: Electromagnetic Fields and Interactions. Blaisdell, New York (1964)
2. Seitz, F.: The Modern Theory of Solids. Dover, Mineola (1987) (reprint)
3. Holian, B.D., Hoover, W.G., Posch, H.A.W.: Resolution of loschmidt paradox: the origin of irreversible behavior in reversible atomistic dynamics. Phys. Rev. Lett. **59**, 10–13 (1987)
4. Evans, D.J., Morriss, G.P.: Statistical Mechanics of Nonequilibrium Fluids. Academic Press, New York (1990)
5. Gallavotti, G.: Topics in chaotic dynamics. In: Garrido-Marro (ed.) Lecture Notes in Physics, vol. 448, pp. 271–311 (1995)
6. Gallavotti, G.: Equivalence of dynamical ensembles and navier stokes equations. Phys. Lett. A **223**, 91–95 (1996)
7. Gallavotti, G., Cohen, E.G.D.: Dynamical ensembles in stationary states. J. Stat. Phys. **80**, 931–970 (1995)
8. Gallavotti, G.: New methods in nonequilibrium gases and fluids. Open Syst. Inf. Dyn. **6**, 101–136 (1999) (preprint chao-dyn/9610018)
9. Gallavotti, G.: Statistical Mechanics. A Short Treatise. Springer, Berlin (2000)
10. Ruelle, D.: A remark on the equivalence of isokinetic and isoenergetic thermostats in the thermodynamic limit. J. Stat. Phys. **100**, 757–763 (2000)
11. Evans, D.J., Sarman, S.: Equivalence of thermostatted nonlinear responses. Phys. Rev. E **48**, 65–70 (1993)
12. Feitosa, K., Menon, N.: A fluidized granular medium as an instance of the fluctuation theorem. Phys. Rev. Lett. **92**, 164301 (+4) (2004)
13. Bonetto, F., Gallavotti, G., Giuliani, A., Zamponi, F: Fluctuations relation and external thermostats: an application to granular materials. J. Stat. Mech. **2006**(5), P05009 (2006)
14. Puglisi, A., Visco, P., Barrat, A., Trizac, E,. van Wijland, F: Fluctuations of internal energy flow in a vibrated granular gas. Phys. Rev. Lett. **95**, 110202 (+4) (2005). cond-mat/0509105

15. Grossman, E.L., Tong Zhou, E.L., Ben-NaimE.: Towards granular hydrodynamics in two-dimensions (1996). cond-mat/9607165
16. Brey, J., Ruiz-Montero, M.J., Moreno, F.: Boundary conditions and normal state for a vibrated granular fluid. Phys. Rev. E **62**, 5339–5346 (2000)
17. Bonetto, F., Chernov, N., Lebowitz, J.L.: (global and local) fluctuations in phase-space contraction in deteministic stationary nonequilibrium. Chaos **8**, 823–833 (1998)
18. Bonetto, F., Lebowitz, J.L.: Thermodynamic entropy production fluctuation in a two-dimensional shear flow model. Phys. Rev. E **64**, 056129 (2001)
19. Zamponi, F., Ruocco, G., Angelani, L.: Fluctuations of entropy production in the isokinetic ensemble. J. Stat. Phys. **115**, 1655–1668 (2004)
20. Giuliani, A., Zamponi, F., Gallavotti, G.: Fluctuation relation beyond linear response theory. J. Stat. Phys. **119**, 909–944 (2005)
21. Visco, P., Puglisi, A., Barrat, A., Trizac, E., van Wijland, F.: Fluctuations of power injection in randomly driven granular gases. J. Stat. Phys. **125**, 529–564 (2005)
22. Jepps, O., Evans, D., Searles, D.: The fluctuation theorem and lyapunov weights. Physica D **187**, 326–337 (2004)
23. Lanford, O.: Time evolution of large classical systems. Dynamical systems, theory and applications. In: Moser, J. (ed.) Lecture Notes in Physics, vol. 38, pp. 1–111. Springer Verlag, New York, Heidelberg (1974)
24. She, Z.S., Jackson, E.: Constrained euler system for navier-stokes turbulence. Phys. Rev. Lett. **70**, 1255–1258 (1993)
25. Sagaut, P.: Large Eddy Simulation for Incompressible Flows: Scientific Computation. Springer, Berlin (2006)
26. Germano, M., Piomell, U., Moin, P., Cabot, W.H.: A dynamic subgridscale eddy viscosity model. Phys Fluids A **3**, 1760–1766 (1991)
27. Gallavotti, G.: Dynamical ensembles equivalence in fluid mechanics. Physica D **105**, 163–184 (1997)
28. Gallavotti, G.: Foundations of Fluid Dynamics, (2nd printing). Springer, Berlin (2005)
29. Presutti, E.: Scaling limits in Statistical Mechanics and Microstructures in Continuum Mechanics. Springer, Berlin (2009)
30. de Groot, S., Mazur, P.: Non-equilibrium Thermodynamics. Dover, Mineola (1984)
31. Gallavotti, G.: Intermittency and time arrow in statistical mechanics and turbulence. Fields Inst. Commun. **31**, 165–172 (2002)
32. Gallavotti, G.: Microscopic chaos and macroscopic entropy in fluids. J. Stat. Mech. (JSTAT) **2006**, P10011 (+9) (2006)
33. Kurchan, J.: Fluctuation theorem for stochastic dynamics. J Phys A **31**, 3719–3729 (1998)
34. Lebowitz, J., Spohn, H.: A gallavotti-cohen type symmetry in large deviation functional for stochastic dynamics. J. Stat. Phys. **95**, 333–365 (1999)
35. Maes, C.: The fluctuation theorem as a gibbs property. J. Stat. Phys. **95**, 367–392 (1999)
36. Kernigham, B.W., Ritchie, D.M.: The C Programming Language. Prentice Hall Software Series. Prentice Hall, Engelwood Cliffs, N.J. (1988)
37. Bonetto, F., Gallavotti, G., Garrido, P.: Chaotic principle: an experimental test. Physica D **105**, 226–252 (1997)
38. Bonetto, F., Gallavotti, G., Gentile, G.: A fluctuation theorem in a random environment. Ergodic Theor. Dyn. Syst. **28**, 21–47 (2008)
39. Bricmont, J., Kupiainen, A.: Diffusion in energy conserving coupled maps. Commun. Math. Phys. **321**, 311–369 (2013)
40. Collet, P., Epstein, H., Gallavotti, G.: Perturbations of geodesic flows on surfaces of constant negative curvature and their mixing properties. Commun. Math. Phys. **95**, 61–112 (1984)
41. Bonetto, F., Gallavotti, G., Giuliani, A., Zamponi, F.: Chaotic hypothesis, fluctuation theorem and singularities. J. Stat. Phys. **123**, 39–54 (2006)
42. Van Zon, R., Cohen, E.G.D.: Extension of the fluctuation theorem. Phys. Rev. Lett. **91**, 110601 (+4) (2003)

43. Van Zon, R., Cohen, E.G.D.: Extended heat-fluctuation theorems for a system with deterministic and stochastic forces. Phys. Rev. E **69**, 056121 (+14) (2004)

44. Joubaud, S., Garnier, N.B., Ciliberto, S.: Fluctuation theorems for harmonic oscillators. J. Stat. Mech. **2007**(9), P09018 (2007). cond-mat/0703798

45. Evans, D.J., Searles, D.J., Rondoni, L.: Application of the Gallavotti-Cohen fluctuation relation to thermostated steady states near equilibrium. Phys. Rev. E **71**, 056120 (+12) (2005)

46. Zamponi, F.: Is it possible to experimentally verify the fluctuation relation? a review of theoretical motivations and numerical evidence. J. Stat. Mech. **2–7**, P02008 (2007)

47. Cugliandolo, L.F., Kurchan, J., Peliti, L.: Energy flow, partial equilibration, and effective temperatures in systems with slow dynamics. Phys. Rev. E **55**, 2898–3914 (1997)

48. Crisanti, A., Ritort, F.: Violation of the fluctuation-dissipation theorem in glassy systems: basic notions and the numerical evidence. J. Phys. A **36**, R181–R290 (2003)

49. Nakabeppu, O., Suzuki, T.: Microscale temperature measurement by scanning thermal microscopy. J. Therm. Anal. Calorim. **69**, 727–737 (2002)

50. Smith, D.T.: A square root circuit to linearize feedback in temperature controllers. J. Phys. E Sci. Instrum. **5**, 528 (1972)

51. Gallavotti, G.: Nonequilibrium statistical mechanics (stationary): overview. In: Françoise, J.P., Naber, G.L., Tsun, T.S. (eds.) Encyclopedia of Mathematical Physics, vol. 3, pp. 530–539 (2006)

52. Ruelle, D.: Positivity of entropy production in nonequilibrium statistical mechanics. J. Stat. Phys. **85**, 1–25 (1996)

53. Ruelle, D.: Entropy production in nonequilibrium statistical mechanics. Commun. Math. Phys. **189**, 365–371 (1997)

54. Ruelle, D.: Differentiation of SRB states. Commun. Math. Phys. **187**, 227–241 (1997)

55. Goldstein, S., Lebowitz, J.L.: On the (Boltzmann) entropy of nonequilibrium systems. Physica D **193**, 53–66 (2004)

56. Bonetto, F., Gentile, G., Mastropietro, V.: Electric fields on a surface of constant negative curvature. Ergodic Theor. Dyn. Syst. **20**, 681–686 (2000)

Chapter 6
Historical Comments

6.1 Proof of the Second Fundamental Theorem

Partial translation and comments of L. Boltzmann, *Über die mechanische Bedeutung des zweiten Haupsatzes der Wägrmetheorie*, Wien. Ber. **53**, 195–220, 1866. Wissenshaftliche Abhanlunger, Vol. **1**, p. 9–33, #2, [1].

[*The distinction between the "second theorem" and "second law" is important: the first is $\oint \frac{dQ}{T} = 0$ in a reversible cycle, while the second is the inequality $\oint \frac{dQ}{T} \leq 0$. For a formulation of the second law here intend the Clausius formulation, see footnote at p. 19. The law implies the theorem but not viceversa.*]

[*In Sect. I Boltzmann's aim is to explain the mechanical meaning of temperature. Two bodies in contact will be in thermal equilibrium if the average kinetic energy (or any average property) of the atoms of each will not change in time: the peculiarity of the kinetic energy is that it is conserved in the collisions. Let two atoms of masses m, M and velocities $\mathbf{v}, \mathbf{V}$ collide and let $\mathbf{c}, \mathbf{C}$ be the velocities after collision. The kinetic energy variation of the atom m is $\ell = \frac{1}{2}mv^2 - \frac{1}{2}mc^2$. Choosing as z-axis (called G) so that the momentum $m\mathbf{v}$ and $m\mathbf{c}$ have the same projection on the xy plane it follows from the collision conservation rules (of kinetic energy and momentum) that if φ, Φ are inclinations of $\mathbf{v}, \mathbf{V}$ over the z-axis (and likewise φ', Φ' are inclinations of $\mathbf{c}, \mathbf{C}$ over the z-axis) it is*]

$$\ell = \frac{2mM}{(m+M)^2}(MC^2 \cos^2 \Phi - mc^2 \cos^2 \varphi + (m-M)cC \cos \varphi \cos \Phi)$$

which averaged over time and over all collisions between atoms of the two bodies yield an average variation of ℓ which is $L = \frac{4mM}{3(m+M)^2}\left(\frac{MC^2}{2} - \frac{mc^2}{2}\right)$ so that the average kinetic energies of the atoms of the two species have to be equal hence $T = A\left(\frac{1}{2}mc^2\right) + B$. The constant $B = 0$ if T is identified with the absolute temperature of a perfect gas. The identification of the average kinetic energy with the absolute temperature was already a well established fact, based on the theory of the perfect

G. Gallavotti, *Nonequilibrium and Irreversibility*,
Theoretical and Mathematical Physics, DOI: 10.1007/978-3-319-06758-2_6,
© Springer International Publishing Switzerland 2014

gas, see Sect. 1.2. The analysis is somewhat different from the one presented by Maxwell few years earlier in [2, p. 383].]

[In Sect. II it is shown, by similar arguments, that the average kinetic energy of the center of mass of a molecule is the same as the average kinetic energy of each of its atoms.]

[In Sect. III the laws called of Ampère-Avogadro, of Dulong-Petit and of Neumann for a free gas of molecules are derived.]

Section IV: Proof of the second theorem of the mechanical theory of heat

The just clarified meaning of temperature makes it possible to undertake the proof of the second fundamental theorem of heat theory, and of course it will be entirely coincident with the form first exposed by Clausius.

$$\int \frac{dQ}{T} \leq 0 \qquad (20)$$

To avoid introducing too many new quantities into account, we shall right away treat the case in which actions and reactions during the entire process are equal to each other, so that in the interior of the body either thermal equilibrium or a stationary heat flow will always be found. Thus if the process is stopped at a certain instant, the body will remain in its state.[1] In such case in Eq. (20) the equal sign will hold. Imagine first that the body, during a given time interval, is found in a state at temperature, volume and pressure constant, and thus atoms will describe curved paths with varying velocities.

We shall now suppose that an arbitrarily selected atom runs over every site of the region occupied by the body in a suitable time interval (no matter if very long), of which the instants t_1 and t_2 are the initial and final times, at the end of it the speeds and the directions come back to the original value in the same location, describing a closed curve and repeating, from this instant on, their motion,[2] possibly not exactly

---

[1] Here B. means a system in thermal equilibrium evolving in a "reversible" way, i.e. "quasi static" in the sense of thermodynamics, and performing a cycle (the cyclicity condition is certainly implicit even though it is not mentioned): in this process heat is exchanged, but there is no heat flow in the sense of modern nonequilibrium thermodynamics (because the process is quasi static); furthermore the process takes place while every atom follows approximate cycles with period $t_2 - t_1$, of duration possibly strongly varying from atom to atom, and the variations induced by the development of the process take place over many cycles. Eventually it will be assumed that in solids the cycles period is a constant to obtain, as an application, theoretical evidence for the Dulong-Petit and Neumann laws.

[2] This is perhaps the first time that what will become the *ergodic hypothesis* is formulated. It is remarkable that an analogous hypothesis, more clearly formulated, can be found in the successive paper by Clausius [3, 1.8, p. 438] (see the following Sect. 6.4), which Boltzmann criticized as essentially identical to Sect. IV of the present paper: this means that already at the time the idea of recurrence and ergodicity must have been quite common. Clausius imagines that the atoms follow closed paths, i.e. he conceives the system as "integrable", while Boltzmann appears to think, at least at first, that the entire system follows a single closed path. It should however be noticed that later, concluding Sect. IV, Boltzmann will suppose that every atom will move staying within a small volume element, introduced later and denoted dk, getting close to Clausius' viewpoint.

equal[3] nevertheless so similar that the average kinetic energy over the time $t_2 - t_1$ can be seen as the atoms average kinetic energy during an arbitrarily long time; so that the temperature of every atom is[4]

$$T = \frac{\int_{t_1}^{t_2} \frac{mc^2}{2} dt}{t_2 - t_1}.$$

Every atom will vary [*in the course of time*] its energy by an infinitely small quantity ε, and certainly every time the work produced and the variation of kinetic energy will be in the average redistributed among the various atoms, without contributing to a variation of the average kinetic energy. If the exchange was not eventually even, it would be possible to wait so long until the thermal equilibrium or stationarity will be attained[5] and then exhibit, as work performed in average, an average kinetic energy increase greater by what was denoted ε, per atom.

At the same time suppose an infinitely small variation of the volume and pressure of the body.[6] Manifestly the considered atom will follow one of the curves, infinitely close to each other.[7] Consider now the time when the atom is on a point of the new

---

[3] Apparently this contradicts the preceding statement: because now he thinks that motion does not come back exactly on itself; here B. rather than taking into account the continuity of space (which would make impossible returning exactly at the initial state) refers to the fact that his argument does not require necessarily that every atom comes back exactly to initial position and velocity but it suffices that it comes back "infinitely close" to them, and actually only in this way it is possible that a quasi static process can develop.

[4] Here use is made of the result discussed in the Sect. I of the paper which led to state that every atom, and molecule alike, has equal average kinetic energy and, therefore, it makes sense to call it "temperature" of the atom.

[5] A strange remark because the system is always in thermodynamic equilibrium: but it seems that it would suffice to say "it would be possible to wait long enough and then exhibit ...". A faithful literal translation is difficult. The comment should be about the possibility that diverse atoms may have an excess of kinetic energy, with respect to the average, in their motion: which however is compensated by the excesses or defects of the kinetic energies of the other atoms. Hence ε has zero average because it is the variation of kinetic energy when the system is at a particular position in the course of a quasi static process (hence it is in equilibrium). However in the following paragraph the same symbol ε indicates the kinetic energy variation due to an infinitesimal step of a quasi static process, where there is no reason why the average value of ε be zero because the temperature may vary. As said below the problem is that this quantity ε seems to have two meanings and might be one of the reasons of Clausius complaints, see footnote at p. 157.

[6] Clausius' paper is, however, more clear: for instance the precise notion of "variation" used here, with due meditation can be derived, as proposed by Clausius and using his notations in which ε has a completely different meaning, as the function with two parameters $\delta i, \varepsilon$ changing the periodic function $x(t)$ with period $i \equiv t_2 - t_1$ into $x'(t)$ with $x'(t) = x(it/(i + \delta i)) + \varepsilon \xi(it/(i + \delta i))$ periodic with period $i + \delta i$ which, to first order in $\delta x = -\dot{x}(t) \frac{\delta i}{i} t + \varepsilon \xi(t)$.

[7] Change of volume amounts at changing the external forces (volume change is a variation of the confining potential): but no mention here is made of this key point and no trace of the corresponding extra terms appears in the main formulae. Clausius *essential critique* to Boltzmann, in the priority dispute, is precisely that no account is given of variations of external forces. Later Boltzmann recognizes that he has not included external forces in his treatment, see p. 159, without mentioning this point.

path from which the position of the atom at time t_1 differs infinitely little, and let it be t_1' and denote t_2' the instant in which the atom returns in the same position with the same velocity; we shall express the variation of the value of T via the integrals

$$\int_{t_1}^{t_2} \frac{mc^2}{2}\, dt = \frac{m}{2} \int_{s_1}^{s_2} c\, ds$$

where ds is the line element values of the mentioned arc with as extremes s_1 and s_2 the positions occupied at the times t_1 and t_2. The variation is

$$\frac{m}{2} \delta \int_{s_1}^{s_2} c\, ds = \frac{m}{2} \int_{s_1'}^{s_2'} c'\, ds - \frac{m}{2} \int_{s_1}^{s_2} c\, ds$$

where the primed quantities refer to the varied curve and s_1', s_2' are the mentioned arcs of the new curve at the times t_1', t_2'. To obtain an expression of the magnitude of the variation we shall consider also ds as variable getting

$$\frac{m}{2} \delta \int_{s_1}^{s_2} c\, ds = \frac{m}{2} \int_{s_1}^{s_2} (\delta c\, ds + c\, \delta ds); \tag{21}$$

It is $\frac{m}{2} \int_{s_1}^{s_2} \delta c\, ds = \int_{t_1}^{t_2} \frac{dt}{2} \delta \frac{mc^2}{2}$, and furthermore, if X, Y, Z are the components on the coordinate axes of the force acting on the atom, it follows:

$$d\frac{m\,c^2}{2} = X\, dx + Y\, dy + Z\, dz$$

$$d\delta\frac{mc^2}{2} = \delta X\, dx + \delta Y\, dy + \delta Z\, dz + X\, \delta dx + Y\, \delta dy + Z\, \delta dz$$

$$= d(X\delta x + Y\delta Y + z\delta Z)$$
$$+ (\delta X dx - dX\delta x + \delta Y dy - dY\delta y + \delta Z dz - dZ\delta z).$$

Integrate then,[8] considering that for determining the integration constant it must be $\delta\frac{mc^2}{2} = \varepsilon$ when the right hand side vanishes,[9] one gets the

---

[8] This point, as well as the entire argument, may appear somewhat obscure at first sight: but it arrives at the same conclusions that 4 years later will be reached by Clausius, whose derivation is instead very clear; see the sections of the Clausius paper translated here in Sect. 6.4 and the comment on the action principle below.

[9] The initial integration point is not arbitrary: it should rather coincide with the point where the kinetic energy variation equals the variation of the work performed, in average, on the atom during the motion.

$$\delta \frac{m\,c^2}{2} - \varepsilon = (X\delta x + Y\delta Y + z\delta Z)$$

$$+ \int (\delta X dx - dX\delta x + \delta Y dy - dY\delta y + \delta Z dz - dZ\delta z).$$

Here the term on the left contains the difference of the kinetic energies, the expression on the right contains the work made on the atom, hence the integral on the r.h.s. expresses the kinetic energy communicated to the other atoms. The latter certainly is not zero at each instant, but its average during the interval $t_2 - t_1$ is, in agreement with our assumptions, $=0$ because the integral is extended to this time interval. Taking into account these facts one finds

$$\int\limits_{t_1}^{t_2} \frac{dt}{2} \delta \frac{mc^2}{2} = \frac{t_2 - t_1}{2}\varepsilon + \frac{1}{2}\int\limits_{t_1}^{t_2} (X\delta x + Y\delta Y + z\delta Z)dt$$

$$= \frac{t_2 - t_1}{2}\varepsilon + \frac{m}{2}\int\limits_{t_1}^{t_2} \left(\frac{d^2x}{dt^2}\delta x + \frac{d^2y}{dt^2}\delta y + \frac{d^2z}{dt^2}\delta z\right)dt,$$

(22)

a formula which, by the way,[10] also follows because ε is the sum of the increase in average of the kinetic energy of the atom and of the work done in average on the atom.[11] If $ds = \sqrt{dx^2 + dy^2 + dz^2}$ is set and $c = \frac{ds}{dt}$,

$$\frac{m}{2}\int\limits_{s_1}^{s_2} c\,\delta ds = \frac{m}{2}\int\limits_{t_1}^{t_2} \left(\frac{dx}{dt}d\delta x + \frac{dy}{dt}d\delta y + \frac{dz}{dt}d\delta z\right).$$

(23)

Inserting Eqs. (22) and (23) in Eq. (21) follows[12]:

---

[10] Here too the meaning of ε is not clear. The integral from t_1 to t_2 is a line integral of a differential $d(X\delta x + \cdots)$ but does not vanish because the differential is not exact, as $\delta x, \delta y, \delta z$ is not parallel to the integration path.

[11] In other words this is the "vis viva" theorem because the variation of the kinetic energy is due to two causes: namely the variation of the motion, given by ε, and the work done by the acting (internal) forces because of the variation of the trajectory, given by the integral. Clausius considered the statement in need of being checked.

[12] This is very close to the least action principle according to which the difference between average kinetic energy and average potential energy is stationary within motions with given extremes. Here the condition of fixed extremes does not apply and it is deduced that the action of the motion considered between t_1 and t_2 has a variation which is a boundary term; precisely $\{m\mathbf{v}\cdot\mathbf{x}\}_{t_1}^{t_2}$ (which is 0) is the difference ε between the average kinetic energy variation $m\delta\int_{s_1}^{s_2} c\,ds$ and that of the average potential energy. Such formulation is mentioned in the following p. 139.

$$\frac{m}{2}\delta\int_{s_1}^{s_2} c\,ds = \frac{t_2 - t_1}{2}\varepsilon + \int_{t_1}^{t_2} d\left(\frac{dx}{dt}\delta x + \frac{dy}{dt}\delta y + \frac{dz}{dt}\delta z\right)$$

$$= \frac{t_2 - t_1}{2}\varepsilon + \left\{\frac{m}{2}\left(\frac{dx}{dt}\delta x + \frac{dy}{dt}\delta y + \frac{dz}{dt}\delta z\right)\right\}\Big|_{t_1}^{t_2}.$$

However since the atom at times t_1 and t_1' is in the same position with the same velocity as at the times t_2 and t_2', then also the variations at time t_1 have the same value taken at time t_2, so in the last part both values cancel and remains

$$\varepsilon = \frac{m\delta\int_{t_1}^{t_2} c\,ds}{t_2 - t_1} = \frac{2\delta\int_{t_1}^{t_2}\frac{mc^2}{2}dt}{t_2 - t_1},\tag{23a}$$

which, divided by the temperature, yield[13]:

$$\frac{\varepsilon}{T} = \frac{2\delta\int_{t_1}^{t_2}\frac{mc^2}{2}dt}{\int_{t_1}^{t_2}\frac{mc^2}{2}dt} = 2\delta\log\int_{t_1}^{t_2}\frac{mc^2}{2}dt.$$

Suppose right away that the temperature at which the heat is exchanged during the process is the same everywhere in the body, realizing in this way the assumed hypothesis. Hence the sum of all the ε equals the amount of heat transferred inside the body measured in units of work. Calling the latter δQ, it is:

$$\delta Q = \sum\varepsilon = 2\sum\frac{\delta\int_{t_1}^{t_2}\frac{mc^2}{2}dt}{t_2 - t_1}$$
$$\frac{\delta Q}{T} = \frac{1}{T}\sum\varepsilon = 2\delta\sum\log\int_{t_1}^{t_2}\frac{mc^2}{2}dt.\tag{24}$$

If now the body temperature varies from place to place, then we can subdivide the body into volume elements dk so small that in each the temperature and the heat transfer can be regarded as constant; consider then each of such elements as external and denote the heat transferred from the other parts of the body as $\delta Q \cdot dk$ and, as before,

$$\frac{\delta Q}{T} dk = 2\delta\sum\log\int_{t_1}^{t_2}\frac{mc^2}{2}dt,$$

---

[13] It should be remarked that physically the process considered is a reversible process in which no work is done: therefore the only parameter that determines the macroscopic state of the system, and that can change in the process, is the temperature: so strictly speaking Eq. (24) might be not surprising as also Q would be function of T. Clausius insists that this is a key point which is discussed in full detail in his work by allowing also volume changes and more generally action of external forces, see p. 159.

if the integral as well as the sum runs over all atoms of the element dk. From this it is clear that the integral

$$\int \int \frac{\delta Q}{T} dk$$

where one integration yields the variation δ of what Clausius would call entropy, with the value

$$2 \sum \log \int_{t_1}^{t_2} \frac{mc^2}{2} dt + C,$$

[and the integration] between equal limit vanishes, if pressure and counter-pressure remain always equal.[14]

Secondly if this condition was not verified it would be possible to introduce all along a new force to restore the equality. The heat amount, which in the last case, through the force added to the ones considered before, must be introduced to obtain equal variations of volumes and temperatures in all parts of the body, must be such that the equation

$$\int \int \frac{\delta Q}{T} dk = 0$$

holds; however the latter [heat amount] is necessarily larger than that [the heat amount] really introduced, as at an expansion of the body the pressure had to overcome the considered necessary positive force[15]; at a compression, in the case of equal pressures, a part of the compressing force employed, and hence also the last heat generated, must always be considered.[16] It yields also for the necessary heat supplied no longer the equality, instead it will be[17]:

$$\int \int \frac{\delta Q}{T} dk < 0$$

---

[14] I.e. in a cycle in which no work is done. This is criticized by Clausius.

[15] The pressure performs positive work in an expansion.

[16] In a compression the compressing force must exceed (slightly) the pressure, which therefore performs a negative work. In other words in a cycle the entropy variation is 0 but the Clausius integral is <0.

[17] The latter comments do not seem to prove the inequality, unless coupled with the usual formulation of the second law (e.g. in the Clausius form, as an inequality). On the other hand this is a place where external forces are taken into account: but in a later letter to Clausius, who strongly criticized his lack of consideration of external forces, Boltzmann admits that he has not considered external forces, see p. 159, and does not refer to his comments above. See also comment at p. 134.

[*The following page deals with the key question of the need of closed paths: the lengthy argument concludes that what is really needed is that in arbitrarily long time two close paths remain close. The change of subject is however rather abrupt.*]

I will first of all consider times t_1, t_2, t_1' and t_2', and the corresponding arcs, also in the case in which the atom in a given longer time does not describe a closed path. At first the times t_1 and t_2 must be thought as widely separated from each other, as well separated as wished, so that the average "vis viva" during $t_2 - t_1$ would be the true average "vis viva". Then let t_1' and t_2' be so chosen that the quantity

$$\frac{dx}{dt}\delta x + \frac{dy}{dt}\delta y + \frac{dz}{dt}\delta z \tag{25}$$

assumes the same value at both times. One easily convinces himself that this quantity equals the product of the atom speed times the displacement $\sqrt{\delta x^2 + \delta y^2 + \delta z^2}$ times the cosine of the angle between them. A second remark will also be very simple, if s_1' and s_2' are the corresponding points, which lie orthogonally to the varied trajectory across the points s_1 and s_2 on the initial trajectory, then the quantity (25) vanishes for both paths. This condition on the variation of the paths, even if not be satisfied, would not be necessary for the vanishing of the integrals difference, as it will appear in the following. Therefore from all these arguments, that have been used above on closed paths, we get

$$\int\int \frac{\delta Q}{T} dk = 2\sum \log \frac{\int_{t_1}^{t_2} \frac{mc^2}{2} dt}{\int_{\tau_1}^{\tau_2} \frac{mc^2}{2} dt},$$

if τ_1 and τ_2 are limits of the considered [*varied*] path, chosen in correspondence of the integral on the left. It is now possible to see that the value of this integral taken equally on both paths does not vanish since, if one proceeds in the above described way, the normal plane at s_1 at its intersection with the next path is again a normal plane and in the end it is not necessary a return on the same curve again at the same point s_1; only the point reached after a time $t_2 - t_1$ will be found at a finite not indefinitely increasing distance from s_1, hence

$$\int_{t_1}^{t_2} \frac{mc^2}{2} dt \quad \text{and} \quad \int_{\tau_1}^{\tau_2} \frac{mc^2}{2} dt$$

now differ by finite amounts and the more the ratio

$$\frac{\int_{t_1}^{t_2} \frac{mc^2}{2} dt}{\int_{\tau_1}^{\tau_2} \frac{mc^2}{2} dt}$$

is close to unity the more its logarithm is close to zero[18]; the more $t_2 - t_1$ increases the more both integrals increase, and also more exactly the average kinetic energy takes up its value; subdivide then both domains of the integrals $\int \int \frac{\delta Q}{T} dk$, so that one of the integrals differs from the other by a quantity in general finite, thus the ratio and therefore the logarithm does not change although it is varied by infinitely many increments.

This argument, together with the mathematical precision of the statement, is not correct in the case in which the paths do not close in a finite time, unless they could be considered closed in an infinite time [italics added in the translation, p. 30].[19]

[Having completed the very important discussion, which Clausius may have overlooked, see Sect. 6.7, on the necessity of periodicity of the motion, Boltzmann returns to the conceptual analysis of the results.]

It is easily seen that our conclusion on the meaning of the quantities that intervene here is totally independent from the theory of heat, and therefore the second fundamental theorem is related to a theorem of pure mechanics to which it corresponds just as the "vis viva" principle corresponds to the first principle; and, as it immediately follows from our considerations, it is related to the least action principle, in form somewhat generalized about as follows:

If a system of point masses under the influence of forces, for which the principle of the "vis viva" holds, performs some motion, and if then all points undergo an infinitesimal variation of the kinetic energy and are constrained to move on a path infinitely close to the precedent, then $\delta \sum \frac{m}{2} \int c\,ds$ equals the total variation of the kinetic energy multiplied by half the time interval during which the motion develops, when the sum of the product of the displacements of the points times their speeds and the cosine of the angles on each of the elements are equal, for instance the points of the new elements are on the normal of the old paths.

This proposition gives, for the kinetic energy transferred and if the variation of the limits of integration vanishes, the least action principle in the usual form.

It is also possible to interpret the problem differently; if the second theorem is already considered sufficiently founded because of experiment credit or other, as done by Zeuner [4], in his new monograph on the mechanical theory of heat, and temperature is defined as the integrating divisor of the differential quantity dQ, then the derivation exposed here implies that the reciprocal of the value of the average kinetic energy is the integrating factor of δQ, hence temperature equals the product of this average kinetic energy time an arbitrary function of entropy. Such entirely arbitrary function must be fixed in a way similar to what done in the quoted case: it is then clear that it will never be possible to separate the meaning of temperature from the second theorem.

Finally I will dedicate some attention to the applicability of Eq. (24) to the determination of the heat capacity.

---

[18] The role of this particular remark is not really clear (to me).

[19] I.e. the distance between the points corresponding to s_2' and s_2 remains small forever: in other words, we would say, if no Lyapunov exponent is positive, i.e. the motion is not chaotic.

Differentiation of the equality $T = \frac{\int_{t_1}^{t_2} \frac{mc^2}{2} dt}{t_2 - t_1}$ leads to

$$\delta T = \frac{\delta \int_{t_1}^{t_2} \frac{mc^2}{2} dt}{t_2 - t_1} - \frac{\int_{t_1}^{t_2} \frac{mc^2}{2} dt}{t_2 - t_1} \cdot \frac{\delta (t_2 - t_1)}{t_2 - t_1};$$

and we shall look for the heat δH spent to increase temperature by δT of all atoms

$$\delta H = \sum \frac{\delta \int_{t_1}^{t_2} \frac{mc^2}{2} dt}{t_2 - t_1} - \sum \frac{\int_{t_1}^{t_2} \frac{mc^2}{2} dt}{t_2 - t_1} \cdot \frac{\delta (t_2 - t_1)}{t_2 - t_1};$$

and combining with Eq. (24) it is found

$$\delta Q = 2 \delta H + 2 \sum \frac{\int_{t_1}^{t_2} \frac{mc^2}{2} dt}{t_2 - t_1} \cdot \frac{\delta (t_2 - t_1)}{t_2 - t_1};$$

and the work performed, both internal and external[20]

$$\delta L = \delta H + 2 \sum \frac{\int_{t_1}^{t_2} \frac{mc^2}{2} dt}{t_2 - t_1} \cdot \frac{\delta (t_2 - t_1)}{t_2 - t_1} \tag{25a}$$
$$= \sum \frac{\delta \int_{t_1}^{t_2} \frac{mc^2}{2} dt}{t_2 - t_1} + \sum \frac{\int_{t_1}^{t_2} \frac{mc^2}{2} dt}{t_2 - t_1} \cdot \frac{\delta (t_2 - t_1)}{t_2 - t_1}$$

and the quantity

$$\delta Z = \int \frac{\delta L}{T} dh = \sum \frac{\delta \int_{t_1}^{t_2} \frac{mc^2}{2} dt}{\int_{t_1}^{t_2} \frac{mc^2}{2} dt} + \sum \frac{\delta (t_2 - t_1)}{t_2 - t_1};$$

called by Clausius "disgregation" integral[21] has therefore the value

$$Z = \sum \log \int_{t_1}^{t_2} \frac{mc^2}{2} dt + \sum \log(t_2 - t_1) + C. \tag{25b}$$

In the case when $t_2 - t_1$, which we can call period of an atom, does not change it is: $\delta (t_2 - t_1) = 0$, $\delta Q = 2\delta H$, $\delta L = \delta H$; i.e. the heat transferred is divided in two parts, one for the heating and the other as work spent.

Suppose now that the body has everywhere absolutely the same temperature and also that it is increased remaining identical everywhere, thus $\frac{\int_{t_1}^{t_2} \frac{mc^2}{2} dt}{t_2 - t_1}$ and $\delta \frac{\int_{t_1}^{t_2} \frac{mc^2}{2} dt}{t_2 - t_1}$

---

[20] See the Clausius' paper where this point is clearer; see also the final comment.

[21] It is the free energy.

are equal for all atoms and the heat capacity γ is expressed by $\frac{\delta Q}{p\delta T}$, if heat and temperature are expressed in units of work and p is the weight of the body:

$$\gamma = \frac{\delta Q}{p\delta T} = \frac{2\delta \int_{t_1}^{T_2} \frac{mc^2}{2} dt}{\frac{p}{N}\left[\delta \int_{t_1}^{T_2} \frac{mc^2}{2} dt - \frac{\int_{t_1}^{t_2} \frac{mc^2}{2} dt}{t_2-t_1} \cdot \frac{\delta(t_2-t_1)}{t_2-t_1}\right]}$$

where N is the number of atoms of the body and, if a is the atomic number or, in composed bodies, the total molecular weight and n the number of molecules in the atom, it will be $\frac{p}{N} = \frac{a}{n}$. In the case $\delta(t_2 - t_1) = 0$ it will also be $\frac{a\gamma}{n} = 2.^{22}$Therefore the product of the specific heat and the atomic weight is twice that of a gas at constant volume, which is $=1$. This law has been experimentally established by Masson for solids (see the published paper "*Sur la correlation ...*", Ann. de. Chim., Sect. III, vol. 53), [5]; it also implies the isochrony of the atoms vibrations in solids; however it is possibly a more complex question and perhaps I shall come back another time on the analysis of this formula for solids; in any event we begin to see in all the principles considered here a basis for the validity of the Dulong-Petit's and Neumann's laws.

6.2 Collision Analysis and Equipartition

Translation and comments of: L. Boltzmann, *Studien über das Gleichgewicht der lebendigen Kraft zwischen bewegten materiellen Punkten*, Wien. Ber., **58**, 517–560, 1868, Wissenschaftliche Abhandlungen, ed. F. Hasenöhrl, **1**, #5, (1868).

All principles of analytic mechanics developed so far are limited to the transformation of a system of point masses from a state to another, according to the evolution laws of position and velocity when they are left unperturbed in motion for a long time and are concerned, with rare exceptions, with theorems of the ideal, or almost ideal, gas. This might be the main reason why the theorems of the mechanical theory of heat which deal with the motions so far considered are so uncorrelated and defective. In the following I shall treat several similar examples and finally I shall establish a general theorem on the probability that the considered point masses occupy distinct locations and velocities.

I. The case of an infinite number of point masses
Suppose we have an infinite number of elastic spheres of equal mass and size and constrained to keep the center on a plane. A similar more general problem has been solved by Maxwell (Phil. Mag. march 1868); however partly because of the non complete exposition partly also because the exposition of Maxwell in its broad lines is difficult to understand, and because of a typo (in formulae (2) and (21) on the quantities called dV^2 and dV) will make it even more difficult, I shall treat here the problem again from the beginning.

---

[22] The hypothesis $\delta(t_2 - t_1)$ looks "more reasonable" in the case of solid bodies in which atoms can be imagined bounded to periodic orbits around the points of a regular lattice.

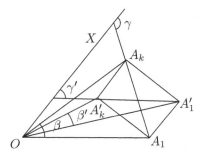

Fig. 6.1 Geometric illustration of the quantities entering in the analysis

It is by itself clear that in this case every point of the plane is a possibly occupied location of the center of one of the elastic spheres and every direction has equal probability, and only the speeds remain to determine. Let $\varphi(c)dc$ be the sum of the time intervals during which the speed of one of the spheres happens to have a value between c and $c + dc$ divided by such very large time: which is also the probability that c is between c and $c + dc$ and let N be the average number of the spheres whose center is within the unit of surface where the velocities are between c and $c + dc$.

Consider now a sphere, that I call I, with speed c_1 towards OA_1, Fig. 6.1, represented in size and direction, and let OX the line joining the centers at the impact moment and let β be the angle between the velocities c_1 and c_k, so that the velocities components of the two spheres orthogonal to OX stay unchanged, while the ones parallel to OX will simply be interchanged, as the masses are equal; hence let us determine the velocities before the collision and let $A_1 A_1'$ be parallel to OX and let us build the rectangle $A_1 A_1' A_k A_k'$; $O A_1'$ and $O A_k'$ be the new velocities and β' their new angle. Consider now the two diagonals of the rectangle $A_1 A_k$ and $A_1' A_k'$ which give the relative velocities of the two spheres, g before and g' after the collision, and call γ the angle between the lines $A_1 A_k$ and OX, and γ' that between the lines $A_1' A_k'$ and OX; so it is easily found:

$$g^2 = c_1^2 + c_2^2 - 2c_1 c_k \sin \gamma \cdot \gamma \sin \beta \tag{1}$$
$$c_1'^2 = c_1^2 \sin^2 \gamma + c_k^2 \cos^2 \gamma - 2c_1 c_k \sin \gamma \cos \gamma \sin \beta$$
$$c_2'^2 = c_1^2 \cos^2 \gamma + c_k^2 \sin^2 \gamma + 2c_1 c_k \sin \gamma \cos \gamma \sin \beta$$
$$\tan \beta' = \frac{(c_1^2 - c_k^2) \sin \gamma \cos \gamma - c_1 c_k (\cos^2 \gamma - \sin^2 \gamma) \sin \beta}{c_1 c_k \cos \beta}$$
$$= \frac{\sqrt{c_1'^2 c_k'^2 - c_1^2 c_k^2 \cos^2 \beta}}{c_1 c_k \cos \beta}$$
$$c_1 c_k \cos \beta = c_1' c_k' \cos \beta'; \quad \gamma' = \pi - \gamma.$$

We immediately ask in which domains c_1' and c_k' are, given c_1, c_k. For this purpose ....

[*A long analysis follows about the relation between the area elements $d^2 \mathbf{c}_1 d^2 \mathbf{c}_k$ and the corresponding $d^2 \mathbf{c}_1' d^2 \mathbf{c}_k'$. Collisions occur with an angle, between the* collision direction and the line connecting the centers, between β and $\beta + d\beta$ with probability proportional to $\sigma(\beta)d\beta$ where $\sigma(\beta)$ is the cross section (equal to $\sigma(\beta) = \frac{1}{2}r\sin\beta$ in the case, studied here, of disks of radius r). Then the density $f(\mathbf{c})d^2\mathbf{c}$ must have the property $\varphi(\mathbf{c}_1)f(\mathbf{c}_k)\sigma(\beta)d\beta = \varphi(\mathbf{c}_1')f(\mathbf{c}_k')\,\sigma(\pi - \beta)d\beta \cdot J$ where J is the ratio $\frac{d^2\mathbf{c}_1'd^2\mathbf{c}_k'}{d^2\mathbf{c}_1d^2\mathbf{c}_k}$, if the momentum and kinetic energy conservation relations hold: $\mathbf{c}_1 + \mathbf{c}_k = \mathbf{c}_1' + \mathbf{c}_k'$ and $\mathbf{c}_1^2 + \mathbf{c}_k^2 = \mathbf{c}_1'^2 + \mathbf{c}_k'^2$ and if the angle between $\mathbf{c}_1 - \mathbf{c}_k$ and $\mathbf{c}_k' - \mathbf{c}_1'$ is β.

The analysis leads to the conclusion, well known, that $J = 1$ and therefore it must be $f(\mathbf{c}_1)f(\mathbf{c}_k) = f(\mathbf{c}_1')f(\mathbf{c}_k')$ for all four velocities that satisfy the conservation laws of momentum and energy: this implies that $f(\mathbf{c}) = const\, e^{-hc^2}$. Boltzmann uses always the directional uniformity supposing $f(\mathbf{c}) = \varphi(c)$ and therefore expresses the probability that the modulus of the velocity is between c and $c + dc$ as $\varphi(c)cdc$ and therefore the result is expressed by $\varphi(c) = b\,e^{-hc^2}$, with $b = 2h$ a normalization constant (keeping in mind that the 2-dimensional case is considered).

In reality if systematic account was taken of the volume preserving property of canonical transformations (i.e. to have Jacobian 1) the strictly algebraic part of evaluating the Jacobian, would save several pages of the paper. It is interesting that Boltzmann had to proceed to rediscover this very special case of a general property of Hamiltonian mechanics.

Having established this result for planar systems of elastic disks (the analysis has been very verbose and complicated and B. admits that "Maxwell argument was simpler but he has chosen on purpose a direct approach based on very simple examples", p. 58), Boltzmann considers the 3-dimensional case in which the interaction is more general than elastic collision between rigid spheres, and admits that it is described by a potential $\chi(r)$, with short range. However he says that, since Maxwell has treated completely the problems analogous to the ones just treated in the planar case, he will study a new problem. Namely:

"Along a line OX an elastic ball of mass M is moving attracted by O with a force depending only on the distance. Against it is moving other elastic balls of mass m and their diverse velocities during disordered time intervals dart along the same line, so that if one considers that all flying balls have run towards O long enough on the line OX without interfering with each other, the number of balls with velocity between c and $c + dc$, which in average are found in the unit length, is a given function of c, i.e. $N\varphi(c)dc$.

The potential of the force, which attracts M towards O be $\chi(x)$, hence as long as the motion does not encounter collisions it will be

$$\frac{MC^2}{2} = \chi(x) + A \tag{9}$$

where C is the speed of the ball M and x the distance between its center and O. Through the three quantities x, A and c the kind of collision is fixed. The fraction of time during which the constant A of Eq. (9) will be between A and $A + dA$ be $\Phi(A)dA$. The time during which again x is between the limits x and $x + dx$ behaves as $\frac{dx}{C}$, and we shall call $t(A)$, as it is a function of A, the fraction of time during which the segment dx is run and x grows from its smallest value to its largest. Consider a variation of the above values that is brought about by the collisions, we want to compare the time interval between two collisions with $t(A)$; this time is

$$\frac{\Phi(A)dA\,dx}{Ct(A)} \cdots$$

The discussion continues (still very involved) to determine the balance between collisions that change A, x, c into A', x', c': it very much resembles Maxwell's treatment in [6, XXVIII, vol. 2] and is a precursor of the later development of the Boltzmann's equation, [7, #22], and follows the same path. The analysis will reveal itself very useful to Boltzmann in the 1871 "trilogy" and then in the 1872 paper because it contains all technical details to be put together to obtain the Boltzmann's equation. The result is

$$\varphi(c) = be^{-h\cdot\frac{mc^2}{2}}, \quad \frac{\Phi(A)dA\,dx}{Ct(A)} = 2Be^{h[\chi(x)-\frac{MC^2}{2}]}$$

with $2B$ a normalization, and it has to be kept in mind that only events on the line OX are considered so that the problem is essentially 1-dimensional.

The 3-dimensional corresponding problem it treated in the rest of Sect. I, subsection 3, and a new related problem is posed and solved in subsections 4 (p. 70) and 5 (p. 73). There a point mass, named I with mass M, is imagined on the on a line OX attracted by O and a second kind point masses, named II, with mass m, interacting with I via a potential with short range ℓ. It is supposed that the fraction of time the point II has speed between c and $c + dc$ (the problem is again 1-dimensional) is $N\varphi(c)dc$ and that events in which two or more particles II come within ℓ of I can be neglected. The analysis leads to the "same" results of subsections 2 and 3 respectively for the 1 and 3 dimensional cases.]

II. On the equipartition of the "vis viva" for a finite number of point masses (p. 80)

In a very large, bounded in every direction, planar region let there be n point masses, of masses $m_1, m_2, \ldots, m_n$ and velocity $c_1, c_2, \ldots, c_n$, and between them act arbitrary forces, which *just begin to act at a distance which vanishes compared to their mean distance* [italics added].[23] Naturally all directions in the plane are equally probable for such velocities. But the probability that the velocity of a point be within assigned limits and likewise that the velocity of the others be within other limits, will certainly not be the product of the single probabilities; the others will mainly depend on the value chosen for the velocity of the first point. The velocity of the last point

---

[23] Often it is stated that Boltzmann does not consider cases in which particles interact: it is here, and in the following, clear that he assumes interaction but he also assumes that the average distance between particles is very large compared to the range of interaction. This is particularly important also in justifying the later combinatorial analysis. See also below.

depends from that of the other $n - 1$, because the entire system must have a constant amount of "vis viva".

I shall identify the fraction of time during which the velocities are so partitioned that c_2 is between c_2 and $c_2 + dc_2$, likewise c_3 is between c_3 and $c_3 + dc_3$ and so on until c_n, with the probability $\varphi_1(c_2, c_3, \ldots, c_n) dc_2 \, dc_3 \ldots dc_n$ for this velocity configuration.

The probability that c_1 is between c_1 and $c_1 + dc_1$ and the corresponding velocities different from c_2 are between analogous limits be $\varphi_2(c_1, c_3, \ldots, c_n) \cdot dc_1 \, dc_3 \ldots dc_n$, etc..

Furthermore let

$$\frac{m_1 c_1^2}{2} = k_1, \quad \frac{m_2 c_2^2}{2} = k_2, \ldots, \quad \frac{m_n c_n^2}{2} = k_n$$

be the kinetic energies and let the probability that k_2 is between k_2 and $k_2 + dk_2$, k_3 is between k_3 and $k_3 + dk_3 \ldots$ until k_n be $\psi_1(k_2, k_3, \ldots, k_n) \, dk_2 \, dk_3 \ldots dk_n$. And analogously define $\psi_2(k_1, k_3, \ldots, k_n) \, dk_1 \, dk_3 \ldots dk_n$ etc., so that

$$m_2 c_2 \cdot m_3 c_3 \cdots m_n c_n \, \psi_1 \left(\frac{m_2 c_2^2}{2}, \frac{m_3 c_1^3}{2}, \ldots, \frac{m_n c_n^2}{2} \right) = \varphi_1(c_2, c_3, \ldots, c_n) \quad \text{or}$$

$$\varphi_1(c_2, c_3, \ldots, c_n) = 2^{\frac{n-1}{2}} \sqrt{m_2 m_3 \cdots m_n} \sqrt{k_2 k_3 \cdots k_n} \, \psi_1(k_2, k_3, \ldots, k_n)$$

and similarly for the remaining φ and ψ.

Consider a collision involving a pair of points, for instance m_r and m_s, which is such that c_r is between c_r and $c_r + dc_r$, and c_s is between c_s and $c_s + dc_s$. Let the limit values of these quantities after the collision be between c_r' and $c_r' + dc_r'$ and c_s' be between c_s' and $c_s' + dc_s'$.

It is now clear that the equality of the "vis viva" will remain valid always in the same way when many point, alternatively, come into collision and are moved within different limits, as well as the other quantities whose limits then can be remixed, among which there are the velocities of the remaining points. [*Here it seems that the constancy of the total kinetic energy is claimed to be clear: which seems strange since at the same time a short range interaction is now present. The reason behind this assumption seems that, as B. says at the beginning of Sect. II, (p. 80), the range of the forces is small compared to the mean interparticle distance.*]

The number of points that are between assigned limits of the velocities, which therefore have velocities between c_2 and $c_2 + dc_2 \ldots$, are different from those of the preceding problems because instead of the product $\varphi(c_r) dc_\varphi(c_s) dc_s$ appears the function $\varphi_1(c_2, c_3, \ldots, c_n) dc_2 \, dc_3 \ldots dc_n$. This implies that instead of the condition previously found

$$\frac{\varphi(c_r) \cdot \varphi(c_s)}{c_r \cdot c_r} = \frac{\varphi(c_r') \cdot \varphi(c_s')}{c_r' \cdot c_r'}$$

the new condition is found:

$$\frac{\varphi_1(c_2, c_3, \ldots, c_n)}{c_r \cdot c_r} = \frac{\varphi_1(c_2, \ldots, c'_r, \ldots, c'_s, \ldots, c_n)}{c_r \cdot c_s}$$

The same holds, of course, for $\varphi_2, \varphi_3, \ldots$. If the function φ is replaced by ψ it is found, equally,

$$\psi_1(k_2, k_3, \ldots, k_n) = \psi_1(k_2, k_3, \ldots, k'_r, \ldots, k'_s, \ldots, k_n), \quad \text{if } k_r + k_s = k'_r + k'_s.$$

Subtract the differential of the first of the above relations $\frac{d\psi_1}{dk_r}dk_r + \frac{d\psi_1}{dk_s}dk_s = \frac{d\psi_1}{dk'_r}dk'_r + \frac{d\psi_1}{dk'_s}dk'_s$ that of the second $[dk_r + dk_s = dk'_r + dk'_s]$ multiplied by λ and set equal to zero the coefficient of each differential, so that it is found:

$$\lambda = \frac{d\psi_1}{dk_r} = \frac{d\psi_1}{dk_s} = \frac{d\psi_1}{dk'_r} = \frac{d\psi_1}{dk'_s}.$$

I.e., in general, $\frac{d\psi_1}{dk_2} = \frac{d\psi_1}{dk_3} = \frac{d\psi_1}{dk_4} = \cdots \frac{d\psi_1}{dk_n}$, hence ψ_1 is function of $k_2 + \cdots + k_n$. Therefore we shall write $\psi_1(k_2, \ldots, k_n)$ in the form $\psi_1(k_2 + k_3 + \cdots + k_n)$. We must now find the meaning of the equilibrium about m_1 and the other points. And we shall determine the full ψ_1.

It is obtained simply with the help of the preceding ψ of which of course the ψ_1 must be a sum. But these are all in reciprocal relations. If in fact the total "vis viva" of the system is $n\kappa$, it is

$$k_1 + k_2 + \cdots + k_n = n\kappa$$

It follows that $\psi_1(k_2 + k_3 + \cdots + k_n)dk_2dk_3 \ldots dk_n$ can be expressed in the new variables[24]

$$k_3, k_4, \ldots, n\kappa - k_1 - k_3 - \cdots - k_n = k_2$$

and it must be for $\psi_2(k_1 + k_3 + \cdots + k_n)dk_1dk_3 \ldots dk_n$. Hence $\psi_1(k_2 + k_3 + \cdots + k_n)$ can be converted in $\psi_1(n\kappa - k_1)$ and $dk_2dk_3 \ldots dk_n$ in $dk_1dk_3 \ldots dk_n$. Hence also

$$\psi_1(n\kappa - k_1) = \psi_2(k_1 + k_3 + \cdots + k_n) = \psi_2(n\kappa - k_2) = \psi_2(n\kappa - k_2)$$

for all k_1 and k_2, therefore all the ψ are equal to the same constant h. This is also the probability that in equal time intervals it is k_1 between k_1 and $k_1 + dk_1$, k_2 is between k_2 and $k_2 + dk_2$ etc.., thus for a suitable h, it is $h\, dk_1 dk_2 \ldots dk_n$ were again the selected differential element must be absent. Of course this probability that at a given instant $k_1 + k_2 + k_3 + \cdots$ differs from $n\kappa$ is immediately zero.

The probability that c_2 is between c_2 and $c_2 + dc_2$, c_3 between c_3 and $c_3 + dc_3 \ldots$ is given by

---

[24] In the formula k_2 and k_1 are interchanged.

$$\varphi_1(c_2, c_3, \ldots, c_n)\, dc_2\, dc_3 \ldots dc_n = m_2 m_3 \ldots m_n \cdot h \cdot c_2 c_3 \ldots c_n dc_2\, dc_3 \ldots dc_n.$$

Therefore the point c_2 is in an annulus of area $2\pi c_2 dc_2$, the point c_3 in one of area $2\pi c_3 dc_3$ etc.., that of c_1 on the boundary of length $2\pi c_1$ of a disk and all points have equal probability of being in such annuli.

Thus we can say: the probability that the point c_2 is inside the area $d\sigma_2$, the point c_3 in $d\sigma_3$ etc.., while c_1 is on a line element $d\omega_1$, is proportional to the product

$$\frac{1}{c_1}\, d\omega_1\, d\sigma_2\, d\sigma_3 \ldots d\sigma_n,$$

if the mentioned locations and velocities, while obeying the principle of conservation of the "vis viva", are not impossible.

We must now determine the fraction of time during which the "vis viva" of a point is between given limits k_1 and $k_1 + dk_1$, without considering the "vis viva" of the other points. For this purpose subdivide the entire "vis viva" in infinitely small equal parts (p), so that if now we have two point masses, for $n = 2$ the probability that k_1 is in one of the p intervals $\left[0, \frac{2\kappa}{p}\right], \left[\frac{2\kappa}{p}, \frac{4\kappa}{p}\right], \left[\frac{4\kappa}{p}, \frac{6\kappa}{p}\right]$ etc. is equal and the problem is solved.

For $n = 3$ if k_1 is in $\left[(p-1)\frac{3\kappa}{p}, p\frac{3\kappa}{p}\right]$, then k_2 and k_3 must be in the interior of the p intervals. If k_1 is in the next to the last interval, i.e. if

$$(p-2)\frac{3\kappa}{p} \le k_1 \le (p-1)\frac{3\kappa}{p}$$

two cases are possible ....

[*Here follows the combinatorial calculation of the number of ways to obtain the sum of n multiples $p_1, \ldots, p_n$ of a unit ε and $p_1\varepsilon = k_1$ such that $\sum_{i=2}^{n-1} p_i\varepsilon = n\kappa - p_1\varepsilon$, and B. chooses $\varepsilon = \frac{2\kappa}{p}$ with p "infinitely large": i.e.*]

$$\sum_{p_2=0}^{n\kappa/\varepsilon - p_1} \sum_{p_3=0}^{n\kappa/\varepsilon - p_1 - p_2} \cdots\cdots \sum_{p_{n-1}=0}^{n\kappa/\varepsilon - p_1 - \cdots - p_{n-2}} 1$$

the result is obtained by explicitly treating the cases $n = 2$ and $n = 3$ and inferring the general result in the limit in which $\varepsilon \to 0$.

The ratio between this number and the same sum performed also on p_1 is, fixing $p_1 \in [k_1/\varepsilon, (k_1 + dk_1)/\varepsilon]$,

$$\frac{dk_1 \int_0^{n\kappa - k_1} dk_2 \int_0^{n\kappa - k_1 - k_2} dk_3 \cdots \int_0^{n\kappa - k_1 - k_2 - \cdots - k_{n-2}} dk_{n-1}}{\int_0^{n\kappa} dk_1 \int_0^{n\kappa - 1} dk_2 \int_0^{n\kappa - k_1 - k_2} dk_3 \cdots \int_0^{n\kappa - k_1 - k_2 - \cdots - k_{n-2}} dk_{n-1}}$$

$$= \frac{(n-1)(n\kappa - k_1)^{n-2} dk_1}{(n\kappa)^{n-1}},$$

This is, however, remarked after the explicit combinatorial analysis of the case $n = 2$ *and* $n = 3$ *from which the last equality is inferred in general (for* $\varepsilon \to 0$*).*

Hence the "remark" is in reality a proof simpler than the combinatorial analysis of the number of ways to decompose the total energy as sum of infinitesimal energies. The choice of B. is certainly a sign of his preference for arguments based on a discrete view of physical quantities. And as remarked in [8] this remains, whatever interpretation is given to it, and important analysis of B.

In the successive limit, $n \to \infty$*, the Maxwell's distribution is obtained.*

$$\frac{1}{\kappa} e^{-k_1/\kappa} dk_1$$

concluding the argument.

In the next subsection 7B. repeats the analysis in the 3-dimensional case obtaining again the Maxwellian distribution for the speed of a single particle in a system of n *point masses in a finite container with perfectly elastic walls.*

Finally in Sect. III.8 the case is considered in which also an external force acts whose potential energy is χ *(not to be confused with the interparticle potential energy, also present and denoted with the same symbol; and which is always considered here as acting instantaneously at the collision instant, as assumed at the beginning of Sect. II).*

The Sect. III is concluded as follows.]

p. 96. As special case from the first theorem it follows, as already remarked in my paper on the mechanical interpretation of the second theorem, that the "vis viva" of an atom in a gas is equal to that of the progressive motion of the molecule.[25] The latter demonstration also contains the solution of others that were left incomplete; it tells us that for such velocity distributions the kinetic energies are balanced in a way that could not take place otherwise.

An exception to this arises when the variables $x_1, y_1, z_1, x_2, \ldots, v_n$ are not independent of each other. This can be the case of all systems of points in which the variables have special values, which are constrained by assigned relations that remain unaltered in the motions, but which under perturbations can be destroyed (weak balance of the "vis viva"), for instance when all points and the fixed centers are located on a mathematically exact line or plane.

A stable balance is not possible in this case, when the points of the system are so assigned that the variables for all initial data within a certain period come back to the initial values, without having consequently taken all values compatible with the 'principle of the "vis viva". Therefore such way of achieving balance is always so infinitely more possible that immediately the system ends up in the set of domains discussed above when, for instance, a perturbation acts on a system of points which

---

[25] To which the atom belongs.

evolves within a container with elastic walls or, also, on an atom in free motion which collides against elastic walls.[26]

[*Boltzmann is assuming that the potential energy could be given its average value, essentially 0, and be replaced by a constant as a mean field: because he, Sect. II, (p. 80), assumes that the range of the forces is small compared to the mean interparticle distance and (tacitly here, but explicitly earlier in subsection 4, p. 144) that multiple collisions can be neglected. The fraction of time in which the potential energy is sizable is supposed "by far" too small to affect averages: this is not so in the case of gases consisting of polyatomic molecules as he will discuss in detail in the first paper of the trilogy, [9, #17]. The analysis of the problem in modern terms would remain essentially the same even not neglecting that the total kinetic energy is not constant if the interaction between the particles is not pure hard core: in modern notations it would mean studying (in absence of external forces, for simplicity)*

$$\frac{\int \delta(\sum_j \frac{1}{2}mc_j^2 + \sum_{i,j} \chi(x_i - x_j) - Nu)d^{3N-1}\mathbf{c}d^{3N}\mathbf{x}}{\int \delta(\sum_j \frac{1}{2}mc_j^2 + \sum_{i,j} \chi(x_i - x_j) - Nu)d^{3N}\mathbf{c}d^{3N}\mathbf{x}}$$

where u is the specific total energy, if the pair interaction is short range and stable (in the sense of existence of a constant B such that for all N it is $\sum_{i,j}^{N} \chi(x_i - x_j) > -BN$) and the integral in the numerator is over all velocity components but one: the analysis would then be easily reduced to the case treated here by B.

In [8] the question is raised on whether Boltzmann would have discovered the Bose-Einstein distribution before Planck referring to the way he employs the discrete approach to compute the number of ways to distribute kinetic energy among the various particles, after fixing the value of that of one particle, in [10, #5, p. 84, 85]. This is an interesting and argumented view, however Boltzmann considered here the discrete view a "fiction", see also [11, #42, p. 167] and Sect. 6.12, and the way the computation is done would not distinguish whether particles were considered distinguishable or not: the limiting case of interest would be the same in both cases (while it would be quite different if the continuum limit was not taken, leading to a Bose-Einstein like distribution). This may have prevented him to be led to the extreme consequence of considering the difference physically significant and to compare the predictions with those that follow in the continuum limit with the distribution found with distinguishable particles, discussed later in [11], see Sect. 6.12 and also [12, Sects. 2.2, 2.6].]

---

[26] NoA: This last paragraph seems to refer to lack of equipartition in cases in which the system admits constants of motion due to symmetries that are not generic and therefore are destroyed by "any" perturbation.

6.3 Dense Orbits: An Example

Comments on: L. Boltzmann, *Lösung eines mechanischen Problems*, Wissenschaftliche Abhandlungen, ed. F. Hasenöhrl, **1**, #6, 97–105, (1868), [13].

The aim of this example it to exhibit a simple case in which the difficult problem of computing the probability of finding a point mass occupying a given position with given velocity.

Here B. presents an example where the ideas of the previous work, Sect. 6.2, can be followed via exact calculations. A point mass [mass $= 1$] subject to a central gravitational force but with a centrifugal barrier augmented by a potential $+\frac{\beta}{2R^2}$ and which is reflected by an obstacle consisting in a straight line, e.g. $x = \gamma > 0$.

The discussion is an interesting example of a problem in ergodic theory for a two variables map. Angular momentum a is conserved between collisions and there motion is explicitly reducible to an elementary quadrature which yields a function (here $r \equiv R$ and A is a constant of motion equal to twice the total energy, constant because collisions with the line are supposed elastic):

$$F(r, a, A) \overset{def}{=} \frac{a}{\sqrt{a^2 + \beta}} \arccos\left(\frac{\frac{2(a+\beta)}{r} - \alpha}{\sqrt{\alpha^2 + 4A(a^2 + \beta)}}\right)$$

such that the polar angle at time t is $\varphi(t) - \varphi(0) = F(r(t), a_0, A) - F(r(0), a_0, A)$. Let $\varepsilon_0 \overset{def}{=} \varphi(0) - F(r(0), a_0, A)$, then if φ_0, a_0 are the initial polar angle and the angular momentum of a motion that comes out of a collision at time 0 then $r(0) \cos\varphi_0 = \gamma$ and $\varphi(t) - \varepsilon_0 = F(r(t), a_0, A)$ until the next collision. Which will take place when $\varphi_1 - \varepsilon_0 = F\left(\frac{\gamma}{\cos\varphi_1}, a_0, A\right)$ and if a_1 is the outgoing angular momentum from then on $\varphi(t) - \varepsilon_1 = F(r(t), a_1, A)$ with $\varepsilon_1 \overset{def}{=} \varphi_1 - F\left(\frac{\gamma}{\cos\varphi_1}, a_1, A\right)$.

Everything being explicit B. computes the Jacobian of the just defined map S : $(a_0, \varepsilon_0) \to (a_1, \varepsilon_1)$ and shows that it is 1 (which is carefully checked, without reference to the canonicity of the map). The map is supposed to exist, i.e. that the Poincaré's section defined by the timing event "hit of the fixed line" is transverse to the solution flow (which may be not true, for instance if $A < 0$ and γ is too large). Hence the observations timed at the collisions has an invariant measure $d\varepsilon da$: if the allowed values of a, ε vary in a bounded set (which certainly happens if $A < 0$) the measure $\frac{d\varepsilon da}{\int d\varepsilon da}$ is an invariant probability measure, i.e. the microcanonical distribution, which can be used to compute averages and frequency of visits to the points of the plane ε, a. The case $\beta = 0$ would be easy but in that case it would also be obvious that there are motions in which ε, a does not roam on a dense set, so it is excluded here.

The interest of B. in the example seems to have been to show that, unless the interaction was very special (e.g. $\beta = 0$) the motion would invade the whole energy surface, in essential agreement with the idea of ergodicity. In reality *neither density nor ergodicity is proved*. It is likely that the confined motions of this system are quasi periodic unless A has special (dense) values corresponding to "resonant" (i.e. periodic) motions. B. does not make here comments about the possible exceptional ("resonant") values of E; assuming that he did not even think to such possibilities, it is clear that he would not have been shocked by their appearance: at least for many value of E (i.e. but for a zero measure set of E' 's) the system would admit only one invariant measure and that would be the microcanonical one, and this would still have been his point.

6.4 Clausius' Version of Recurrence and Periodicity

Translation and comments on Sect. 10 of: R., *Ueber die Zurückführung des zweites Hauptsatzes der mechanischen Wärmetheorie und allgemeine mechanische Prinzipien*, Annalen der Physik, **142**, 433–461, 1871.

Section 10 deals with the necessity of closed atomic paths in the derivation of the second theorem of thermodynamics from mechanics.

10. So far we considered the simple case of an isolated point moving on a closed path and we shall now consider more complicated cases.

We want to consider a very large number of point masses, interacting by exercising a reciprocal force as well as subject to an external force. It will be supposed that the points behave in a stationary way under the action of such force. Furthermore it will be supposed that the forces have an *ergale*,[27] i.e. that the work, performed by all forces upon an infinitesimal displacement be the differential, with sign changed, of a function of all coordinates. If the initial stationary motion is changed into a varied stationary motion, still the forces will have an "ergale" [potential energy], which does not depend only on the changed position of the point, but which also can depend from other factors. The latter can be thought, from a mathematical viewpoint, by imagining that the ergale [the potential energy] is a quantity that in every stationary motion is constant, but it has a value that can change from a stationary motion to another.

Furthermore we want to set up an hypothesis which will clarify the following analysis and correlates its contents which concern the motion that we call heat, that the system consists in only one chemical species, and therefore all its atoms are equal, or possibly that it is composed, but every species contains a large number of atoms. It is certainly not necessary that all these atoms are of the same species. If for instance the body is an aggregate of different substances, atoms of a part move as those of the others. Then, after all, we can suppose that every possible motion will take place as one of those followed by the large number of atoms subject to equal forces, proceeding in the same way, so that also the different phases[28] of such motions will be realized. What said means that we want to suppose that in our system of point masses certainly we can find among the large number of the same species a large number which goes through the same motion under the action of equal forces and with different phases.

---

[27] I.e. a potential energy.

[28] Here it is imagined that each atom moves on a possible orbit but different atoms have different positions on the orbit, at any given time, which is called its "phase".

Finally temporarily and for simplicity we shall assume, as already done before, that all points describe closed trajectories, For such points about which we are concerned and which move in the same way we suppose, more in particular, that they go through equal trajectories with equal periods. If the stationary motion is transformed into another, hence on a different trajectory with different period, nevertheless it will still follow a closed trajectories each of which will be the same for a large number of points.[29]

6.5 Clausius' Mechanical Proof of the Heat Theorem

Translation of §13, §14, §15, and comments: R. Clausius, *Ueber die Zurückführung des zweites Hauptsatzes der mechanischen Wärmetheorie und allgemeine mechanische Prinzipien*, Annalen der Physik, **142**, 433–461, 1871.

[*The translation is here because I consider several sentences in it interesting to appear within their context: the reader interested to a rapid self contained summary of Clausius' proof is referred to Sect. 1.4 and to Appendix A.*]

13. In the present work we have supposed until now that all points move on closed paths. We now want to set aside also this assumption and concentrate on the hypothesis that the motion is stationary.

For motions that do not run over closed trajectories the notion of recurrence is no longer usable in a literal sense, therefore it is necessary to talk about them in another sense. Consider therefore right away motions that have a given component in a given direction, for instance the x direction in our coordinates system. It is then clear that motions go back and forth alternatively, and also for the elongation, speed and return time, as it is proper for stationary motion, the same form of motion is realized. The time interval within which each group of points which behave, approximately, in the same way admits an average value.

Denote with i this time interval, so that without doubt we can consider valid, also for this motion, the Eq. 28), [i.e. *the equality, in average, of the two sides of*]

$$-\sum m \frac{d^2x}{dt^2} \delta x = \sum \frac{m}{2} \delta (\frac{dx}{dt})^2 + \sum m (\frac{dx}{dt})^2 \delta \log i$$

[here δx denotes a variation of a quantity x between two infinitely close stationary states].[30] The above equation can also be written for the components y, z, and of course we shall suppose that the motions in the different directions behave in the

---

[29] The assumption differs from the ergodic hypothesis and it can be seen as an assumption that all motions are quasi periodic and that the system is integrable: it is a view that *mutatis mutandis* resisted until recent times both in celestial mechanics, in spite of Poincaré's work, and in turbulence theory as in the first few editions of Landau-Lifschitz' treatise on fluid mechanics, [14].

[30] In Clausius $\delta x = x'(i'\varphi) - x(i\varphi)$, $t' = i'\varphi$ and $t = i\varphi$ is defined much more clearly than in Boltzmann, through the notion of *phase* $\varphi \in [0, 1]$ assigned to a trajectory, and calculations are performed up to infinitesimals of order higher than δx and $\delta i = (i' - i)$.

same way and that, for each group of points, the quantity $\delta \log i$ assumes the same value for the three coordinates.

If then with the three equations so obtained we proceed as above for the Eqs. (28), (28b), (28c), we obtain the Eq. (31)[31]:

$$\delta L = \sum \frac{m}{2} \delta v^2 + \sum m v^2 \delta \log i$$

14. To proceed further to treat such equations a difficulty arises because the velocity v, as well as the return time interval i, may be different from group to group and both quantities under the sum sign cannot be distinguished without a label. But imagining it the distinction will be possible allowing us to keep the equation in a simpler form.

Hence the different points of the system, acting on each other, interact so that the kinetic energy of a group cannot change unless at the concomitant expenses of another, while always a balance of the kinetic energies of the different points has to be reached, before the new state can be stationary. We want to suppose, for the motion that we call heat, that a balance is established between the kinetic energies of the different points and a relation is established, by which each intervening variation reestablishes the kinetic energy balance. Therefore the average kinetic energy of each point can be written as mcT, where m is the mass of the points and c another constant for each point, while T denotes a variable quantity equal for all points.

Inserting this in place of $\frac{m}{2} v^2$ the preceding equation becomes:

$$\delta L = \sum m c \, \delta T + \sum 2m c T \, \delta \log i \qquad (32)$$

Here the quantity T can become a common factor of the second sum. We can, instead, leave the factor δT inside the first sum. We get

$$\begin{aligned}
\delta L &= \sum m c \, \delta T + T \sum 2 m c \, \delta \log i \\
&= T \left(\sum m c \, \tfrac{\delta T}{T} + \sum 2 m c \, \delta \log i \right) \qquad (33) \\
&= T \left(\sum m c \, \delta \log T + \sum 2m c \, \delta \log i \right)
\end{aligned}$$

or, merging into one both sums and extracting the symbol of variation

$$T \delta \sum (m c \, \log T + 2 \, \log i)$$

from which we finally can write

$$\delta L = T \delta \sum m c \, \log(T i^2) \qquad (34)$$

---

[31] δL is the work in the process. It seems that here the integration of both sides is missing, or better the sign of average over v^2, which instead is present in the successive Eq. (32).

15. The last equation entirely agrees, intending for T the absolute temperature, with Eq. (1) for the heat

$$dL = \frac{T}{A} dZ$$

making clear its foundation on mechanical principles. The quantity denoted Z represents the *disgregation* [free energy] of the body which after this is represented as

$$A \sum m c \log T i^2$$

And it is easy also to check its agreement with another equation of the mechanical theory of heat.

Imagine that our system of moving point masses has a kinetic energy which changes because of a temporary action of a force and returns to the initial value. In this way the kinetic energy so communicated in part increments the kinetic energy content and in part it performs mechanical work.

If δq is the communicated average kinetic energy and h is the kinetic energy available in the system, it will be possible to write:

$$\delta q = \delta h + \delta L = \delta \sum m c T + \delta L = \sum m c \delta T + \delta L$$

and assigning to δL its value Eq. (33), it is found

$$\begin{aligned}
\delta q &= \sum 2m c \delta T + T \sum 2m c \delta \log i \\
&= T \left(\sum 2m c \delta \log T + \sum 2m c \delta \log i \right) \\
&= T \sum 2m c \log(T i)
\end{aligned}$$

i.e. also

$$\delta q = T \delta \sum 2 m c \log(T i) \tag{35}$$

This equation appears as the Eq. (59) of my 1865 paper [15]. Multiply, in fact, both sides of the preceding equation by A (the caloric equivalent of the work) and interpret the product $A \delta q$ as the variation of the kinetic energy spent to increment the quantity of heat transferred and let it be δQ, defining the quantity S by

$$S = A \sum 2m c \log(T i) \tag{36}$$

so the equation becomes

$$\delta Q = T \delta S \tag{37}$$

where the quantity S introduced here is the one I called *entropy*.

In the last equation the signs of variation can be replaced by signs of differentiation because both are auxiliary to the argument (the variation between a stationary motion transient to another) and the distinction between such two symbols will not be any longer necessary because the first will no longer intervene. Dividing again the equation by T, we get

$$\frac{dQ}{T} = dS$$

Imagine to integrate this relation over a cyclic process, and remark that at the end S comes back to the initial value, so we can establish:

$$\int \frac{dQ}{T} = 0 \tag{38}$$

This is the equation that I discovered for the first time as an expression of the second theorem of the mechanical theory of heat for reversible cyclic processes.[32] At the time I set as foundation *that heat alone cannot be trasferred from a colder to a warmer body*. Later[33] I derived the same equation in a very different way, i.e. based on the preceding law *that the work that the heat of a body can perform in a transformation is proportional to the absolute temperature and does not depend on its composition*. I treated, in this way, the fact that in other way it can be proved the equation as a key consequence of each law. The present argument tells us, as well, that each of these laws and with them the second theorem of the mechanical theory of heat can be reduced to general principles of mechanics.

6.6 Priority Discussion of Boltzmann (vs. Clausius)

Partial translation and comments: L. Boltzmann, *Zur priorität der auffindung der beziehung zwischen dem zweiten hauptsatze der mechanischen wärmetheorie und dem prinzip der keinsten wirkung*, Pogg. Ann. **143**, 211–230, 1871, [18], and Wissenschaftliche Abhandlungen, ed. F. Hasenöhrl, **1**, #17 p. 228–236.

Hrn. Clausius presented, at the meeting of 7 Nov. 1870 of the "Niederrheinischen Gesellschaft fr Natur und Heilkunde vorgetragenen" and in Pogg. Ann. 142, S. 433, [19], a work where it is proved that the second fundamental theorem of the mechanical theory of heat follows from the principle of least action and that the corresponding arguments are identical to the ones implying the principle of least action. I have already treated the same question in a publication of the Wien Academy of Sciences

---

[32] NoA: Pogg. Ann. **93**, 481 (1854) and Abhandlungen über die mechanische Wärmetheorie, **I**, 127, [16, p. 460].

[33] NoA: "Ueber die Anwendung des Satzes von der Aequivalenz der Verwandlungen auf die innere Arbeit", Pogg. Ann. **116**, 73–112 (1862) and Abhandlungen über die mechanische Wärmetheorie, **I**, 242–279, [17].

of 8 Feb. 1866, printed in the volume 53 with the title *On the mechanical meaning of the second fundamental theorem of the theory of heat*, [[1, #2] and Sect. 6.6]; and I believe I can assert that the fourth Section of my paper published 4 years earlier is, in large part, identical to the quoted publication of Hr. Clausius. Apparently, therefore, my work is entirely ignored, as well as the relevant part of a previous work by Loschmidt. It is possible to translate the notations of Hr. Clausius into mine, and via some very simple transformation make the formulae identical. I claim, to make a short statement, that given the identity of the subject nothing else is possible but some agreement. To prove the claim I shall follow here, conveniently, the fourth section of my work of 8 Feb. 1866, of which only the four formulae Eqs. (23a), (24a), (25a) and (25b), must be kept in mind.[34]

6.7 Priority Discussion: Clausius' Reply

Translation and comments: R. Clausius *Bemerkungen zu der prioritätreclamation des Hrn. Boltzmann*, Pogg. Ann. **144**, 265–274, 1871.

In the sixth issue of this Ann., p. 211, Hr. Boltzmann claims to have already in his 1866 paper reduced the second main theorem of the mechanical theory of heat to the general principles of mechanics, as I have discussed in a short publication. This shows very correctly that I completely missed to remark his paper, therefore I can now make clear that in 1866 I changed twice home and way of life, therefore naturally my attention and my action, totally involuntarily, have been slowed and made impossible for me to follow regularly the literature. I regret overlooking this all the more because I have subsequently missed the central point of the relevant paper.

It is plain that, in all point in which his work overlaps mine, the priority is implicit and it remains only to check the points that agree.

In this respect I immediately admit that his expressions about disgregation [*free energy*] and of entropy overlap with mine on two points, about which we shall definitely account in the following; but his mechanical equations, on which such expressions are derived are not identical to mine, of which they rather are a special case.

We can preliminarily limit the discussion to the simplest form of the equations, which govern the motion of a single point moving periodically on a closed path.

Let m be the mass of the point and let i its period, also let its coordinates at time t be x, y, z, and the acting force components be X, Y, Z and v its velocity. The latter quantities as well as other quantities derived from them, vary with the motion, and we want to denote their average value by over lining them. Furthermore we think that near the initially considered motion there is another one periodic and infinitely little

---

[34] Clausius answer, see Sect. 6.7, was to apologize for having been unaware of Boltzmann's work but rightly pointed out that Boltzmann's formulae became equal to his own after a suitable interpretation, absent from the work of Boltzmann; furthermore his version was more general than his: certainly, for instance, his analysis takes into account the action of external forces. As discussed, the latter is by no means a minor remark: it makes Clausius and Boltzmann results deeply different. See also p. 134 and 159.

different, which follows a different path under a different force. Then the difference between a quantity relative to the first motion and the one relative to the varied motion will be called "variation of the quantity", and it will be denoted via the symbol δ. And my equation is written as:

$$- \overline{X \delta x + Y \delta y + Z \delta Z} = \frac{m}{2} \overline{\delta v^2} + m \overline{v^2} \delta \log i \tag{I}$$

or, if the force acting on the point admits and ergale [*potential*], that we denote U, for the initial motion,[35]

$$\delta \overline{U} = \frac{m}{2} \overline{\delta v^2} + m \overline{v^2} \delta \log i \tag{Ia}$$

Boltzmann now asserts that these equations are identical to the equation that in his work is Eq. (22), if elaborated with the help of the equation denoted (23a). Still thinking to a point mass moving on a closed path and suppose that it is modified in another for which the point has a kinetic energy infinitely little different from the quantity ε, then Boltzmann's equation, after its translation into my notations, is

$$\frac{m}{2} \overline{\delta v^2} = \varepsilon + \overline{X \delta x + Y \delta y + Z \delta Z} \tag{1}$$

and thanks to the mentioned equation becomes:

$$\varepsilon = \frac{\delta i}{i} m \overline{v^2} + m \delta \overline{v^2} \tag{2}$$

The first of these Boltzmann's equations will be identical to my Eq. (I), if the value assigned to ε can be that of my equation.[36]

I cannot agree on this for two reasons.

The first is related to a fact, that already Boltzmann casually mentions, as it seems to me, to leave it aside afterwards. In his equations both quantities $\overline{\delta v^2}$ and $\delta \overline{v^2}$ (i.e. the average value of the variation δv^2 and the variation of the average value of v^2) are fundamentally different from each other, and therefore it happens that his and my

---

[35] For Clausius' notation used here see Sect. 1.4. Here an error seems present because the (I) implies that in the following (Ia) there should be $\overline{\delta U}$: but it is easy to see, given the accurate definition of variation by Clausius, see Eq. (1.4.2) and Appendix A for details, that the following (Ia) is correct because $\overline{\delta U} = \delta \overline{U}$. In reality the averages of the variations are quantities not too interesting physically because they depend on the way followed to establish the correspondence between the points of the initial curve and the points of its variation, and an important point of Clausius's paper is that it established a notion of variation that implies that the averages of the variations, in general of little interest because quite arbitrary, coincide with the variations of the averages.

[36] I.e. to obtain the identity, as Clausius remarks later, it is necessary that $\delta \overline{U} = \varepsilon - \frac{m}{2} \overline{\delta v^2}$ which is obtained if ε is interpreted as conservation of the total average energy, as in fact Boltzmann uses ε after his Eq. (23a): *but* instead in Boltzmann ε is introduced, and used first, as variation of the average kinetic energy. The problem is, as remarked in Sect. 6.1 that in Boltzmann ε does not seem clearly defined.

equations cannot be confronted.[37] Hence I have dedicated, in my research, extreme care to avoid leaving variations vaguely defined. And I use a special treatment of the variations by means of the notion of *phase*. This method has the consequence that for every varied quantity the average of the variation is the variation of the average, so that the equations are significantly simple and useful. Therefore I believe that the introduction of such special variations is essential for the subsequent researches, and do not concern a point of minor importance.

If now my variations are inserted in Boltzmann's Eq. (1) the following is deduced:

$$\frac{m}{2}\overline{\delta v^2} = \varepsilon + \overline{X\,\delta x + Y\delta\,y + Z\,\delta Z} \tag{1a}$$

and if next we suppose tat the force acting on the point has an ergale [*potential*], which we denote U, the equation becomes $\frac{m}{2}\overline{\delta v^2} = \varepsilon - \delta\overline{U}$, alternatively written as

$$\varepsilon = \frac{m}{2}\overline{\delta v^2} + \delta\overline{U}. \tag{1b}$$

If the value of ε is inserted in Eq. (2) my Eqs. (I), (Ia) follow. In spite of the changes in Eq. (1a, 1b) Boltzmann's equations so obtained are not identical to mine for a second and very relevant reason.

I.e. it is easy to recognize that both Boltzmannian equations and Eqs. (1) and (2) hold under certain limiting conditions, which are not necessary for the validity of mine. To make this really evident, we shall instead present the Boltzmannian equations as the most general equations, not subject to any condition. Therefore we shall suppose more conveniently that they take the form taken when the force acting on the point has an ergale [*potential energy*].

Select, in some way, on the initial trajectory a point as initial point of the motion, which starts at time t_1 as in Boltzmann, and denote the corresponding values of v and U with v_1 and U_1. Then during the entire motion the equation

$$\frac{m}{2}v^2 + U = \frac{m}{2}v_1^2 + U_1 \tag{3}$$

will hold; thus, likewise, we can set for the average values:

$$\frac{m}{2}\overline{v}^2 + \overline{U} = \frac{m}{2}v_1^2 + \overline{U}_1 \tag{4}$$

About the varied motion suppose that it starts from another point, with another initial velocity and takes place under the action of other forces. Hence we shall suppose that the latter have an ergale $U + \mu V$, where V is some function of the coordinates and μ an infinitesimal constant factor. Consider now again the two specified on the initial

---

[37] Indeed if in (1) ε is interpreted as what it should really be according to what follows in Boltzmann, i.e. $\varepsilon = (\delta(\overline{U} + \overline{K}))$ Eq. (I) becomes a trivial identity while Eq. (Ia) is non trivial. However it has to be kept in mind that Eq. (I) is not correct!

trajectory and on the varied one, so instead of v^2 we shall have in the varied motion the value $v^2 + \delta v^2$ and instead of U the value $U + \delta U + \mu(V + \delta V)$; therefore, since $\mu \, \delta V$ is a second order infinitesimal, this can be written $U + \delta U + \mu V$. Hence for the varied motion Eq. (3) becomes:

$$\frac{m}{2}v^2 + \frac{m}{2}\delta v^2 + U + \delta U + \mu V = \frac{m}{2}v_1^2 + \frac{m}{2}\delta v_1^2 + U_1 + \delta U_1 + \mu V_1 \quad (5)$$

so that my calculation of the variation leads to the equation:

$$\frac{m}{2}\overline{v^2} + \frac{m}{2}\overline{\delta v^2} + \overline{U} + \overline{\delta U} + \mu \overline{V} = \frac{m}{2}v_1^2 + \frac{m}{2}\delta v_1^2 + U_1 + \delta U_1 + \mu V_1. \quad (6)$$

Combining the last equation with the Eq. (4) it finally follows

$$\frac{m}{2}\delta v_1^2 + \delta U_1 + \mu(V_1 - \overline{V}) = \frac{m}{2}\overline{\delta v^2} + \overline{\delta U}. \quad (7)$$

This the equation that in a more general treatment should be in place of the different Boltzmannian Eq. (1b). Thus instead of the Boltzmannian Eq. (2) the following is obtained

$$\frac{m}{2}\delta v_1^2 + \delta U_1 + \mu(V_1 - \overline{V}) = \frac{\delta i}{i}m\overline{v^2} + m\overline{\delta v^2}. \quad (8)$$

As we see, since such new equations are different from the Boltzmannian ones, we must treat more closely the incorrect quantity ε. As indicated by the found infinitesimal variation of the "vis viva", due to the variation of the motion, it is clear that in the variation ε of the "vis viva" at the initial time one must understand, and hence set:

$$\varepsilon = \frac{m}{2}\delta v_1^2.$$

Hence the Boltzmannian equations of the three terms, that are to the left in Eqs. (7) and (8), should only contain the first.

Hr. Boltzmann, whose equations incompleteness I have, in my view, briefly illustrated, pretends a wider meaning for ε in his reply, containing at the same time the "vis viva" of the motion and the work, and consequently one could set

$$\varepsilon = \frac{m}{2}\delta v_1^2 + \delta U_1.$$

But I cannot find that this is said anywhere, because in the mentioned places where the work can be read it seems to me that there is a gain that exchanges the "vis viva" with another property of the motion that can transform it into work, which is not in any way understandable, and from this it does not follow that the varied original trajectory could be so transformed that it has no point in common with it and also in the transformation the points moved from one trajectory to the other could be moved without spending work.

Hence if one wishes to keep the pretension that the mentioned meaning of ε, then always two of the three terms appearing in Eqs. (7) and (8) are obtained, *the third of them, i.e. $\mu(V_1 - \overline{V})$ no doubt is missing in his equations.*

On this point he writes: "The term $\mu(\overline{V} - V_1)$ is really missing in my equations, because I have not explicitly mentioned the possibility of the variation of the ergale. Certainly all my equations are so written that they remain correct also in this case. The advantage, about the possibility of considering some small variation of the ergale and therefore to have at hand the second independent variable in the infinitesimal δU exists and from now on it will not be neglected...".

I must strongly disagree with the remark in the preceding reply, that all his equations are written so that also in the case in which the ergale varies still remain valid. The above introduced Eqs. (1) and (2), even if the quantity ε that appears there receives the extended meaning $\frac{m}{2}\delta v_1^2 + \delta U_1$, are again false in the case in which by the variation of the motion of a point the ergale so changes that the term $\mu(\overline{V}_1 - V)$ has an intrinsic value [*see p. 134.*].

It cannot be said that my Eq. (I) is *implicitly* contained in the Boltzmannian work, but the relevant equations of his work represent, also for what concerns my method of realizing the variations, only a special case of my equations.

Because I must remark that the development of the treatment of the case in which the ergale so changes is not almost unessential, but for researches of this type it is even necessary.

It is in fact possible to consider a body as an aggregate of very many point masses that are under the influence of external and internal forces. The internal forces have an ergale, depending only on the points positions, but in general it stays unchanged in all states of the body; on the contrary this does not hold for the external forces. If for instance the body is subject to a normal pressure p and later its volume v changes by dv, then the external work $p\,dv$ will be performed. This term, when p is varied independently of v, is not an exact differential and the work of the external force cannot, consequently, be representable as the differential of an ergale. The behavior of this force can be so represented. For each given state of the body in which its components are in a state of stationary type it is possible to assign an ergale also to the external forces which, however, does not stay unchanged, unlike that of the internal forces, but it can undergo variations while the body evolved into another state, independent of the change of position of the points.

Keep now in mind the equations posed in the thermology of the changes of state to build their mechanical treatment, which have to be reconsidered to adapt them to the case in which the ergale changes.

I can say that I looked with particular care such generalizations. Hence it would not be appropriate to treat fully the problem, but I obtained in my mechanical equations the above mentioned term $\mu(V_1 - \overline{V})$, for which the corresponding term cannot be found in the mechanical equations. I must now discuss the grounds for this difference and under which conditions such term could vanish. I find that this term will be obtained if the ergale variation is not instantaneous happening at a given moment, but gradual and uniform while an entire cycle takes place, and at the same time I claim that the same result is obtained if it is supposed that we do not deal with a single moving point but with *very large numbers of equal points*, all moving in the same way but with different phases, so that at every moment the phases are uniformly distributed and this suffices for each quantity to be evaluated at a point where it assumes a different value thus generating the average value. The latter case arises in the theory of heat, in which the motions, that we call heat, are of a type in which the quantities that are accessible to our senses are generated by many equal points in the same way, Hence the preceding difficulty is solved, but I want to stress that such solution appears well simpler when it is found than when it is searched.

The circumstance that for the motions that we call heat those terms disappear from the averages had as a result that Boltzmann could obtain for the digregation [*free energy*] and the entropy, from his more restricted analysis, results similar to those that I obtained with a more general analysis; but it will be admitted that the real and complete foundation of this solution can only come from the more general treatment.

The validity condition of the result, which remains hidden in the more restricted analyses, will also be evident.

In every case B. restricts attention to motions that take place along closed trajectories. Here we shall consider motions on non closed curves, hence it now becomes necessary a special argument.[38]

Here too I undertook another way with respect to Boltzmann, and this is the first of the two points mentioned above, in which Boltzmann's result on disgregation and entropy differ. In his method taking into account of time is of the type that I called *characteristic time of the period of a motion*, essentially different. The second point of difference is found in the way in which we defined temperature. The special role of these differences should be followed here in detail, but I stop here hoping to come back to it elsewhere.[39]

Finally it will nor be superfluous to remark that in another of my published works the theorem whereby in every stationary motion *the average "vis viva" equals the virial* remains entirely outside of the priority question treated here. This theorem, as far as I know, has not been formulated by anyone before me.

6.8 On the Ergodic Hypothesis (Trilogy: #1)

Partial translation and comments: L. Boltzmann, a *Über das Wärmegleichgewicht zwischen mehratomigen Gasmolekülen*, 1871, in Wissenschaftliche Abhandlungen, ed. F. Hasenöhrl, **1**, #18, 237–258, [9].

---

[38] The case of motions taking place on non closed trajectories is, *however*, treated by Boltzmann, as underlined in p. 138 of Sect. 6.1, quite convincingly.

[39] NoA: While this article was in print I found in a parallel research that the doubtful expression, to be correct in general, requires a change that would make it even more different from the Boltzmannian one.

[This work is remarkable particularly for its Sect. II where the Maxwell's distribution is derived as a consequence of the of the assumption that, because of the collisions with other molecules, the atoms of the molecule visit all points of the molecule phase space. It is concluded that the distribution of the atoms inside a molecule is a function of the molecule energy. So the distribution of the coordinates of the body will depend on the total energy (just kinetic as the distance between the particles in very large compared with the interaction range). Furthermore the form of the distribution is obtained by supposing the particles energies discretized regularly and using a combinatorial argument and subsequently by passing to the limit as N →∞ and the level spacing → 0. The question of uniqueness of the microcanonical distribution is explicitly raised. Strictly speaking the results do not depend on the ergodic hypothesis. The relation of the "Trilogy" papers with Einstein's statistical mechanics is discussed in [20].]

According to the mechanical theory of heat every molecule of gas is in motion while, in its motion, it does not experience, by far for most of the time, any collision; and its baricenter proceeds with uniform rectilinear motion through space. When two molecules get very close, they interact via certain forces, so that the motion of each feels the influence of the other.

The different molecules of the gas take over all possible configurations[40] and it is clear that it is of the utmost importance to know the probability of the different states of motion.

We want to compute the average kinetic energy, the average potential energy, the mean free path of a molecule etc., and, furthermore, also the probability of each of their values. Since the latter value is not known we can then at most conjecture the most probable value of each quantity, as we cannot even think of the exact value.

If every molecule is a point mass, Maxwell provides the value of the probability of the different states (Phil. Mag., March[41] 1868), [6, 21]. In this case the state of a molecule is entirely determined as soon as the size and direction of its velocity are known. And certainly every direction, in space, of the velocity is equally probable, so that it only remains to determine the probability of the different components of the velocity.

If we denote N the number of molecules per unit volume, Maxwell finds that the number of molecules per unit volume and speed between c and $c + dc$, equals, [21, Eq. (26), p. 187]:

$$4\sqrt{\frac{h^3}{\pi}} N e^{-hc^2} c^2 \, dc,$$

---

[40] In this paper B. imagines that a molecule of gas, in due time, goes through all possible states, but this is not yet the ergodic hypothesis because this is attributed to the occasional interaction of the molecule with the others, see p. 164. The hypothesis is used to extend the hypothesis formulated by Maxwell for the monoatomic systems to the case of polyatomic molecules. For these he finds the role of the internal potential energy of the molecule, which must appear together with the kinetic energy of its atoms in the stationary distribution, thus starting what will become the theory of statistical ensembles, and in particular of the canonical ensemble.

[41] Maybe February?

where h is a constant depending on the temperature. We want to make use of this expression: through it the velocity distribution is defined, i.e. it is given how many molecules have a speed between 0 and dc, how many between dc and $2dc$, $2dc$ and $3dc$, $3dc$ and $4dc$, etc., up to infinity.

Natural molecules, however, are by no means point masses. We shall get closer to reality if we shall think of them as systems of more point masses (the so called atoms), kept together by some force. Hence the state of a molecule at a given instant can no longer be described by a single variable but it will require several variables. To define the state of a molecule at a given instant, think of having fixed in space, once and for all, three orthogonal axes. Trace then through the point occupied by the baricenter three orthogonal axes parallel to the three fixed directions and denote the coordinates of the point masses of our molecule, on every axis and at time t, with $\xi_1, \eta_1, \zeta_1, \xi_2, \eta_2, \zeta_2, \ldots, \xi_{r-1}, \eta_{r-1}, \zeta_{r-1}$. The number of point masses of the molecule, that we shall always call atoms, be r. The coordinate of the r-th atom be determined besides those of the first $r-1$ atoms from the coordinates of the baricenter. Furthermore let c_1 be the velocity of the atom 1, u_1, v_1, w_1 be its components along the axes; the same quantities be defined for the atom 2, c_2, u_2, v_2, w_2; for the atom 3 let them be c_3, u_3, v_3, w_3 etc. Then the state of our molecule at time t is given when the values of the $6r-3$ quantities $\xi_1, \eta_1, \zeta_1, \xi_2, \ldots, \zeta_{r-1}, u_1, v_1, w_1, u_2, \ldots, w_r$ are known at this time. The coordinates of the baricenter of our molecule with respect to the fixed axes do not determine its state but only its position.

We shall say right away, briefly, that a molecule is at a given place when its baricenter is there, and we suppose that in the whole gas there is an average number N of molecules per unit volume. Of such N molecules at a given instant t a much smaller number dN will be so distributed that, at the same time, the coordinates of the atom 1 are between ξ_1 and $\xi_1 + d\xi_1$, η_1 and $\eta_1 + d\eta_1$, ζ_1 and $\zeta_1 + d\zeta_1$, those of the atom 2 are between ξ_2 and $\xi_2 + d\xi_2$, η_2 and $\eta_2 + d\eta_2$, ζ_2 and $\zeta_2 + d\zeta_2$, and those of the $r-1$-th between ξ_{r-1} and $\xi_{r-1} + d\xi_{r-1}$, η_{r-1} and $\eta_{r-1} + d\eta_{r-1}$, ζ_{r-1} and $\zeta_{r-1} + d\zeta_{r-1}$, while the velocity components of the atom 1 are between u_1 and $u_1 + du_1$, v_1 and $v_1 + dv_1$, w_1 and $w_1 + dw_1$, those of the atom 2 are between u_2 and $u_2 + du_2$, v_2 and $v_2 + dv_2$, w_2 and $w_2 + dw_2$, and those of the $r-1$-th are between u_{r-1} and $u_{r-1} + du_{r-1}$, v_{r-1} and $v_{r-1} + dv_{r-1}$, w_{r-1} and $w_{r-1} + dw_{r-1}$.

I shall briefly say that the so specified molecules are in the domain (A). Then it immediately follows that

$$dN = f(\xi_1, \eta_1, \zeta_1, \ldots, \zeta_{r-1}, u_1, v_1, \ldots, w_r) d\xi_1 d\eta_1 d\zeta_1 \ldots d\zeta_{r-1} du_1 dv_1 \ldots dw_r.$$

I shall say that the function f determines a distribution of the states of motion of the molecules at time t. The probability of the different states of the molecules would be known if we knew which values has this function for each considered gas when it is left unperturbed for a long enough time, at constant density and temperature. For monoatomic molecules gases Maxwell finds that the function f has the value

$$4\sqrt{\frac{h^3}{\pi}} N e^{-hc^2} c^2 \, dc.$$

The determination of this function for polyatomic molecules gases seems very difficult, because already for a three atoms complex it is not possible to integrate the equations of motion. Nevertheless we shall see that just from the equations of motion, without their integration, a value for the function f is found which, in spite of the motion of the molecule, will not change in the course of a long time and therefore, represents, at least, a possible distribution of the states of the molecules.[42]

That the value pertaining to the function f could be determined without solving the equations of motion is not so surprising as at first sight seems. Because the great regularity shown by the thermal phenomena induces to suppose that f be almost general and that it should be independent from the properties of the special nature of every gas; and also that the general properties depend only weakly from the general form of the equations of motion, except when their complete integration presents difficulties not unsurmountable.[43]

Suppose that at the initial instant the state of motion of the molecules is entirely arbitrary, i.e. think that the function f has a given value.[44] As time elapses the state of each molecule, because of the motion of its atoms while it follows its rectilinear motion and also because of its collisions with other molecules, becomes constant; hence the form of the function f will in general change, until it assumes a value that in spite of the motion of the atoms and of the collisions between the molecules will no longer change.

When this will have happened we shall say that the states of the molecules are distributed in *thermal equilibrium*. From this immediately the problem is posed to find for the function f a value that will not any more change no matter which collisions take place. For this purpose we shall suppose, to treat the most general case, that we deal with a mixture of gases. Let one of the kinds of gas (the kind G) have N molecules per unit volume. Suppose that at a given instant t there are dN molecules whose state is in the domain (a). Then as before

$$dN = f(\xi_1, \eta_1, \zeta_1, \ldots, \zeta_{r-1}, u_1, v_1, \ldots, w_r) d\xi_1 d\eta_1 d\zeta_1 \ldots d\zeta_{r-1} du_1 dv_1 \ldots dw_r.$$
$$(1)$$

The function f gives us the complete distribution of the states of the molecules of the gas of kind G at the instant t. Imagine that a certain time δt elapses. At time $t + \delta t$ the distribution of the states will in general have become another, hence the function f becomes different, which I denote f_1, so that at time $t + \delta t$ the number of molecules per unit volume whose state in the domain (A) equals:

---

[42] Remark the care with which the possibility is not excluded of the existence invariant distributions different from the one that will be determined here.

[43] Here B. seems aware that special behavior could show up in integrable cases: he was very likely aware of the theory of the solution of the harmonic chain of Lagrange, [22, Vol. I].

[44] This is the function called "empirical distribution", [23, 24].

$$f_1(\xi_1, \eta_1, \ldots, w_r) \, d\xi_1 \, dh_1 \ldots dw_r. \tag{2}$$

§I. Motion of the atoms of a molecule

[*Follows the analysis of the form of f in absence of collisions: via Liouville's theorem it is shown that if f is invariant then it has to be a function of the coordinates of the molecules through the integrals of motion. This is a wide extension of the argument by Maxwell for monoatomic gases, [21].*]

§II. Collisions between molecules

[*It is shown that to have a stationary distribution also in presence of binary collisions it must be that the function f has the form $Ae^{-h\varphi}$ where φ is the total energy, sum of the kinetic energy and of the potential energy of the atoms of the molecule. Furthermore if the gas consists of two species then h must be the same constant for the distribution of either kinds of molecules and it is identified with the inverse temperature. Since a gas, monoatomic or not, can be considered as a giant molecule it is seen that this is the derivation of the canonical distribution. The kinetic energies equipartition and the ratios of the specific heats is deduced. It becomes necessary to check that this distribution "of thermal equilibrium" generates average values for observables compatible with the heat theorem: this will be done in the successive papers. There it will also be checked that the ergodic hypothesis in the form that each group of atoms that is part of a molecule passes through all states compatible with the value of the energy (possibly with the help of the collisions with other molecules) leads to the same result if the number of molecules is infinite or very large. The question of the uniqueness of the equilibrium distribution is however left open as explicitly stated at p. 225.*]

p. 255, (line 21) Against me is the fact that, until now, the proof that these distributions are the only ones that do not change in presence of collisions is not complete. Nevertheless remains the fact that [the distribution shows] that the same gas with equal temperature and density can be in many states, depending on the given initial conditions, a priori improbable and which will even never be observed in experiments.

[*This paper is also important as it shows that Boltzmann was well aware of Maxwell's paper, [21]: in which a key argument towards the Boltzmann's equation is discussed in great detail. One can say that Maxwell's analysis yields a form of "weak Boltzmann's equation", namely several equations which can be seen as equivalent to the time evolution of averages of one particle observable with what we call now the one particle distribution of the particles. Boltzmann will realize, [7, #22], that the one particle distribution itself obeys an equation (the Boltzmann equation) and obtain in this way a major conceptual simplification of Maxwell's approach and derive the H-theorem.*]

6.9 Canonical Ensemble and Ergodic Hypothesis (Trilogy: #2)

Partial translation and comments of: L. Boltzmann, *Einige allgemeine sätze über Wärmegleichgewicht*, (1871), in Wissenschaftliche Abhandlungen, ed. F. Hasenöhrl, **1**, 259–287, #19, [25].

§I. *Correspondence between the theorems on the polyatomic molecules behavior and Jacobi's principle of the last multiplier.*[45]

The first theorem that I found in my preceding paper *Über das Wärmegleichgewicht zwischen mehratomigen Gasmolekülen*, 1871, [9, #17], is strictly related to a theorem, that at first sight has nothing to do with the theory of gases, i.e. with Jacobi's principle of the last multiplier.

To expose the relation, we shall leave aside the special form that the mentioned equations of the theory of heat have, whose relevant developments will be generalized here later.

Consider a large number of systems of point masses (as in a gas containing a large number of molecules of which each is a system of point masses). The state of a given system of such points at a given time is assigned by n variables $s_1, s_2, \ldots, s_n$ for which we can pose the following differential equations:

$$\frac{ds_1}{dt} = S_1, \frac{ds_2}{dt} = S_2, \ldots, \frac{ds_n}{dt} = S_n.$$

Let the $S_1, S_2, \ldots, S_n$ be functions of the $s_1, s_2, \ldots, s_n$ and possibly of time. Through these equations and the initial value of the n variables $s_1, s_2, \ldots, s_n$ are known the values of such quantities at any given time. To arrive to the principle of the last multipliers, we can use many of the conclusions reached in the already quoted paper; hence we must suppose that between the point masses of the different systems of points never any interaction occurs. As in the theory of gases the collisions between molecules are neglected, also in the present research the interactions will be excluded.

[*Follows a discussion on the representation of a probability distribution giving the number of molecules in a volume element of the phase space with $2n$ dimensions. The Liouville's theorem is proved for the purpose of obtaining an invariant distribution in the case of equations of motion with vanishing divergence.*

Subsequently it is discussed how to transform this distribution into a distribution of the values of n constants of motion, supposing their existence; concluding that the distribution is deduced by dividing the given distribution by the Jacobian determinant of the transformation expressing the coordinates in terms of the constants of motion and of time: the last multiplier of Jacobi is just the Jacobian determinant of the change of coordinates.

Taking as coordinates $n - 1$ constants of motion and as n-th the s_1 it is found that a stationary distribution is such that a point in phase space spends in a volume element,

---

[45] This section title is quotes as such by Gibbs in the introduction to his *Elementary principles in statistical mechanics*, [26], thus generating some confusion because, of course, this title is not found in the list of publications by Boltzmann.

in which the $n - 1$ constants of motion $\varphi_2, \varphi_3, \ldots, \varphi_n$ have a value in the set D of the points where $\varphi_2, \varphi_3, \ldots$ are between φ_2 and $\varphi_2 + d\varphi_2$, φ_3 and $\varphi_3 + d\varphi_3 \ldots$, and s_1 is between s_1 and $s_1 + ds_1$, a fraction of time equal to the fraction of the considered volume element with respect to the total volume in which the constants have value in D and the n-th has an arbitrary value.

The hypothesis of existence of n constants of motion is not realistic in the context in which it is assumed that the motion is regulated by Hamiltonian differential equations. It will become plausible in the paper of 1877, [11, #42], where a discrete structure is admitted for the phase space and time.]

§II. *Thermal equilibrium for a finite number of point masses.*
[In this section the method is discussed to compute the average kinetic energy and the average potential energy in a system with n constants of motion.]
§III. p. 284 *Solution for the thermal equilibrium for the molecules of a gas with a finite number of point masses under an hypothesis.*

Finally from the equations derived we can, under an assumption which it does not seem to me of unlikely application to a warm body, directly access to the thermal equilibrium of a polyatomic molecule, and more generally of a given molecule interacting with a mass of gas. The great chaoticity of the thermal motion and the variability of the force that the body feels from the outside makes it probable that the atoms get in the motion, that we call heat, all possible positions and velocities compatible with the equation of the "vis viva", and even that the atoms of a warm body can take all positions and velocities compatible with the last equations considered.[46]

---

[46] Here comes back the ergodic hypothesis in the form saying that not only the atoms of a single molecule take all possible positions and velocities but also that the atoms of a "warm body" with which a molecule is in contact take all positions and velocities.
This is essentially the ergodic hypothesis. The paper shows how, through the ergodic hypothesis assumed for the whole gas it is possible to derive the canonical distribution for the velocity and position distribution both of a single molecule and of an arbitrary number of them. It goes beyond the preceding paper deducing the *microcanonical* distribution, on the assumption of the ergodic hypothesis which is formulated here for the first time as it is still intended today, and finding as a consequence the *canonical* stationary distribution of the atoms of each molecule or of an arbitrary number of them by integration on the positions and velocities of the other molecules.
This also *founds the theory of the statistical ensembles*, as recognized by Gibbs in the introduction of his treatise on statistical mechanics, [26]. Curiously Gibbs quotes this paper of Boltzmann attributing to it a title which, instead, is the title of its first Section. The Jacobi's principle, that B. uses in this paper, is the theorem that expresses the volume element in a system of coordinates in terms of that in another through a "final multiplier", that today we call "Jacobian determinant" of the change of coordinates. B. derives already in the preceding paper what we call today "Liouville's theorem" for the conservation of the volume element of phase space and here he gives a version that takes into account the existence of constants of motion, such as the energy. From the uniform distribution on the surface of constant total energy (suggested by the ergodic hypothesis) the canonical distribution of subsystems (like molecules) follows by integration and use of the formula $(1 - \frac{c}{\lambda})^\lambda = e^{-c}$ if λ (total number of molecules) is large.
Hence imagining the gas large the canonical distribution follows for every finite part of it, be it constituted by 1 or by 10^{19} molecules: a finite part of a gas is like a giant molecule.

Let us accept this hypothesis, and thus let us make use of the formulae to compute the equilibrium distribution between a gas in interaction with a body supposing that only r of the mentioned λ atoms of the body interact with the mass of gas.

Then χ [*potential energy*] has the form $\chi_1 + \chi_2$ where χ_1 is a function of the coordinates of the r atoms, χ_2 is a function of the coordinates of the remaining $\lambda - r$. Let us then integrate formula (24) for dt_4 [*which is*

$$dt_4 = \frac{(a_n - \chi)^{\frac{3\lambda}{2}-1} dx_1 \, dy_1 \, \dots \, dz_\lambda}{\int \int (a_n - \chi)^{\frac{3\lambda}{2}-1} dx_1 \, dy_1 \, \dots \, dz_\lambda} \tag{24}$$

expressing the time during which, in average, the coordinates are between x_1 *and* $x_1 + dx_1 \dots z_\lambda$ *and* $z_\lambda + dz_\lambda$] *over all values of* $x_{r+1}, y_{r+1}, \dots, z_\lambda$ *obtaining for the time during which certain* $x_1, y_1, \dots, z_r$ *are, in average, between* x_1 *and* $x_1 + dx_1$ etc.; hence for the average time that the atom m_1 spends in the volume element $dx_1 dy_1 dz_1$, m_2 spends in $dx_2 dy_2 dz_2 \dots$, the value is found of

$$dt_5 = \frac{dx_1 \, dy_1 \, \dots \, dz_r \int \int \dots (a_n - \chi_1 - \chi_2)^{\frac{3\lambda}{2}-1} dx_{r+1} dy_{r+1} \dots dz_\lambda}{\int \int \dots (a_n - \chi_1 - \chi_2)^{\frac{3\lambda}{2}-1} dx_1 dy_1 \dots dz_\lambda}$$

If the elements $dx_1 dy_1 dz_1, dx_2 dy_2 dz_2 \dots$ were chosen so that $\chi_1 = 0$ gave the true value of dt_5 it would be

$$dt_6 = \frac{dx_1 \, dy_1 \, \dots \, dz_r \int \int \dots (a_n - \chi_2)^{\frac{3\lambda}{2}-1} dx_{r+1} dy_{r+1} \dots dz_\lambda}{\int \int \dots (a_n - \chi_1 - \chi_2)^{\frac{3\lambda}{2}-1} dx_1 dy_1 \dots dz_\lambda}$$

And then the ratio is

$$\frac{dt_5}{dt_6} = \frac{\int \int \dots (a_n - \chi_1 - \chi_2)^{\frac{3\lambda}{2}-1} dx_1 dy_1 \dots dz_\lambda}{\int \int \dots (a_n - \chi_2)^{\frac{3\lambda}{2}-1} dx_1 dy_1 \dots dz_\lambda}.$$

The domain of the integral in the denominator is, because of the unchanged presence of the function χ_2, dependent from a_n. The domain of the integral in the numerator, in the same way, which does not contain the variable on which the integral has to be made. The last integral is function of $(a_n - \chi_1)$. Let $\frac{a_n}{\lambda} = \varrho$, where λ is the number, naturally constant, of atoms, thus also the integral in the denominator is a function of ϱ, that we shall denote $F(\varrho)$; the integral in the numerator is the same function of $\varrho - \frac{\chi_1}{\lambda}$, hence equal to $F(\varrho - \frac{\chi_1}{\lambda})$ and therefore

$$\frac{dt_5}{dt_6} = \frac{F\left(\varrho - \frac{\chi_1}{\lambda}\right)}{F(\varrho)}.$$

Let now λ be very large, Hence also r can be very large; we now must eliminate λ. If dt_5/dt_6 is a finite and continuous function of ϱ and χ_1 then ϱ and $\frac{\chi_1}{r}$ have the order of magnitude of the average "vis viva" of an atom. Let $\frac{dt_5}{dt_6} = \psi(\varrho, \chi_1)$, then

$$\psi(\varrho, \chi_1) = \frac{F\left(\varrho - \frac{\chi_1}{\lambda}\right)}{F(\varrho)} \tag{28}$$

Hence

$$\frac{F\left(\varrho - \frac{2\chi_1}{\lambda}\right)}{F\left(\varrho - \frac{\chi_1}{\lambda}\right)} = \psi\left(\varrho - \frac{\chi_1}{\lambda}\right) = \psi_1$$

$$\frac{F\left(\varrho - \frac{3\chi_1}{\lambda}\right)}{F\left(\varrho - \frac{2\chi_1}{\lambda}\right)} = \psi\left(\varrho - \frac{2\chi_1}{\lambda}\right) = \psi_2$$

$$\cdots\cdots$$

$$\frac{F\left(\varrho - \frac{\mu\chi_1}{\lambda}\right)}{F\left(\varrho - \frac{\mu-1\chi_1}{\lambda}\right)} = \psi\left(\varrho - \frac{(\mu-1)}{\lambda}, \chi_1\right) = \psi_{\mu-1}.$$

Multiplying all these equations yields

$$\log F(\varrho - \frac{\mu\chi_1}{\lambda}) - \log F(\varrho) = \log \psi + \log \psi_1 + \cdots + \log \psi_{\mu-1}$$

Let now $\mu\chi_1 = \chi_3$ then

$$\log F(\varrho - \chi_3) - \log(\varrho) = \lambda \log \Psi(\varrho, \chi_3)$$

where Ψ is again finite and continuous and if ϱ and $\frac{\chi_3}{r}$ are of the order of the average "vis viva" also $\frac{\lambda}{\mu}$ is finite. It is also:

$$F(\varrho - \chi_3) = F(\varrho) \cdot [\Psi(\varrho, \chi_3)]^{\lambda}.$$

Let us now treat ϱ as constant and χ_3 as variable and set

$$\varrho - \chi_3 = \sigma, \quad F(\varrho) = C, \quad \Psi(\varrho, \varrho - \sigma) = f(\sigma)$$

thus the last formula becomes $F(\sigma) = C \cdot [f(\sigma)]^{\lambda}$, and therefore formula (28) becomes

$$\psi(\varrho, \chi_1) = \left[\frac{f(\varrho - \frac{\chi_1}{\lambda})}{f(\varrho)}\right]^{\lambda} = \left[1 - \frac{f'(\varrho)}{f(\varrho)} \cdot \frac{\chi_1}{\lambda}\right]^{\lambda} = e^{\frac{f'(\varrho)}{f(\varrho)} \cdot \chi_1}$$

and if we denote $\frac{f'(\varrho)}{f(\varrho)}$ with h

$$dt_5 = C'e^{-h\chi_1}dx_1dy_1 \ldots dz_r.$$

Exactly in the same way the time can be found during which the coordinates of the r atoms are between x_1 and $x_1 + dx_1 \ldots$ and their velocities are between c_1 and $c_1 + dC_1 \ldots$. It is found to be equal to

$$C''e^{-h\left(\chi_1+\sum \frac{mc^2}{2}\right)}c_1^2c_2^2 \ldots c_r^2dx_1dy_2 \ldots dc_r.$$

These equations must, under our hypothesis, hold for an arbitrary body in a mass of gas, and therefore also for a molecule of gas. In the considered case it is easy to see that this agrees with the formulae of my work *Über das Wärmegleichgewicht zwischen mehratomigen Gasmolekülen*, 1871, [9, #17]. We also arrive here in a much easier way to what found there. Since however the proof, that in the present section makes use of the hypothesis about the warm body, which certainly is acceptable but it had not yet been proposed, thus I avoided it in the quoted paper obtaining the proof in a way independent from that hypothesis.[47]

6.10 Heat Theorem Without Dynamics (Trilogy: #3)

Comment: L. Boltzmann, *Analytischer Beweis des zweiten Hauptsatzes der mechanischen Wärmetheorie aus den Sätzen über das Gleichgewicht des lebendigen Kraft*, 1871, Wissenschaftliche Abhandlungen, ed. F. Hasenöhrl, **1**, 288–308, #20, [27].

Here it is shown how the hypothesis that, assuming that the equilibrium distribution of the entire system is the microcanonical one, then defining the heat dQ received by the body as the variation of the total average energy dE plus the work dW done by the system on the outside (average variation in time of the potential energy due to a change of the values of the external parameters) it follows that $\frac{dQ}{T}$ is an exact differential if T is proportional to the average kinetic energy. This frees (apparently) equilibrium statistical mechanics from the ergodic hypothesis and will be revisited in the paper of 1884, [28, #73] see also Sect. 6.13, with the general theory of statistical ensembles and of the states of thermodynamic equilibrium. Here dynamics enters only through the conservation laws and the hypothesis (molecular chaos, see the first trilogy paper [9, #17]) that never a second collision (with a given molecule) takes place when one is still taking place: properties before and after the collision are only used to infer again the canonical distribution, which is then studied as the source of the heat theorem. The hypothesis of molecular chaos preludes to the paper, following this a little later, that will mark the return to a detailed dynamical

---

[47] He means that he proved in the quoted reference the invariance of the canonical distribution (which implies the equidistribution) without the present hypothesis. However even that was not completely satisfactory as he had also stated in the quoted paper that he had not been able to prove the uniqueness of the solution found there (that we know today to be not true in general).

analysis with the determination, for rarefied gases, of the time scales needed to reach equilibrium, based on the Boltzmann's equation, [7, #22].

6.11 Irreversibility: Loschmidt and "Boltzmann's Sea"

Partial translation and comments: L. Boltzmann, *Bemerkungen über einige Probleme der mechanischen Wärmetheorie*, 1877, in Wissenschaftliche Abhandlungen, ed. F. Hasenöhrl, **2**, 112–148, #39, [29].

§II. On the relation between a general mechanics theorem and the second main theorem[48] of the theory of heat (p. 116)

In his work on the states of thermal equilibrium of a system of bodies, with attention to the force of gravity, Loschmidt formulated an opinion, according to which he doubts about the possibility of an entirely mechanical proof of the second theorem of the theory of heat. With the same extreme sagacity suspects that for the correct understanding of the second theorem an analysis of its significance is necessary deeper than what appears indicated in my philosophical interpretation, in which perhaps various physical properties are found which are still difficult to understand, hence I shall immediately here undertake their explanation with other words.

We want to explain in a purely mechanical way the law according to which all natural processes proceed so that

$$\int \frac{dQ}{T} \leq 0$$

and so, therefore, behave bodies consistent of aggregates of point masses. The forces acting between these point masses are imagined as functions of the relative positions of the points. If they are known as functions of these relative positions we say that the interaction forces are known. Therefore the real motion of the point masses and also the transformations of the state of the body will be known once given the initial positions and velocities of the generic point mass. We say that the initial conditions must be given.

We want to prove the second theorem in mechanical terms, founding it on the nature of the interaction laws and without imposing any restriction on the initial conditions, knowledge of which is not supposed. We look also for the proof that, provided initial conditions are similar, the transformations of the body always take place so that

$$\int \frac{dQ}{T} \leq 0.$$

---

[48] In this paper the discussion is really about the second law rather than about the second main theorem, see the previous sections.

Suppose now that the body is constituted by a collection of point like masses, or virtually such. The initial condition be so given that the successive transformation of the body proceed so that

$$\int \frac{dQ}{T} \leq 0$$

We want to claim immediately that, provided the forces stay unchanged, it is possible to exhibit another initial condition for which it is

$$\int \frac{dQ}{T} \geq 0.$$

Because we can consider the values of the velocities of all point masses reached at a given time t_1. We now want to consider, instead of the preceding initial conditions, the following: at the beginning all point masses have the same positions reached, starting form the preceding initial conditions, in time t_1 but with the all velocities inverted. We want in such case to remark that the evolution of the state towards the future retraces exactly that left by the preceding evolution towards the time t_1.

It is clear that the point masses retrace the same states followed by the preceding initial conditions, but in the opposite direction. The initial state that before we had at time 0 we see it realized at time t_1 [*with opposite velocities*]. Hence if before it was

$$\int \frac{dQ}{T} \leq 0$$

we shall have now ≥ 0.

On the sign of this integral the interaction cannot have influence, but it only depends on the initial conditions. In all processes in the world in which we live, experience teaches us this integral to be ≤ 0, and this is not implicit in the interaction law, but rather depends on the initial conditions. If at time 0 the state [*of the velocities*] of all the points of the Universe was opposite to the one reached after a very long time t_1 the evolution would proceed backwards and this would imply

$$\int \frac{dQ}{T} \leq 0$$

Every experimentation on the nature of the body and on the mutual interaction law, without considering the initial conditions, to check that

$$\int \frac{dQ}{T} \leq 0$$

would be vain. We see that this difficulty is very attractive and we must consider it as an interesting sophism. To get close to the fallacy that is in this sophism we shall immediately consider a system of a finite number of point masses, which is isolated from the rest of the Universe.

We think to a very large, although finite, number of elastic spheres, which are moving inside a container closed on every side, whose walls are absolutely still and perfectly elastic. No external forces be supposed acting on our spheres. At time 0 the distribution of the spheres in the container be assigned as non uniform; for instance the spheres on the right be denser than the ones on the left and be faster if higher than if lower and of the same order of magnitude. For the initial conditions that we have mentioned the spheres be at time t_1 almost uniformly mixed. We can then consider instead of the preceding initial conditions, the ones that generate the inverse motion, determined by the initial conditions reached at time t_1. Then as time evolves the spheres come back; and at time t_1 will have reached a non uniform distribution although the initial condition was almost uniform. We then must argue as follows: a proof that, after the time t_1 the mixing of the spheres must be with absolute certainty uniform, whatever the initial distribution, cannot be maintained. This is taught by the probability itself; every non uniform distribution, although highly improbable, is not absolutely impossible. It is then clear that every particular uniform distribution, that follows an initial datum and is reached in a given time is as improbable as any other even if not uniform; just as in the lotto game every five numbers are equally probable as the five 1, 2, 3, 4, 5. And then the greater or lesser uniformity of the distribution depends on the greater size of the probability that the distribution becomes uniform, as time goes.

It is not possible, therefore, to prove that whatever are the initial positions and velocities of the spheres, after a long enough time, a uniform distribution is reached, nevertheless it will be possible to prove that the initial states which after a long enough time evolve towards a uniform state will be infinitely more than those evolving towards a nonuniform state, and even in the latter case, after an even longer time, they will evolve towards a uniform state.[49]

Loschmidt's proposition teaches also to recognize the initial states that really at the end of a time t_1 evolve towards a very non uniform distribution; but it does not imply the proof that the initial data that after a time t_1 evolve into uniform distributions are infinitely many more. Contrary to this statement is even the proposition itself which enumerates as infinitely many more uniform distributions than non uniform, for which the number of the states which, after a given time t_1 arrive to uniform distribution must also be as numerous as those which arrive to nonuniform distributions, and these are just the configurations that arise in the initial states of Loschmidt, which become non uniform at time t_1.

It is in reality possible to calculate the ratio of the numbers of the different initial states which determines the probabilities, which perhaps leads to an interesting method to calculate the thermal equilibria.[50] Exactly analogous to the one that leads to the second theorem. It is at least in some special cases successfully checked, when a system undergoes a transformation from a nonuniform state to a uniform one, that $\int \frac{dQ}{T}$ will be intrinsically negative, while it will be positive in the inverse case. Since there are infinitely more uniform then nonuniform distributions of the states,

---

[49] Today this important discussion is referred to as the argument of the *Boltzmann's sea*, [30].
[50] Boltzmann will implement the idea in [11, #42], see also Sect. 6.12.

therefore the last case will be extremely improbable: and in practice it could be considered impossible that at the beginning a mixture of oxygen and nitrogen are given so that after one month the chemically pure oxygen is found in the upper part and that the nitrogen in the lower, an event that probability theory states as improbable but not as absolutely impossible.

Nevertheless it seems to me that the Loschmidtian theorem has a great importance, since it tells us how intimately related are the second principle and the calculus of probabilities. For all cases in which $\int \frac{dQ}{T}$ can be negative it is also possible to find an initial condition very improbable in which it is positive. It is clear to me that for closed atomic trajectories $\int \frac{dQ}{T}$ must always vanish. For non closed trajectories it can also be negative. Now a peculiar consequence of the Loschmidtian theorem which I want to mention here, i.e. that the state of the Universe at an infinitely remote time, with fundamentally equal confidence, can be considered with large probability both as a state in which all temperature differences have disappeared, and as the state in which the Universe will evolve in the remote future.[51]

This is analogous to the following case: if we want that, in a given gas at a given time, a non uniform distribution is realized and that the gas remains for a very long time without external influences, then we must think that as the distribution of the states was uniform before so it will become entirely uniform.

In other words: as any nonuniform distribution evolves at the end of a time t_1 towards a uniform one the latter if inverted as the same time t_1 elapses again comes back to the initial nonuniform distribution (precisely for the said inversion). The [new] but inverted initial condition, chosen as initial condition, after a time t_1 similarly will evolve to a uniform distribution.[52]

But perhaps such interpretation relegates in the domain of probability theory the second principle, whose universal use appears very questionable, and nevertheless just because of the theory of probability it will be realized in every laboratory experimentation.

[*§III, p. 127 and following: a check of the heat theorem is presented in the case of a central motion, which will be revisited in the papers of 1884 by v. Helmholtz and Boltzmann.*]

Let M be a point mass, whose mass will be denoted m, and let $OM = r$ its distance from a fixed point O. The point M is pushed towards O by a force $f(r)$. We suppose that the work

$$\varphi(r) = \int_r^\infty f(r)dr,$$

necessary to bring the point M to an infinite distance from O, is finite, so that it is a function φ whose negative derivative in any direction gives the force acting in the

---

[51] Reference to the view of Clausius which claims that in the remote future the Universe will be in an absolutely uniform state. Here B. says that the same must have happened, with equal likelihood in the remote past.

[52] I.e. if once having come back we continue the evolution for as much time again a uniform distribution is reached.

same direction, also called the force function [*minus the potential*]. Let us denote by v the velocity of M and with θ the angle that the radius vector OM form with an arbitrarily fixed line in the plane of the trajectory, thus by the principle of the "vis viva":

$$\frac{m}{2}\left(\frac{dr}{dt}\right)^2 + \frac{mr^2}{2}\left(\frac{d\theta}{dt}\right)^2 = \alpha - \varphi(r), \tag{1}$$

and by the area law

$$r^2\frac{d\theta}{dt} = \sqrt{\beta} \tag{2}$$

α and β remain constant during the whole motion. From these equations follows

$$dt = \frac{dr}{\sqrt{\frac{2\alpha}{m} - \frac{2\varphi(r)}{m} - \frac{\beta}{r^2}}}, \tag{3}$$

We define for the average value of the force function the term $\overline{\varphi} = \frac{z}{n}$ with

$$z = \int \frac{\varphi(r)\,dr}{\sqrt{\frac{2\alpha}{m} - \frac{2\varphi(r)}{m} - \frac{\beta}{r^2}}}$$

$$n = \int \frac{dr}{\sqrt{\frac{2\alpha}{m} - \frac{2\varphi(r)}{m} - \frac{\beta}{r^2}}}$$

The integration is from a minimum to a maximum of the radius vector r, or also it is extended from a smaller to the next larger positive root of the equation

$$\alpha - \varphi(r) - \frac{\beta m}{2r^2}$$

the average "vis viva" T_1 is $\alpha - \overline{\varphi}$. The polynomial equation must obviously have two positive roots. The total sum of the "vis viva" and of the work, which must be done on the point mass to lead it on a path to infinity and its velocity to a standstill is α. The force function be unchanged, so also the work δQ necessary for somehow moving the point mass from a path into another is equal to the increment $\delta\alpha$ of the quantity α. Change now the force function, e.g. because of the intervening constant parameters $c_1, c_2, \ldots$, so under the condition that the change of the nature of the force function demands no work, $\delta\alpha$ is the difference of the works, which are necessary for bringing the point mass standing and again at infinite distance on the other path deprived of the mentioned velocity. The amount of "vis viva" and work necessary to move from an initial site M of the old path to the new initial site M' of the varied path as well as its velocity in M to the one in M' differs from $\delta\alpha$ for the extra work resulting from the change of the force function in the varied state over the one originally necessary to bring the point mass to an infinite distance from the initial M. It is therefore

$$\delta\alpha + \sum \frac{\partial\varphi(r)}{\partial c_k}\delta c_k$$

where for r one has to set the distance of the initial location M from O. The extra work, which is due to the change of the force function because of the required infinitely small displacement, is now infinitely small of a higher order. According to Clausius' idea the average work in the change from one to the other paths is $dQ = \delta\alpha + \frac{\zeta}{n}$, with

$$\zeta = \int \frac{\left[\sum\right]\frac{\partial\varphi}{\partial c_k}\delta c_k\, dr}{\sqrt{\frac{2\alpha}{m} - \frac{2\varphi(r)}{m} - \frac{\beta}{r^2}}}$$

and n is the above defined value.[53] To let the integration easier to perform, one sets

$$\varphi(r) = -\frac{a}{r} + \frac{b}{2r^2}$$

with an attraction of intensity $a/r^2 - n/r^3$ expressed in the distance r, a and b play here the role of the constants c_k. By the variation of the motion let α, β, a and b respectively change by $\delta\alpha, \delta\beta, \delta a$ and δb, then it is:

$$\sum \frac{\partial\varphi(r)}{\partial c_k}\delta c_k = -\frac{d\alpha}{r} + \frac{\delta b}{2r^2}.$$

And also δQ is the average value of

$$\delta\alpha = -\frac{\delta a}{r} + \frac{\delta b}{2r^2}$$

These can also be compared with the results found above [*in the previous text*]. The quantity denoted before with $\delta_2 V$ is in this case

$$\delta_2\left(\alpha - \frac{a}{r} + \frac{b}{2r^2}\right) = \delta\alpha - \frac{\delta a}{r} + \frac{\delta b}{2r^2}$$

We now set, for brevity, formally

$$\varrho = r^2, \ s = \frac{1}{r}, \text{ and } \sigma = \frac{1}{r^2}$$

Hence let the trajectory be real so its endpoints must be the maximum and the minimum of the radius vector r or the pair of roots of the equation

---

[53] Here it seems that there is a sign incorrect as ζ should have a minus sign. *I have not modified the following equations*; but this has to be kept in mind; in Appendix D the calculation for the Keplerian case is reported in detail. See also the footnote to p. 129 of [27, #19].

$$\frac{2\alpha}{m} - \frac{2\varphi(r)}{m} - \frac{\beta}{r^2} = \frac{2\alpha}{m} - \frac{2a}{m}\frac{1}{r} - \frac{b+m\beta}{m}\frac{1}{r^2} = 0$$

must be positive, and the polynomial must be for the r, which are between the pair of roots, likewise positive, and also for r infinitely large negative; i.e. α must be negative and positive at the Cartesian coordinates a and $b + b$. We want always to integrate from the smallest to the largest value of r; then dr, $d\varrho$ and

$$\sqrt{\frac{2\alpha}{m} - \frac{2a}{m}\frac{1}{r} - \frac{b+m\beta}{m}\frac{1}{r^2}}$$

on the contrary ds and $d\sigma$ are negative. Remark that

$$\int_{w_1}^{w_2} \frac{dx}{\sqrt{A + Bx + Cx^2}} = \frac{\pi}{\sqrt{-C}},$$

$$\int_{w_1}^{w_2} \frac{x\,dx}{\sqrt{A + Bx + Cx^2}} = -\frac{\pi B}{2C\sqrt{-C}},$$

if w_1 is the smaller, w_2 the larger root of the equation $A + Bx + Cx^2 = 0$, so for the chosen form of the force function it is found

$$z = -a \int \frac{dr}{\sqrt{\frac{2\alpha}{m}r^2 - \frac{2a}{m}r - \frac{b+m\beta}{m}\frac{1}{r^2}}} + \frac{b}{2} \int \frac{-ds}{\sqrt{\frac{2\alpha}{m} + \frac{2a}{m}s - \frac{b+m\beta}{m}s^2}}$$

$$= -a \cdot \pi \sqrt{\frac{m}{-2\alpha}} + \frac{b}{2}\pi \sqrt{\frac{m}{b+m\beta}},$$

$$\zeta = -\delta\alpha\,\pi \sqrt{\frac{m}{-2\alpha}} + \frac{\delta b}{2} \cdot \pi \sqrt{\frac{m}{b+m\beta}}$$

$$n = \int \frac{dr}{\sqrt{\frac{2\alpha}{m}r^2 - \frac{2a}{m}r - \frac{b+m\beta}{m}\frac{1}{r^2}}} = -\frac{\pi}{2}\frac{a}{\alpha}\sqrt{\frac{m}{-2\alpha}}$$

As α is negative, so to have all integrals essentially positive all roots must be taken positive. It is

$$\overline{\varphi} = \frac{z}{n} = 2\alpha - \frac{b\alpha}{a}\sqrt{\frac{-2\alpha}{b+m\beta}},$$

$$T_1 = \alpha - \overline{\varphi} = \frac{b\alpha}{a}\sqrt{\frac{-2\alpha}{b+m\beta}} - \alpha$$

$$dQ = \delta\alpha + \frac{\zeta}{n} = \delta\alpha + 2\alpha\frac{\delta a}{a} - \frac{\alpha\delta b}{a}\sqrt{\frac{-2\alpha}{b+m\beta}}$$

it is built here the term $\delta Q/T_1$, so it is immediately seen that it is not an exact differential, since $\delta\beta$ does not appear; and, furthermore, if a and b and also the force function stay constant also it is not an exact differential. On the contrary if $b = \delta b = 0$ the trajectory is closed and $\delta Q/T_1$ is an exact differential.[54]

As a second case, in which the integration is much easier, is if

$$\varphi(r) = -ar^2 + \frac{b}{2r^2}$$

[*The analysis follows the same lines and the result is again that $\frac{\delta Q}{T_1}$ is an exact differential only if $\delta b = b = 0$. It continues studying various properties of central motions of the kinds considered so far as the parameters vary, without reference to thermodynamics. The main purpose and conclusion of Sect. III seems however to be that when there are other constants of motion it cannot be expected that the average kinetic energy is an integrating factor for $dQ = dE - \langle\partial_{\mathbf{c}}\varphi \cdot \delta\mathbf{c}\rangle$. The Newtonian potential is a remarkable exception, see Appendix D. Other exceptions are the 1-dimensional systems, obtained as special cases of the central potentials cases with zero area velocity, $\beta = 0$. However, even for the one dimensional case, only special cases are considered here: the general 1-dimensional case was discussed a few years later by v. Helmoltz, [31, 32], and immediately afterwards by Boltzmann, [28, #73], see Sect. 6.13.*]*

6.12 Discrete Phase Space, Count of Its Points and Entropy

Partial translation and comments: L. Boltzmann, Über die Beziehung zwischen dem zweiten Hauptsatze der mechanischen Wärmetheorie und der Wahrscheinlichkeitsrechnung, respektive den Sätzen über das Wärmegleichgewicht, in Wissenschaftliche Abhandlungen, ed. F. Hasenöhrl, Vol. 2, #42 p. 164–233, [11].

p. 166 …We now wish to solve the problem which on my above quoted paper *Bemerkungen über einige Probleme der mechanischen Wärmetheorie*, [29, #39], I have already formulated clearly, i.e. the problem of determining the "ratios of the number of different states of which we want to compute the probabilities".

We first want to consider a simple body, i.e. a gas enclosed between absolutely elastic walls and whose molecules are perfect spheres absolutely elastic (or centers of force which, now, interact only when their separation is smaller than a given quantity, with a given law and otherwise not; this last hypothesis, which includes the first as a special case does not at all change the result). Nevertheless in this case

---

[54] The sign error mentioned in the footnote at p. 176 does not affect the conclusion but only some intermediate steps.

the use of probability theory is not easy. The number of molecules is not infinite in a mathematical sense, although it is extremely large. The number of the different velocities that every molecule can have, on the contrary, should be thought as infinite. Since the last fact renders much more difficult the calculations, thus in the first Section of this work I shall rely on easier conceptions to attain the aim, as I often did in previous works (for instance in the *Weiteren Studien*, [7, #22]).

.....

§I. The number of values of the "vis viva" is discrete (p. 167).

We want first to suppose that every molecule can assume a finite number of velocities, for instance the velocities

$$0, \frac{1}{q}, \frac{2}{q}, \frac{3}{q}, \ldots, \frac{p}{q},$$

where p and q are certain finite numbers. At a collision of two molecules will correspond a change of the two velocities, so that the state of each will have one of the above mentioned velocities, i.e.

$$0, \text{ or } \frac{1}{q}, \text{ or } \frac{2}{q}, \text{ etc. until } \frac{p}{q},$$

It is plain that this fiction is certainly not realized in any mechanical problem, but it is only a problem that is much easier to treat mathematically, and which immediately becomes the problem to solve if p and q are allowed to become infinite.

Although this treatment of the problem appears too abstract, also it very rapidly leads to the solution of the problem, and if we think that all infinite quantities in nature mean nothing else than going beyond a bound, so the infinite variety of the velocities, that each molecule is capable of taking, can be the limiting case reached when every molecule can take an always larger number of velocities.

We want therefore, temporarily, to consider how the velocities are related to their "vis viva". Every molecule will be able to take a finite number of values of the "vis viva". For more simplicity suppose that the values of the "vis viva" that every molecule can have form an arithmetic sequence,

$$0, \varepsilon, 2\varepsilon, 3\varepsilon, \ldots, p\varepsilon$$

and we shall denote with P the largest of the possible values of p.

At a collision each of the two molecules involved will have again a velocity

$$0, \text{ or } \varepsilon, \text{ or } 2\varepsilon, \text{ etc.} \ldots, p\varepsilon,$$

and in any case the event will never cause that one of the molecules will end up with having a value of the "vis viva" which is not in the preceding sequence.

Let n be the number of molecules in our container. If we know how many molecules have a "vis viva" zero, how many ε, etc., then we say that the distribution of the "vis viva" between the molecules is given.

If at the initial time a distribution of the molecules states is given, it will in general change because of the collisions. The laws under which such changes take place have been often object of research. I immediately remark that this is not my aim, so that I shall not by any means depend on how and why a change in the distribution takes place, but rather to the probability on which we are interested, or expressing myself more precisely I will search all combinations that can be obtained by distributing $p + 1$ values of "vis viva" between n molecules, and hence examine how many of such combinations correspond to a distribution of states. This last number gives the probability of the relevant distribution of states, precisely as I said in the quoted place of my *Bemerkungen über einige Probleme der mechanischen Wärmetheorie* (p. 121), [29, #39].

Preliminarily we want to give a purely schematic version of the problem to be treated. Suppose that we have n molecules each susceptible of assuming a "vis viva"

$$0, \varepsilon, \ 2\varepsilon, \ 3\varepsilon, \dots, p\varepsilon.$$

and indeed these "vis viva" will be distributed in all possible ways between the n molecules, so that the sum of all the "vis viva" stays the same; for instance is equal to $\lambda\varepsilon = L$.

Every such way of distributing, according to which the first molecule has a given "vis viva", for instance 2ε, the second also a given one, for instance 6ε, etc., until the last molecule, will be called a "complexion", and certainly it is easy to understand that each single complexion is assigned by the sequence of numbers (obviously after division by ε) to which contribute the "vis viva" of the single molecules. We now ask which is the number B of the complexions in which w_0 molecules have "vis viva" 0, w_1 "vis viva" ε, w_2 "vis viva" 2ε, etc., $\dots$ w_p "vis viva" $p\varepsilon$.

We have said before that when is given how many molecules have zero "vis viva", how many ε etc., then the distribution of the states among the molecules is given; we can also say: the number B gives us how many complexions express a distribution of states in which w_0 molecules have zero "vis viva", w_1 "vis viva" ε, etc., or it gives us the probability of every distribution of states. Let us divide in fact the number B by the total number of all possible complexions and we get in this way the probability of such state.

It does not follow from here that the distribution of the states gives which are the molecules which have a given "vis viva", but only how many they are, so we can deduce how many (w_0) have zero "vis viva", and how many (w_1) among them have one unit of "vis viva" ε, etc. All of those values zero, one, etc., we shall call elements of the distribution of the states.

(p. 170) .... (p. 175)

We shall first treat the determination of the number denoted above B for each given distribution of states, i.e. the permutability of such distribution of states. Then denote by J the sum of the permutabilities of all possible distributions of states, and hence the ratio $\frac{B}{J}$ gives immediately the probability of the distribution that from now on we shall always denote W.

We also want right away to calculate the permutability B of the distributions of states characterized by w_0 molecules with zero "vis viva", w_1 with "vis viva" ε etc. Hence evidently

$$w_0 + w_1 + w_2 + \cdots + w_p = n \qquad (1)$$

$$w_1 + 2w_2 + 3w_3 + \cdots + pw_p = \lambda, \qquad (2)$$

and then n will be the total number of molecules and $\lambda\varepsilon = L$ their "vis viva".

We shall write the above defined distribution of states with the method described, so we consider a complexion with w_0 molecules with zero "vis viva", w_1 with "vis viva" unitary etc. We know that the number of permutations of the elements of this complexion, with in total n elements distributed so that among them w_0 are equal between them, and so w_1 are equal between them .... The number of such complexions is known to be[55]

$$B = \frac{n!}{(w_0)! \, (w_1)! \cdots} \qquad (3)$$

The most probable distribution of states will be realized for those choices of the values of $w_0, w_1, \ldots$ for which B is maximal and quantities $w_0, w_1, \ldots$ are at the same time constrained by the conditions (1) and (2). The denominator of B is a product and therefore it will be better to search for the minimum of its logarithm, i.e. the minimum of

$$M = \ell[(w_0)!] + \ell[(w_1)!] + \cdots \qquad (4)$$

where ℓ denotes the natural logarithm.

....

[*Follows the discussion of this simple case in which the energy levels are not degenerate (i.e. this is essentially a 1-dimensional case) ending with the Maxwell distribution of the velocities. In Sect. II, p. 186, B. goes on to consider the case of 2-dimensional and of 3-dimensional cells (of sides da, db or da, db, dc) in the space of the velocities (to be able to take degeneracy into account), treating first the discrete case and then taking the continuum limit: getting the canonical distribution. Sect. III (p. 198) deals with the case of polyatomic molecules and external forces.*

In Sect. IV (p. 204), concludes that it is possible to define and count in other ways the states of the system and discusses one of them.

An accurate analysis of this paper, together with [10], is in [8] where two ways of computing the distribution when the energy levels are discrete are discussed pointing out that unless the continuum limit, as considered by Boltzmann in the two papers, was taken would lead to a distribution of Bose-Einstein type or of Maxwell-Boltzmann type: see the final comment in Sect. 6.2 and also [12, Sects. (2.2), (2.6)].

---

[55] Here particles are considered distinguishable and the total number of complexions is P^n.

In Sect. V, p. 215, the link of the probability distributions found in the previous sections with entropy is discussed, dealing with examples like the expansion of a gas in a half empty container; the example of the barometric formula for a gas is also discussed. On p. 218 the following celebrated statement is made (in italics) about "permutability" (i.e. number of ways in which a given (positions-velocities) distribution can be achieved) and is illustrated with the example of the expansion of a gas in a half empty container:]

Let us think of an arbitrarily given system of bodies, which undergo an arbitrary change of state, without the requirement that the initial or final state be equilibrium states; then always the measure of the permutability of all bodies involved in the transformations continually increases and can at most remain constant, until all bodies during the transformation are found with infinite approximation in thermal equilibrium.[56]

6.13 Monocyclic and Orthodic Systems: Ensembles

Partial translation and comments: L. Boltzmann, "Über die Eigenshaften monozyklischer und anderer damit verwandter Systeme", (1884), Wissenschaftliche Abhandlungen, ed. F. Hasenöhrl, **3,** *122–152, #73, [28].* [57]

The most complete proof of the second main theorem is manifestly based on the remark that, for each given mechanical system, equations that are analogous to equations of the theory of heat hold. [italics added]
Since on the one hand it is evident that the proposition, in this generality, cannot be valid and on the other hand, because of our scarce knowledge of the so called atoms, we cannot establish the exact mechanical properties with which the thermal motion manifests itself, the task arises to search in which cases and up to which point the equations of mechanics are analogous to the ones of the theory of heat. We should not refrain to list the mechanical systems with behavior congruent to the one of the solid bodies, rather than to look for all systems for which it is possible to establish stronger or weaker analogies with warm bodies. The question has been posed in this form by Hrn. von Helmoltz[58] and I have aimed, in what follows, to treat a very special case, before proceeding to general propositions, of the analogy that he discovered between the thermodynamic behavior and that of systems, that he calls monocyclic, and to follow the propositions, to which I will refer, of the mechanical theory of heat intimately related to monocyclic systems.[59]

---

[56] After the last word appears in parenthesis and still in italics *(reversible transformations),* which seems to mean "*or performing reversible transformations*".

[57] The first three paragraphs have been printed almost unchanged in Wien, Ber, **90,** p. 231, 1884;
...

[58] Berl. Ber, 6 and 27 March 1884.

[59] NoA: A very general example of monocyclic system is offered by a current without resistance (see Maxwell, "Treatise on electricity", 579–580, where x and y represent the v. Helmholtzian p_a and p_b).

§1

Let a point mass move, according to Newton's law of gravitation, around a fixed central fixed body O, on an elliptic trajectory. Motion is not in this case monocyclic; but it can be made such with a trick, that I already introduced in the first Section of my work *"Einige allgemeine sätze über Wärmegleichgewicht"*[60] and that also Maxwell[61] has again followed.

Imagine that the full elliptic trajectory is filled with mass, which at every point has a density (of mass per unit length) such that, while time elapses, density in each point of the trajectory remains unchanged. As it would be a Saturn ring thought as a homogeneous flow or as a homogeneous swarm of solid bodies so that, fixed one of the rings among the different possible ones a stationary motion would be obtained. The external force can accelerate the motion or change its eccentricity; this can be obtained increasing or diminishing slowly the mass of the central body, so that external work will be performed on the ring which, by the increase or diminution of the central body mass, in general is not accelerated nor decelerated in the same way. This simple example is treated in my work *Bemerkungen über einige Probleme der mechanischen Wärmetheorie*[62] where in Sect. 3 are derived formulae in which now we must set $b = 0$ and for m we must intend the total mass of the considered ring (there, by the way, the appropriate value of the work is given with wrong signs). I denote always by Φ the total potential energy, with L the total "vis viva" of the system, by dQ the work generated by the increase of the internal motion which, as Hrn. v. Helmholtz. I assume that the external forces always undergo an infinitesimal variation of their value and that are immediately capable to bring back the motion into a stationary state.[63] Let then a/r^2 be the total attraction force that the central body exercises on the mass of the ring if it is at a distance r, let C be an unspecified global constant then it is, [*for more details see Appendix D*],

$$\Phi = C - 2L, \quad dQ = -dL + 2L\frac{da}{a} = L\, d \log \frac{a^2}{L};$$

---

[60] NoA: Wiener. Berl. **63**, 1871, [25, #18], [*see also Sect.* 6.9].

[61] NoA: Cambridge Phil. Trans. **12**, III, 1879 (see also Wiedemanns Beiblätter, **5**, 403, 1881).

[62] NoA: Wien, Ber. **75**, [*see Appendix D and Sect.* 6.11]. See also Clausius, Pogg. Ann. **142**, 433; Math. Ann. von Clebsch, **4**, 232, **6**, 390, Nachricht. d. Gött. Gesellsch. Jahrg. 1871 and 1871; Pogg. nn. **150**, 106, and Ergängzungs, **7**, 215.

[63] The "direct" increase of the internal motion is the amount of work done on the system by the internal and external forces (which in modern language is the variation of the internal energy) summed to the work dW done by the system on the outside: $dQ = dU + dW$; which would be 0 if the system did not absorb heat. If the potential energy W due to the external forces depends on a parameter a then the variation of W changed in sign, $-\partial_a W_a da$, or better its average value in time, is the work that the system does on the outside due to the only variation of W while the energy of the system varies also because the motion changes because of the variation of a. Therefore here it has to be interpreted as the average value of the derivative of the potential energy $W = -a/r$ with respect to a times the variation da of a. Notice that in the Keplerian motion it is $2L = a/r$ and therefore $\langle -\partial_a W\, da \rangle = \langle da/r \rangle = \langle 2L da/a \rangle$, furthermore the total energy is $L + \Phi = -L$ up to a constant and hence $dU = -dL$.

hence L is also the integrating factor of dQ, and the consequent value of the entropy S is $\log a^2/L$ and the consequent value of the characteristic function [*free energy*] is

$$K = \Phi + L - LS = C - L - LS = C - L - L \log \frac{a^2}{L}$$

Let $2L\frac{\sqrt{L}}{a} = q$, $\frac{a}{\sqrt{L}} = s$, so that it is $dQ = qds$ and the characteristic function becomes $H = \Phi + L - qs = C = 3L$ and it is immediately seen that

$$\left(\frac{\partial K}{\partial a}\right)_L = \left(\frac{\partial H}{\partial a}\right)_q = -A$$

is the gain deriving from the increase of a, in the sense that Ada is how much of the internal motion can be converted into work when a changes from a to $a + da$. Thus it is

$$\left(\frac{\partial K}{\partial L}\right)_a = -S, \quad \left(\frac{\partial H}{\partial q}\right)_a = -s$$

The analogy with the monocyclic systems of Hrn. v. Helmholtz' with a single velocity q is also transparent.

....

§2

[*Detailed comparison with the work of Helmoltz follows, in particular the calculation reported in Sect. 1.4, p. 11, is explained. The notion of monocyclic system, i.e. a system whose orbits are all periodic, allows to regard each orbit as a stationary state of the system and, extending Helmotz' conception, as the collection of all its points which is called a "monode", each being a representative of the considered state.*

Varying the parameters of the orbit the state changes (i.e. the orbit changes). Some examples of monocyclic systems are worked out: for all of their periodic orbits is defined the amount of heat dQ that the system receives in a transformation ("work to increase the internal motion" or "infinitesimal direct increment of the internal motion", i.e. the heat acquired by a warm body), and the amount of work dW that the system does against external forces as well as the average kinetic energy L; in all of them it is shown that $\frac{dQ}{L}$ is an exact differential. For a collection of stationary motions of a system this generates the definitions (p. 129–130):]

I would permit myself to call systems whose motion is stationary in this sense with the name *monodes*.[64] They will therefore be characterized by the property that in every point persists unaltered a motion, not function of time as long as the external forces stay unchanged, and also in no point and in any region or through any surface mass or "vis viva" enters from the outside or goes out. If the "vis viva" is the integrating denominator of the differential dQ, which directly gives the work to

---

[64] NoA: With the name "stationary" Hrn. Clausius would denote every motion whose coordinates remain always within a bounded region.

increase the internal motion,[65] then I will say that the such systems are *orthodes*. [*Etymologies: monode*= μόνος + εἶδος = unique + aspect; orthode = ὀρθός + εἶδος = right + aspect.]

...

§3.

After these introductory examples I shall pass to a very general case. Consider an arbitrary system, whose state is characterized by arbitrary coordinates $p_1, p_2, \ldots, p_g$; and let the corresponding momenta be $r_1, r_2, \ldots, r_g$. For brevity we shall denote the coordinates by p_g and the momenta by r_g. Let the internal and external forces be also assigned; the first be conservative. Let ψ be the "vis viva" and χ the potential energy of the system, then also χ is a function of the p_g and ψ is a homogeneous function of second degree of the r_g whose coefficients can depend on the p_g. The arbitrary constant appearing in χ will be determined so that χ vanishes at infinite distance of all masses of the system or even at excessive separation of their positions. We shall not adopt the restrictive hypothesis that certain coordinates of the system be constrained to assigned values, hence also the external forces will not be characterized other than by their almost constancy on slowly varying parameters. The more so the slow variability of the external forces will have to be taken into account either because χ will become an entirely new function of the coordinates p_g, or because some constants that appear in χ, which we shall indicate by p_a, vary slowly.

1. We now imagine to have a large number N of such systems, of exactly identical nature; each system absolutely independent from all the others.[66] The number of such systems whose coordinates and momenta are between the limits p_1 and $p_1 + dp_1$, p_2 and $p_2 + dp_2 \ldots$, r_g and $r_g + dr_g$ be

$$dN = N e^{-h(\chi+\psi)} \frac{\sqrt{\Delta}\, d\sigma\, d\tau}{\int\int e^{-h(\chi+\psi)}\sqrt{\Delta}\, d\sigma\, d\tau},$$

where $d\sigma = \Delta^{-\frac{1}{2}} dp_1 dp_2 \ldots dp_g$, $d\tau = dr_1 dr_2 \ldots dr_g$ (for the meaning of Δ see Maxwell *loc. cit* p. 556).[67]

The integral must be extended to all possible values of the coordinates and momenta. The totality of these systems constitutes a *monode* in the sense of the definition given above (see, here, especially Maxwell *loc. cit*) and I will call

---

[65] in an infinitesimal transformation, i.e. the variation of the internal energy summed to the work that the system performs on the outside, which defines the heat received by the system. The notion of *monode* and *orthode* will be made more clear in the next subsection 3.

[66] In modern language this is an ensemble: it is the generalization of the Saturn ring of Sect. 1: each representative system is like a stone in a Saturn ring. It is a way to realize all states of motion of the same system. Their collection does not change in time and keeps the same aspect, if the collection is stationary, i.e. is a "monode".

[67] In general the kinetic energy is a quadratic form in the r_g and then Δ is its determinant: In the formula for dN the $\sqrt{\Delta}$ should not be there. (if it is in $d\sigma$)

this species of monodes with the name *Holodes*. [*Etymology:* ὅλος + εἶδος *or*
"*global*"+ "*aspect*".[68]]

Each system I will call an *element* of the holode.[69] The total "vis viva" of a
holode is[70]

$$L = \frac{Ng}{2h}.$$

Its potential energy $\varPhi$ equals N times the average value $\overline{\chi}$ of χ, i.e. :

$$\varPhi = N\frac{\int \chi\, e^{-h\chi}\, d\sigma}{\int e^{-h\chi}\, d\sigma}.$$

The coordinates $p_{\mathbf{g}}$ correspond therefore to the v. Helmholtzian $p_{\mathbf{b}}$, which appear in
the "vis viva" ψ and potential energy χ of an element. The intensity of the motion
of the entire ergode[71] and hence also L and $\varPhi$ depend now on h and on $p_{\mathbf{a}}$, as for
Hrn. v. Helmholtz on $q_{\mathbf{b}}$ and $p_{\mathbf{a}}$.

The work provided for a direct increase, see p. 184, of internal motion is:

$$\delta Q = \delta\varPhi + \delta L - N\frac{\int \delta\chi\, e^{-h\chi}\, d\sigma}{\int e^{-h\chi}\, d\sigma}$$

---

[68] Probably because the canonical distribution deals with all possible states of the system and does
not select quantities like the energy or other constants of motion.

[69] Hence a monode is a collection of identical systems called *elements* of the monode, that can be
identified with the points of the phase space. The points are permuted by the time evolution but the
number of them near a phase space volume element remains the same in time, i.e. the distribution
of such points is stationary and keeps the same "unique aspect".

The just given canonical distributions are particular kinds of monodes called holodes. An holode
is therefore an element of a species ("gattung"), in the sense of collection, of monodes that are
identified with the canonical distributions [of a given mechanical system]. A holode will be identified
with a state of thermodynamic equilibrium, because it will be shown to have correct properties. For
its successive use an holode will be intended as a statistical ensemble, i.e. the family of probability
distributions, consisting in the canonical distributions of a given mechanical system: in fact the
object of study will be the properties of the averages of the observables in the holodes as the
parameters that define them change, like h (now β, the inverse temperature) or the volume of the
container.

[70] In the case of a gas the number g must be thought as the Avogadro's number times the number
of moles, while the number N is a number *much larger* and equal to the number of cells which can
be thought to constitute the phase space. Its introduction is not necessary, and Boltzmann already
in 1871 had treated canonical and microcanonical distributions with $N = 1$: it seems that the
introduction of the N copies, adopted later also by Gibbs, intervenes for ease of comparison of the
work of v. Helmholtz with the preceding theory of 1871. Remark that B. accurately avoids to say
too explicitly that the work of v. Helmholz is, as a matter of fact, a different and particular version
of his preceding work. Perhaps this caution is explained by caution of Boltzmann who in 1884 was
thinking to move to Berlin, solicited and supported by v. Helmholtz. We also have to say that the
works of 1884 by v. Helmholtz became an occasion for B. to review and systematize his own works
on the heat theorem which, after the present work, took up the form and the generality that we still
use today as "theory of the statistical ensembles".

[71] This is a typo as it should be holode: the notion of ergode is introduced later in this work.

(see here my work[72] *Analytischer Beweis des zweiten Hauptsatzes der mechanischen Wärmetheorie aus den Sätzen über das Gleichgewicht des lebendigen Kraft*), [27, #19], [*see also Sect.* 6.10]. The amount of internal motion generated by the external work, when the parameter p_a varies[73] by δp_a, is therefore $-P\delta p_a$, with

$$-P = \frac{N \int \frac{\partial \chi}{\partial p_a} e^{-h\chi} d\sigma}{\int e^{-h\chi} d\sigma}$$

The "vis viva" L is the integrating denominator of δQ: all holodes are therefore orthodic, and must therefore also provide us with thermodynamic analogies. Indeed let[74]

$$s = \frac{1}{\sqrt{h}} \left(\int e^{-h\chi} d\sigma \right)^{\frac{1}{g}} e^{\frac{h\chi}{g}} = \sqrt{\frac{2L}{Ng}} \left(\int e^{-h\chi} d\sigma \right)^{\frac{1}{g}} e^{\frac{\Phi}{2L}},$$

$$q = \frac{2L}{s}, \quad K = \Phi + L - 2L \log s, \quad H = \Phi - L,$$

[*the intermediate expression for s is not right and instead of χ in the exponential should have the average $\frac{\Phi}{N}$ of χ.*]

2. Let again be given many (N) systems of the kind considered at the beginning of the above sections; let all be constrained by the constraints

$$\varphi_1 = a_1, \ \varphi_2 = a_2, \ldots, \varphi_k = a_k.$$

These relations must also, in any case, be integrals of the equations of motion. And suppose that there are no other integrals. Let dN be the number of systems whose coordinates and momenta are between p_1 e $p_1 + dp_1$, p_2 and $p_2 + dp_2$, ... r_g

---

[72] NoA: Wien. Ber., **63**, 1871, formula (17).

[73] Here we see that Boltzmann considers among the parameters p_a coordinates such as the dimensions of the molecules container: this is not explicitly said but it is often used in the following.

[74] Here the argument in the original relies to some extent on the earlier paragraphs: a self contained check is therefore reported in this footnote for ease of the reader:

$$F \overset{def}{=} -h^{-1} \log \int e^{-h(\chi+\varrho)} \overset{def}{=} -h^{-1} \log Z(\beta, p_a), \quad T = h^{-1}$$

and remark that

$$dF = (h^{-2} \log Z + h^{-1}(\Phi + L))dh - h^{-1}\partial_{p_a} \log Z \, dp_a$$

Define S via $F \overset{def}{=} U - TS$ and $U = \Phi + L$ then

$$dF = dU - TdS - SdT = -\frac{dT}{T}(-(U - TS) + U) + Pdp_a$$

hence $dU - TdS - SdT = -\frac{dT}{T}TS - Pdp_a$, i.e. $TdS = dU + Pdp_a$ and the factor $T^{-1} = h$ is the integrating factor for $dQ \overset{def}{=} dU + Pdp_a$, see [12, Eq. (2.2.7)].

and $r_g + dr_g$. Naturally here the differentials of the coordinates or momenta that we imagine determined by the equations $\varphi_1 = a_1, \ldots$ will be missing. These coordinates or momenta missing be $p_c, p_d, \ldots, r_f$; their number be k. Then if

$$dN = \frac{\dfrac{N dp_1 dp_2 \cdots dr_g}{\sum \pm \frac{\partial \varphi_1}{\partial p_c} \cdots \frac{\partial \varphi_k}{\partial r_f}}}{\displaystyle\int \int \cdots \frac{dp_1 dp_2 \cdots dr_g}{\sum \pm \frac{\partial \varphi_1}{\partial p_c} \cdot \frac{\partial \varphi_2}{\partial p_d} \cdots \frac{\partial \varphi_k}{\partial r_f}}}$$

the totality of the N systems will constitute a monode, which is defined by the relations $\varphi_1 = a_1, \ldots$. The quantities a can be either constant or subject to slow variations. The functions φ in general may change form through the variation of the p_a, always slowly. Each single system is again called element.

Monodes that are constrained through the only value of the equation of the "vis viva"[75] will be called *ergodes*, while if also other quantities are fixed will be called *subergodes*. The ergodes are therefore defined by

$$dN = \frac{N\, dp_1 dp_2 \cdots dp_g dr_1 \cdots dr_{g-1} \dfrac{\partial \psi}{\partial r_g}}{\displaystyle\int \int \frac{dp_1 dp_2 \cdots dp_g dr_1 \cdots dr_{g-1}}{\frac{\partial \psi}{\partial r_g}}}$$

Hence for the ergodes there is a φ, equal for all the identical systems and which stays, during the motion, equal to the constant energy of each system $\chi + \psi = \frac{1}{N}(\Phi + L)$. Let us set again $\Delta^{-\frac{1}{2}} dp_1 dp_2 \cdots dp_g = d\sigma$, and then (see the works cited above by me and by Maxwell):

$$\Phi = N \frac{\int \chi \psi^{\frac{q}{2}-1} d\sigma}{\int \psi^{\frac{q}{2}-1} d\sigma}, \qquad L = N \frac{\int \psi^{\frac{q}{2}} d\sigma}{\int \psi^{\frac{q}{2}-1} d\sigma},$$

$$\delta Q = N \frac{\int \delta\psi \psi^{\frac{q}{2}-1} d\sigma}{\int \psi^{\frac{q}{2}-1} d\sigma} = \delta(\Phi + L) - N \frac{\int \delta\chi \psi^{\frac{q}{2}-1} d\sigma}{\int \psi^{\frac{q}{2}-1} d\sigma},$$

L is again the integrating factor of δQ,[76] and the entropy thus generated is $\log(\int \psi^{\frac{q}{2}} d\sigma)^{\frac{2}{q}}$, while it will also be $\delta Q = q\, \delta s$ if it will be set:

---

[75] The equation of the "vis viva" is the energy conservation $\varphi = a$ with $\varphi = \psi + \chi$, if the forces are conservative with potential χ.

[76] The (elementary) integrations on the variables r_g with the constraint $\psi + \chi = a$ have been explicitly performed: and the factor $\psi^{\frac{q}{2}-1}$ is obtained, in modern terms, performing the integration $\int \delta(\chi - (a - \psi)) dr_g$ and in the formulae ψ has to be interpreted as $\sqrt{a - \chi}$, as already in the work of 1871.

$$s = \left(\int \psi^{\frac{q}{2}} d\sigma \right)^{\frac{1}{q}}, \quad q = \frac{2L}{s}.$$

Together with the last entropy definition also the characteristic function $\Phi - L$ is generated. The external force in the direction of the parameter p_a is in each system

$$-P = \frac{\int \frac{\partial \chi}{\partial p_a} \psi^{\frac{q}{2}-1} d\sigma}{\int \psi^{\frac{q}{2}-1} d\sigma}.$$

Among the infinite variety of the subergodes I consider those in which for all systems not only is fixed the value of the equation of "vis viva" [*value of the energy*] but also the three components of the angular momentum. I will call such systems *planodes*. Some property of such systems has been studied by Maxwell, *loc. cit.*. Here I mention only that in general they are not orthodic.

The nature of an element of the ergode is determined by the parameters $p_{\mathbf{a}}$.[77] Since every element of the ergode is an aggregate of point masses and the number of such parameters $p_{\mathbf{a}}$ is smaller than the number of all Cartesian coordinates of all point masses of an element, so such $p_{\mathbf{a}}$ will always be fixed as functions of these Cartesian coordinates, which during the global motion and the preceding developments remain valid provided these functions stay constant as the "vis viva" increases or decreases.[78] If there was variability of the potential energy for reasons other than because of the mentioned parameters[79], there would also be a slow variability of these functions, which play the role of the v. Helmholtzian $p_{\mathbf{a}}$, and which here we leave as denoted $p_{\mathbf{a}}$ to include the equations that I obtained previously and the v. Helmholtzians ones.[80] And here is the place of a few considerations.

The formulae, that follow from formulae (18) of my work "*Analytischer Beweis des zweiten Hauptsatzes der mechanischen Wärmetheorie aus den Sätzen über das Gleichgewicht des lebendigen Kraft*", 1871, [27, #19], see also Sect. 6.10, have not been developed in their full generality, in fact there I first speak of *a* system, which goes through all possible configurations compatible with the principle of the "vis viva" and secondly I only use Cartesian coordinates; and certainly this is seen in the very often quoted work of Maxwell "*On the theorem of Boltzmann ...etc.*", [33]. This being said these formulae must also hold for ergodes in any and no matter how generalized coordinates. Let these be, for an element of an ergode, $p_1, p_2, \ldots, p_g,$

---

[77] In the text, however, there is p_b: typo?

[78] Among the $p_{\mathbf{a}}$ we must include the container dimensions a, b, c, for instance: they are functions of the Cartesian coordinates which, however, are *trivial constant functions*. The mention of the variability of the "vis viva" means that the quadratic form of the "vis viva" must not depend on the $p_{\mathbf{a}}$.

[79] I interpret: the parameters controlling the external forces; and the "others" can be the coupling constants between the particles.

[80] It seems that B. wants to say that between the $p_{\mathbf{a}}$ can be included also possible coupling constants that are allowed to change: this permits a wider generality.

and thus it is[81]

$$\frac{d\mathcal{N}}{N} = \frac{\Delta^{-\frac{1}{2}}\psi^{\frac{g}{2}-1}dp_1 \cdots dp_g}{\int\int \cdots \Delta^{-\frac{1}{2}}\psi^{\frac{g}{2}-1}dp_1 \cdots dp_g},$$

where N is the total number of systems of the ergode, $d\mathcal{N}$ the number of such systems whose coordinates are between p_1 and $p_1 + dp_1$, p_2 and $p_2 + dp_2 \ldots p_g$ and $p_g + dp_g$. Let here ψ be the form of the "vis viva" of a system. The relation at the nine-th place of the quoted formula (18) yields.

....

[(p. 136): Follows the argument that shows that the results do not depend on the system of coordinates. Then a few examples are worked out, starting with the case considered by Helmholtz, essentially one dimensional: ergodes "with only one fast variable" (p. 137):]

The monocyclic systems of Hrn. v. Helmholtz with a single velocity are not different from the ergodes with a single rapidly varying coordinate, that will be called p_g which, at difference with respect to the v. Helmhotzian p_a, is not subject to the condition, present in his treatment, of varying very slowly.

Hence the preceding formula is valid equally for monocyclic systems with a unique velocity and for warm bodies, and therefore it has been clarified the mentioned analogy of Hrn. v. Helmholtz between rotatory motions and ideal gases (see Crelles Journal, 07, p. 123,; Berl. Ber. p. 170).

Consider a single system, whose fast variables are all related to the equation of the "vis viva" (*isomonode*), therefore it is $N = 1$, $\psi = L$. For a rotating solid body it is $g = 1$. Let p be the position angle θ and $\omega = \frac{d\theta}{dt}$, then

$$\psi = L = \frac{T\omega^2}{2} = \frac{r^2}{2T}, \quad r = T\omega;$$

where $\Delta = \frac{1}{T}$, and always $\Delta = \mu_1\mu_2\ldots$, while L has the form

$$\frac{\mu_1 r_1^2 + \mu_2 r_2^2 + \cdots}{2}.$$

T is the inertia moment; $\int\int \ldots dp_1 dp_2 \ldots$ is reduced to $\int dp = 2\pi$ and can be treated likewise, so that the preceding general formula becomes $\delta Q = L\,\delta\log(TL)$. If a single mass m rotates at distance ϱ from the axis, we can set p equal to the arc s of the point where the mass is located; then it will be:

---

[81] The dots the follow the double integral signs cannot be understood; perhaps this is an error repeated more times.

$$\psi = L = \frac{mv^2}{2} = \frac{r^2}{2m}, \quad r = mv, \quad \Delta = \frac{1}{m},$$

$$= \int \int \ldots dp_1 dp_2 \ldots = \int dp = 2\pi\varrho,$$

where $v = \frac{ds}{dt}$; therefore the preceding formula follows $\delta Q = L\,\delta \log(mL\varrho^2)$. For an ideal gas of monoatomic molecules it is $N = 1$, $\psi = L$; $p_1, p_2, \ldots, p_g$ are the Cartesian coordinates $x_1, y_1, \ldots, z_n$ of the molecules, hence $g = 3n$, where n is the total number of molecules, v is the volume of the gas and $\int \int \ldots dp_1 dp_2 \ldots$ is v^n, Δ is constant, as long as the number of molecules stays constant; hence the preceding general formula $\delta Q = L\,\delta \log(Lv^{\frac{2}{3}})$ follows, which again is the correct value because in this case the ratio of the specific heats is $\frac{5}{3}$.

....

[p. 138: Follow more examples. The concluding remark (p. 140) in Sect. 3 is of particular interest as it stresses that the generality of the analysis of holodes and ergodes is dependent on the ergodic hypothesis. However the final claim, below, that it applies to polycyclic systems may seem contradictory: it probably refers to the conception of Boltzmann and Clausius that in a system with many degrees of freedom all coordinates had synchronous (B.) or asynchronous (C.) periodic motions, see Sects. 6.1, 6.4, 6.5.]

...

The general formulae so far used apply naturally both to the monocyclic systems and to the polycyclic ones, as long as they are ergodic, and therefore I omit to increase further the number of examples.

[Sects. 4, 5, 6 are not translated.]

6.14 Maxwell 1866

Commented summary of: *On the dynamical theory of gases*, di J.C. Maxwell, *Philosophical Transactions*, **157**, 49–88, 1867, [6, XXVIII, vol. 2].

The statement *Indeed the properties of a body supposed to be a uniform plenum may be affirmed dogmatically, but cannot be explained mathematically*, [6, p. 49], is in the overture of the second main work of Maxwell on kinetic theory.[82]

---

[82] Page numbers refer to the original: the page number of the collected papers, [6], are obtained by subtracting 23.

6.14.1 Friction Phenomenology

The first new statement is about an experiment that he performed on viscosity of almost ideal gases: yielding the result that at pressure p viscosity is independent of density ϱ and proportional to temperature, i.e. to $\frac{p}{\varrho}$. This is shown to be possible if the collisions frequency is also temperature independent or, equivalently, if the collision cross section is independent of the relative speed.

Hence an interaction potential at distance $|x|$ *proportional to* $|x|^{-4}$ is interesting and it might be a key case. The argument on which the conclusion is based is interesting and, as far as I can see, quite an unusual introduction of *viscosity*.

Imagine a displacement S of a body (think of a parallelepiped of sides $a, b = c$ (here $b = c$ for simplicity, *in the original b, c are not set equal*) containing a gas, or a cube of metal and let $S = \delta a$ in the case of stretching or, in the case of deformation *at constant volume*, $S = \delta b^2$). The displacement generates a "stress" F, i.e. a force opposing the displacement that is imagined proportional to S via an elasticity constant E.

If S varies in time then the force varies in time as $\dot{F} = E\dot{S}$: here arises a difference between the iron parallelepiped and the gas one; the iron keeps being stressed as long as $\dot{S}$ stays fixed or varies slowly. On the other hand the gas in the parallelepiped undergoes a stress, i.e. a difference in pressure in the different directions, which goes away, *after some material dependent time*, even if $\dot{S}$ is fixed, because of the equalizing effect of the collisions.

In the gas case (or in general viscous cases) the rate of disappearance of the stress can be imagined proportional to F and F will follow the equation $\dot{F} = E\dot{S} - \frac{F}{\tau}$ where τ is a constant with the dimension of a time as shewn by the solution of the equation $F(t) = E\tau\dot{S} + conste^{-\frac{t}{\tau}}$ and a constant displacement rate results for $t \gg \tau$ in a constant force $E\tau\dot{S}$ therefore $E\tau$ *has the meaning of a viscosity*.

The continuation of the argument is difficult to understand exactly; *my* interpretation is that the variation of the dimension a accompanied by a compensating variation of $b = c$ so that $ab^2 = const$ decreases the frequency of collision with the wall orthogonal to the a-direction by $-\frac{\delta a}{a}$, relative to the initial frequency of collision which is proportional to the pressure; hence the pressure in the direction a undergoes a relative diminution $\frac{\delta p}{p} = -\frac{\delta a}{a} = 2\frac{db}{b}$ [*Maxwell gives instead* $\frac{\delta p}{p} = -2\frac{\delta a}{a}$], and this implies that the force that is generated is $dF = \delta(b^2p)$: a force ("stress" in the above context) called in [6] "linear elasticity" or "rigidity" for changes of form. It disappears upon re-establishment of equal pressure in all directions due to collisions [6]. So that $dF = 2bpdb + b^2dp = 4pbdb = 2pdb^2$ and the rigidity coefficient is $E = 2p$ [*Maxwell obtains p*].

Hence the elasticity constant E is $[2]p$ and by the above general argument the viscosity is $p\tau$: the experiment quoted by Maxwell yields a viscosity proportional to the temperature and independent of the density ϱ, i.e. $p\tau$ proportional to $\frac{p}{\varrho}$ so that τ is temperature independent and inversely proportional to the density.

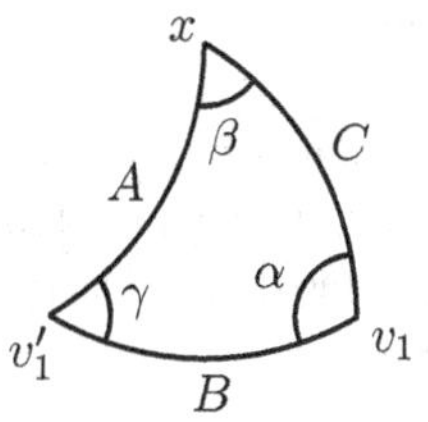

$$\gamma = C, \quad \varphi = \alpha, \quad 2\theta = B, \quad \theta = A$$

$$\cos A = \cos B \cos C + \sin B \sin C \cos \alpha$$

$$\cos \theta' = \cos 2\theta \cos \gamma + \sin 2\theta \sin \gamma \cos \varphi$$

$$v_1' = \frac{M_1 v_1 + M_2 v_2}{M_1 + M_2} + \frac{M_2}{M_1 + M_2}(v_1' - v_2')$$

Fig. 6.2 Spherical triangle for momentum conservation

In an earlier work he had considered the case of a hard balls gas, [2, XX] concluding density independence, proportionality of viscosity to $\sqrt{T}$ and to the inverse r^{-2} of the balls diameter.

6.14.2 Collision Kinematics

The technical work starts with the derivation of the collision kinematics. Calling $v_i = (\xi_i, \eta_i, \zeta_i)$ the velocities of two particles of masses M_i he writes the outcome $v_i' = (\xi_i', \eta_i', \zeta_i')$ of a collision in which particle 1 is deflected by an angle 2θ as:

$$\xi_1' = \xi_1 + \frac{M_2}{M_1 + M_2}\left\{(\xi_2 - \xi_1)2(\sin\theta)^2 + \sqrt{(\eta_2 - \eta_1)^2 + (\zeta_2 - \zeta_1)^2}\,\sin 2\theta \cos\varphi\right\}$$

which follows by considering the spherical triangle with vertices on the x-axis, v_1, v_1', and $\cos\gamma = \frac{(\xi_1 - \xi_2)}{|v_1 - v_2|}$, $\sin\gamma = \frac{\sqrt{(\eta_1 - \eta_2)^2 + (\zeta_1 - \zeta_2)^2}}{|v_1 - v_2|}$, with $|v_1' - v_2'| = |v_1 - v_2|$, [6, p. 59] (Fig. 6.2).

6.14.3 Observables Variation upon Collision

The next step is to evaluate the amount of a quantity $Q = Q(v_1)$ contained per unit (space) volume in a velocity volume element $d\xi_1\, d\eta_1\, d\zeta_1$ around $v_1 = (\xi_1, \eta_1, \zeta_1)$: this is $Q(v_1)dN_1$, with $dN_1 = f(v_1)d^3 v_1$. Collisions at impact parameter b and relative speed $V = |v_1 - v_2|$ occur at rate $dN_1 V b db d\varphi dN_2$ per unit volume. They change the velocity of particle 1 into v_1' hence change the total amount of Q per unit volume by

$$(Q' - Q)\, V b db d\varphi dN_1 dN_2. \tag{2.1}$$

Notice that here independence is assumed for the particles distributions as in the later Boltzmann's *stosszahlansatz*.

Then it is possible to express the variation of the amount of Q per unit volume. Maxwell considers "only" $Q = \xi_1, \xi_1^2, \xi_1(\xi_1^2 + \eta_1^2 + \zeta_1^2)$ and integrates $(Q'-Q)\,Vbdbd\varphi dN_1 dN_2$ over $\varphi \in [0, 2\pi]$, on $b \in [0, \infty)$ and then over dN_1, dN_2. The φ integral yields, setting $s_\alpha =^{def} \sin\alpha$, $c_\alpha =^{def} \cos\alpha$,

$$(\alpha): \int_0^{2\pi} d\varphi(\xi_1' - \xi_1)d\varphi = \frac{M_2}{M_1 + M_2}(\xi_2 - \xi_1)4\pi s_\theta^2$$

$$(\beta): \int_0^{2\pi} (\xi_1'^2 - \xi_1^2)d\varphi = \frac{M_2}{(M_1 + M_2)^2}\Big\{(\xi_2 - \xi_1)(M_1\xi_1 + M_2\xi_2)8\pi s_\theta^2$$
$$+ M_2\Big((\eta_2 - \eta_1)^2 + (\zeta_2 - \zeta_1)^2 - 2(\xi_2 - \xi_1)^2\Big)\pi s_{2\theta}^2\Big\}$$

$$(\beta'): \int_0^{2\pi} (\xi_1'\eta_1' - \xi_1\eta_1)d\varphi = \frac{M_2}{(M_1 + M_2)^2}\Big\{\Big(M_2\xi_2\eta_2 - M_1\xi_1\eta_1$$
$$+ \frac{1}{2}(M_1 - M_2)(\xi_1\eta_2 + \xi_2\eta_1)8\pi s_\theta^2\Big) - 3M_2\Big((\xi_2 - \xi_1)(\eta_2 - \eta_1)\pi s_{2\theta}^2\Big)\Big\}$$

$$(\gamma): \int_0^{2\pi} (\xi_1'V_1'^2 - \xi_1 V_1^2)d\varphi = \frac{M_2}{M_1 + M_2}4\pi s_\theta^2\Big\{(\xi_2 - \xi_1)V_1^2 + 2\xi_1(U - V_1^2)\Big\}$$
$$+ \Big(\frac{M_2}{M_1 + M_2}\Big)^2 \Big((8\pi s_\theta^2 - 3\pi s_{2\theta}^2)2(\xi_2 - \xi_1)(U - V_1^2)$$
$$+ (8\pi s_\theta^2 + 2\pi s_{2\theta}^2\xi_1)V^2\Big) + \Big(\frac{M_2}{M_1 + M_2}\Big)^3 (8\pi s_\theta^2 - 2\pi s_{2\theta}^2)2(\xi_2 - \xi_1)V^2$$

where $V_1^2 =^{def} (\xi_1^2 + \eta_1^2 + \zeta_1^2)$, $U =^{def} (\xi_1\xi_2 + \eta_1\eta_2 + \zeta_1\zeta_2)$, $V_2^2 =^{def} (\xi_2^2 + \eta_2^2 + \zeta_2^2)$ $V^2 =^{def} ((\xi_2 - \xi_1)^2 + (\eta_2 - \eta_1)^2 + (\zeta_2 - \zeta_1)^2)$.

If the interaction potential is $\frac{K}{|x|^{n-1}}$ the deflection θ is a function of b. Multiplying both sides by $Vbdb$ and integrating over b a linear combination with coefficients

$$B_1 = \int_0^\infty 4\pi bs_\theta^2 db = \Big(\frac{K(M_1 + M_2)}{M_1 M_2}\Big)^{\frac{2}{n-1}} V^{\frac{n-5}{n-1}} A_1$$

$$B_2 = \int_0^\infty \pi bs_{2\theta}^2 db = \Big(\frac{K(M_1 + M_2)}{M_1 M_2}\Big)^{\frac{2}{n-1}} V^{\frac{n-5}{n-1}} A_2$$

with A_1, A_2 dimensionless and expressed by a quadrature.

6.14.4 About the "Precarious Assumption"

To integrate over the velocities it is necessary to know the distributions dN_i. The only case in which the distribution has been determined in [6, p. 62] is when the momenta distribution is stationary: this was obtained in [2] under the assumption which "*may appear precarious*" that "the probability of a molecule having a velocity resolved parallel to x lying between given limits is not in any way affected by the knowledge that the molecule has a given velocity resolved parallel to y". Therefore in [6] a different analysis is performed: based on the energy conservation at collisions which replaces the independence of the distribution from the coordinates directions. The result is that (in modern notations), if the mean velocity is 0,

$$dN_1 = f(v_1)d^3v_1 = \frac{N_1}{(2\pi(M_1\beta)^{-1})^{\frac{3}{2}}}e^{-\beta\frac{M_1}{2}v_1^2}d^3v_1$$

where N_1 is the density.

Remark However the velocity distribution is not supposed, in the following, to be Maxwellian but just close to a slightly off center Maxwellian. It will be assumed that the distribution factorizes over the different particles coordinates, and over positions and momenta.

6.14.5 Balance of the Variations of Key Observables

At this point the analysis is greatly simplified if $n = 5$, [6, vol. 2, p. 65–67]. Consider the system as containing two kinds of particles. Let the symbols δ_1 and δ_2 indicate the effect produced by molecules of the first kind and second kind respectively, and δ_3 to indicate the effect of the external forces. Let $\kappa \overset{def}{=} \left(\frac{K}{M_1 M_2(M_1+M_2)}\right)^{\frac{1}{2}}$ and let $\langle\cdot\rangle$ denote the average with respect to the velocity distribution.

$(\alpha):\dfrac{\delta_2\langle\xi_1\rangle}{\delta t} = \kappa\, N_2 M_2 A_1 \langle\xi_2 - \xi_1\rangle$

$(\beta):\dfrac{\delta_2\langle\xi_1^2\rangle}{\delta t} = \kappa\,\dfrac{N_2 M_2}{(M_1 + M_2)}\Big\{2A_1\langle(\xi_2 - \xi_1)(M_1\xi_1 + M_2\xi_2)\rangle$
$\qquad\qquad + M_2 A_2\Big(\langle(\eta_2 - \eta_1)^2 + (z_2 - z_1)^2 - 2(\xi_2 - \xi_1)^2\rangle\Big)\Big\}$

$(\beta'):\dfrac{\delta_2\langle\xi_1\eta_1\rangle}{\delta t} = \kappa\,\dfrac{N_2 M_2}{(M_1 + M_2)}\Big\{A_1\Big(\langle2M_2\xi_2\eta_2 - 2M_1\xi_1\eta_1\rangle$
$\qquad\qquad + (M_1 - M_2)\langle(\xi_1\eta_2 + \xi_2\eta_1)\rangle\Big) - 3A_2 M_2\Big(\langle(\xi_2 - \xi_1)(\eta_2 - \eta_1)\rangle\Big)\Big\}$

$(\gamma):\dfrac{\delta_2\langle\xi_1 V_1\rangle}{\delta t} = \kappa N_2 M_2\Big\{A_1\langle(\xi_2 - \xi_1)V_1^2 + 2\xi_1(U - V_1^2)\rangle\Big\}$

$$+ \frac{M_2}{M_1 + M_2} \Big((2A_1 - 3A_2)2\langle(\xi_2 - \xi_1)(U - V_1^2)\rangle$$

$$+ (2A_1 + 2A_2)\langle\xi_1 V^2\rangle \Big) + \Big(\frac{M_2}{M_1 + M_2}\Big)^2 (2A_1 - 2A_2)2\langle(\xi_2 - \xi_1)V^2\rangle \Big\}$$

More general relations can be found if external forces are imagined to act on the particles. If only one species of particles is present the relations simplify, setting $M = M_1$, $N = N_1$ and $\kappa =^{def} (\frac{K}{2M^3})^{\frac{1}{2}}$, into

$$(\alpha) : \frac{\delta_1\langle\xi\rangle}{\delta t} = 0$$

$$(\beta) : \frac{\delta_1\langle\xi^2\rangle}{\delta t} = \kappa M N A_2 \Big\{ (\langle\eta^2\rangle - \langle\eta\rangle^2) + (\langle\zeta^2\rangle - \langle\zeta\rangle^2) - 2(\langle\xi^2\rangle - \langle\xi\rangle^2) \Big\}$$

$$(\beta') : \frac{\delta_1\langle\xi\eta\rangle}{\delta t} = \kappa M N 3 A_2 \Big\{ \langle\xi\rangle\langle\eta\rangle - \langle\xi\eta\rangle \Big\}$$

$$(\gamma) : \frac{\delta_1\langle\xi_1 V_1\rangle}{\delta t} = \kappa N M 3 A_2 \Big\{ \langle\xi\rangle\langle V_1^2\rangle - \langle\xi V_1^2\rangle \Big\}$$

Adding an external force with components X, Y, Z

$$(\alpha) : \frac{\delta_3\langle\xi\rangle}{\delta t} = X$$

$$(\beta) : \frac{\delta_3\langle\xi^2\rangle}{\delta t} = 2\langle\xi X\rangle$$

$$(\beta') : \frac{\delta_3\langle\xi\eta\rangle}{\delta t} = (\langle\eta X + \xi Y\rangle)$$

$$(\gamma) : \frac{\delta_2\langle\xi_1 V_1\rangle}{\delta t} = 2\langle\xi(\xi X + \eta Y + \zeta Z)\rangle + X\langle V^2\rangle$$

6.14.6 Towards the Continua

Restricting the analysis to the case of only one species the change of the averages due to collisions and to the external force is the sum of $\delta_1 + \delta_3$. Changing notation to denote the velocity $u + \xi$, $v + \eta$, $w + \zeta$, so that ξ, η, ζ have 0 average and "almost Maxwellian distribution" while u, v, w is the average velocity, and if $\varrho =^{def} NM$, see comment following Eq. (56) in [6, XXVIII, vol. 2], it is

$$(\alpha) : \frac{\delta u}{\delta t} = X$$

$$(\beta) : \frac{\delta\langle\xi^2\rangle}{\delta t} = \kappa \varrho A_2 (\langle\eta^2\rangle + \langle\zeta^2\rangle - 2\langle\xi^2\rangle)$$

$$(\beta') : \frac{\delta\langle\xi\eta\rangle}{\delta t} = -3\,\kappa\,\varrho\,A_2\,\langle\xi\eta\rangle$$

$$(\gamma) : \frac{\delta\langle\xi_1 V_1\rangle}{\delta t} = -3\kappa\,\varrho\,A_2\,\langle\xi V_1^2\rangle + X\langle3\xi^2 + \eta^2 + \zeta^2\rangle + 2Y\langle\xi\eta\rangle + 2Z\langle\xi\zeta\rangle$$

Consider a plane moving in the x direction with velocity equal to the average velocity u: then the amount of Q crossing the plane per unit time is $\langle\xi Q\rangle =^{def} \int \xi Q(v_1) \varrho f(v_1)d^3v_1$.

For $Q = \xi$ the momentum in the direction x transferred through a plane orthogonal to the x-direction is $\langle\xi^2\rangle$ while the momentum in the direction y transferred through a plane orthogonal to the x-direction is $\langle\xi\eta\rangle$

The quantity $\langle\xi^2\rangle$ is interpreted as pressure in the x direction and the tensor $T_{ab} = \langle v_a v_b\rangle$ is interpreted as stress tensor.

6.14.7 "Weak" Boltzmann Equation

Supposing the particles without structure and point-like the $Q = \xi v_1^2 \equiv \xi(\xi^2 + \xi\eta^2 + \xi\zeta^2)$ is interpreted as the heat per unit time and area crossing a plane orthogonal to the x-direction.

The averages of observables change also when particles move without colliding. Therefore $\frac{\delta}{\delta t}$ does not give the complete contribution to the variations of the averages of observables. The complete variation is given by

$$\partial_t(\varrho\langle Q\rangle) = \varrho\frac{\delta\langle Q\rangle}{\delta t} - \partial_x\left(\varrho\langle(u+\xi)Q\rangle\right) - \partial_y\left(\varrho\langle(v+\eta)Q\rangle\right) - \partial_z\left(\varrho\langle(w+\zeta)Q\rangle\right)$$

if the average velocity is zero at the point where the averages are evaluated (otherwise if it is $(u', v'w')$ the (u, v, w) should be replaced by $(u - u', v - v', w - w')$). The special choice $Q = \varrho$ gives the "continuity equation".

$$\partial_t\varrho + \varrho(\partial_x u + \partial_y v + \partial_z w) = 0$$

which allows us to write the equation for Q as

$$\varrho\partial_t\langle Q\rangle + \partial_x(\varrho\langle\xi Q\rangle) + \partial_y(\varrho\langle\eta Q\rangle) + \partial_z(\varrho\langle\zeta Q\rangle) = \varrho\frac{\delta Q}{\delta t}$$

Remark Notice that the latter equation is exactly Boltzmann's equation (after multiplication by Q and integration, i.e. in "weak form") for Maxwellian potential $(K|x|^{-4})$ if the expression for $\frac{\delta\langle Q\rangle}{\delta t}$ derived above is used. The assumption on the potential is actually not used in deriving the latter equation as only the stosszahlansatz matters in the expression of $\frac{\delta\langle Q\rangle}{\delta t}$ in terms of the collision cross section (see Eq. (2.1)).

6.14.8 The Heat Conduction Example

The choice $Q = (u + \xi)$ yields

$$\varrho\partial_t u + \partial_x(\varrho\langle\xi^2\rangle) + \partial_y(\varrho\langle\eta^2\rangle) + \partial_z(\varrho\langle\zeta^2\rangle) = \varrho X$$

The analysis can be continued to study several other problems. As a last example the heat conductivity in an external field X pointing in the x-direction is derived by choosing $Q = M(u + \xi)(u^2 + v^2 + w^2 + 2u\xi + 2v\eta + 2w\zeta + \xi^2 + \eta^2 + \zeta^2)$ which yields

$$\varrho\partial_t\langle\xi^3 + \xi\eta^2 + \xi\zeta^2\rangle + \partial_x\varrho\langle\xi^4 + \xi^2\eta^2 + \xi^2\zeta^2\rangle$$
$$= -3\kappa\varrho^2 A_2\langle\xi^3 + \xi\eta^2 + \xi\zeta^2\rangle + 5X\langle\xi^2\rangle$$

having set $(u, v, w) = 0$ and having neglected all odd powers of the velocity fluctuations except the terms multiplied by κ (which is "large": for hydrogen at normal conditions it is $\kappa \sim 1.65 \cdot 10^{13}$, in cgs units, and $\kappa\varrho = 2.82 \cdot 10^9\,s^{-1}$).

In a stationary state the first term is 0 and X is related to the pressure: $X = \partial_x p$ so that

$$\partial_x\varrho\langle(\xi^4 + \xi^2\eta^2 + \xi^2\zeta^2)\rangle - 5\langle\xi^2\rangle\partial_x p = -3\kappa\varrho^2 A_2\langle(\xi^3 + \xi\eta^2 + \xi\zeta^2)\rangle$$

This formula allows us to compute $\frac{1}{2}\varrho\langle(\xi^3 + \xi\eta^2 + \xi\zeta^2)\rangle =^{def} - \chi\partial_x T$ as

$$\chi = \frac{1}{6\kappa\,\varrho\,A_2}\left(\varrho\partial_x\langle(\xi^4 + \xi^2\eta^2 + \xi^2\zeta^2)\rangle - 5\langle x^2\rangle\partial_x p\right)\frac{1}{\partial_x T}$$
$$= \frac{1}{6\kappa A_2}\frac{5\partial_x(\beta^{-2}) - 5\beta^{-1}\partial_x(\beta^{-1})}{M^2\partial_x T} = -\frac{5}{6\kappa A_2}\beta^{-3}\frac{\partial_x\beta}{M^2\partial_x T} = \frac{5k_B}{6\kappa M^2 A_2}k_B T$$

without having to determine the momenta distribution but only assuming that the distribution is close to the Maxwellian. The dimension of χ is $[k_B]\,cm^{-1}s^{-1} = g\,cm\,s^{-3}\,{}^o K^{-1}$. The numerical value is $\chi = 7.2126 \cdot 10^4$ c.g.s. for hydrogen at normal conditions (cgs units).

Remark (1) Therefore the conductivity for Maxwellian potential turns out to be proportional to T rather than to $\sqrt{T}$: in general it will depend on the potential and in the hard balls case it is, according to [2, Eq. (59)], proportional to $\sqrt{T}$. In all cases it is independent on the density. The method to find the conductivity in this paper is completely different from the (in a way elementary) one in [2].

(2) The neglection of various odd momenta indicates that the analysis is a first order analysis in the temperature gradient.

(3) In the derivation density has not been assumed constant: if its variations are taken into account the derivatives of the density cancel if the equation of state is that of a perfect gas (as assumed implicitly). However the presence of the pressure in the

equation for the stationarity of Q is due to the external field X: in absence of the external field the pressure would not appear *but* in such case it would be constant while the density could not be constant; the calculation can be done in the same way and it leads to the *same* conductivity constant.

References

1. Boltzmann, L.: Über die mechanische Bedeutung des zweiten Hauptsatzes der Wärmetheorie. In: Hasenöhrl, F. (ed.) Wissenschaftliche Abhandlungen, vol. 1, #2. Chelsea, New York (1968)
2. Maxwell, J.C.: Illustrations of the dynamical theory of gases. In: Niven, W.D. (ed.) The scientific papers of James Clerk Maxwell, vol. 1. Cambridge University Press, Cambridge (1964)
3. Clausius, R.: Ueber die Zurückführung des zweites Hauptsatzes der mechanischen Wärmetheorie und allgemeine mechanische Prinzipien. Annalen der Physik **142**, 433–461 (1871)
4. Zeuner, G.: Grundgzüge der Mechanischen Wärmetheorie. Buchhandlung J.G. Engelhardt, Freiberg (1860)
5. Masson, M.A.: Sur la corrélation des propriétés physique des corps. Annales de Chimie **53**, 257–293 (1858)
6. Maxwell, JC.: On the dynamical theory of gases. In: Niven, W.D. (ed.) The scientific papers of James Clerk Maxwell, vol. 2. Cambridge University Press, Cambridge (1964)
7. Boltzmann, L.: Weitere Studien über das Wärmegleichgewicht unter Gasmolekülen. In: Hasenöhrl, F. (ed.) Wissenschaftliche Abhandlungen, vol. 1, #22. Chelsea, New York (1968)
8. Bach, A.: Boltzmann's probability distribution of 1877. Arch. Hist. Exact. Sci. **41**, 1–40 (1990)
9. Boltzmann, L.: Über das Wärmegleichgewicht zwischen mehratomigen Gasmolekülen. In: Hasenöhrl, F. (ed.) Wissenschaftliche Abhandlungen, vol. 1, #18 Chelsea, New York (1968)
10. Boltzmann, L.: Studien über das Gleichgewicht der lebendigen Kraft zwischen bewegten materiellen Punkten. In: Hasenöhrl, F. (ed.) Wissenschaftliche Abhandlungen, vol. 1, #5. Chelsea, New York (1968)
11. Boltzmann, L.: Über die Beziehung zwischen dem zweiten Hauptsatze der mechanischen Wärmetheorie und der Wahrscheinlichkeitsrechnung, respektive den Sätzen über das Wärmegleichgewicht. In: Hasenöhrl, F. (ed.) Wissenschaftliche Abhandlungen, vol. 2, #42. Chelsea, New York (1968)
12. Gallavotti, G.: Statistical Mechanics. A Short Treatise. Springer, Berlin (2000)
13. Boltzmann, L.: Lösung eines mechanischen Problems. In: Hasenöhrl, F. (ed.) Wissenschaftliche Abhandlungen, vol. 1, #6. Chelsea, New York (1968)
14. Landau, L.D., Lifschitz, E.M.: Mécanique des Fluides. MIR, Moscow (1971)
15. Clausius, R.: Über einige für Anwendung bequeme formen der Hauptgleichungen der mechanischen Wärmetheorie. Annalen der Physik und Chemie **125**, 353–401 (1865)
16. Clausius, R.: Ueber eine veränderte form des zweiten hauptsatzes der mechanischen wärmetheorie. Annalen der Physik und Chemie **93**, 481–506 (1854)
17. Clausius, R.: On the application of the theorem of the equivalence of transformations to interior work. Phil. Mag. **XXIV**(4):81–201 (1862)
18. Boltzmann, L.: Zur priorität der auffindung der beziehung zwischen dem zweiten hauptsatze der mechanischen wärmetheorie und dem prinzip der keinsten wirkung. In: Hasenöhrl, F. (ed.) Wissenschaftliche Abhandlungen, vol. 1, #17. Chelsea, New York (1968)
19. Clausius, R.: Bemerkungen zu der prioritätreclamation des Hrn. Boltzmann. Annalen der Physik **144**, 265–280 (1872)
20. Renn, J.: Einstein's controversy with Drude and the origin of statistical mechanics: a new glimpse from the "Love Letters". Arch. Hist. Exact Sci. **51**, 315–354 (1997)
21. Maxwell, J.C.: On the dynamical theory of gases. Phil. Mag. **XXXV**:129–145, 185–217 (1868)
22. Lagrange, J.L.: Oeuvres. Gauthiers-Villars, Paris (1867–1892)

23. Goldstein, S., Lebowitz, J.L.: On the (Boltzmann) entropy of nonequilibrium systems. Physica D **193**, 53–66 (2004)

24. Garrido, P. L., Goldstein, S., Lebowitz,J. L.: Boltzmann entropy for dense fluids not in local equilibrium. Phys. Rev. Lett. **92**:050602 (+4) (2005)

25. Boltzmann, L.: Einige allgemeine sätze über Wärmegleichgewicht. In: Hasenöhrl, F. (ed.) Wissenschaftliche Abhandlungen, vol. 1, #19. Chelsea, New York (1968)

26. Gibbs, J.: Elementary Principles in Statistical Mechanics. Schribner, Cambridge (1902)

27. Boltzmann, L.: Analytischer Beweis des zweiten Hauptsatzes der mechanischen Wärmetheorie aus den Sätzen über das Gleichgewicht des lebendigen Kraft. In: Hasenöhrl, F. (ed.) Wissenschaftliche Abhandlungen, vol. 1, #20. Chelsea, New York (1968)

28. Boltzmann, L.: Über die Eigenshaften monozyklischer und anderer damit verwandter Systeme. In: Wissenschaftliche Abhandlungen, vol. 3, #73. Chelsea, New-York, (1968) (1884)

29. L. Boltzmann. Bemerkungen über einige Probleme der mechanischen Wärmetheorie. In: Hasenöhrl, F. (ed.) Wissenschaftliche Abhandlungen, vol. 2, #39. Chelsea, New York (1877)

30. Uhlenbeck, G.E.: An outline of statistical mechanics. In: Cohen, E.G.D. (ed.) Fundamental Problems in Statistical Mechanics, II. North Holland, Amsterdam (1968)

31. Helmholtz, H.: Prinzipien der Statistik monocyklischer Systeme. In: Wissenschaftliche Abhandlungen, vol. III. Barth, Leipzig (1895)

32. Helmholtz, H.: Studien zur Statistik monocyklischer Systeme. In: Wissenschaftliche Abhandlungen, vol. III. Barth, Leipzig (1895)

33. Maxwell, J.C.: On Boltzmann's theorem on the average distribution of energy in a system of material points. Trans. Camb. Phil. Soc. **12**, 547–575 (1879)

34. Gallavotti, G.: Entropy, thermostats and chaotic hypothesis. Chaos **16**:043114 (+6) (2006)

Appendices

Appendix A: Heat Theorem (Clausius Version)

To check Eq. (1.4.3) one computes the first order variation of $\mathcal{A}$ (if $t = i\varphi, t' = i'\varphi$).
Suppose first that the varied motion is subject to the same forces, i.e. the potential
V, sum of the potentials of the internal and external forces, does not vary; then

$$\delta\mathcal{A} = \mathcal{A}(x') - \mathcal{A}(x)$$

$$= \int_0^1 \left[\left(\frac{i'\,m}{2} \dot{x}'(t')^2 - \frac{i\,m}{2} \dot{x}(t)^2 \right) - \left(i'V(x'(t')) - iV(x(t)) \right) \right] d\varphi \qquad (A.1)$$

and, computing to first order in $\delta i = i' - i$ and δx the result is that $\delta\mathcal{A}$ is

$$= \delta i\,(\overline{K} - \overline{V}) + i \int_0^1 \left[\frac{m}{2} (\dot{x}'(t') + \dot{x}(t))(\dot{x}'(t') - \dot{x}(t)) - \partial_x V(x(t))\,\delta x(t) \right] d\varphi$$

$$(A.2)$$

$$= \delta i\,(\overline{K} - \overline{V}) + i \int_0^1 \left[m\dot{x}(t) \frac{d}{d\varphi} \left(\frac{x'(t')}{i'} - \frac{x(t)}{i} \right) - \partial_x V(x(t))\,\delta x(t) \right] d\varphi$$

$$= \delta i\,(\overline{K} - \overline{V}) + i \int_0^1 \left[-i\,m\ddot{x}(t)\delta\left(\frac{x(t)}{i} \right) - \partial_x V(x(t))\,\delta x(t) \right] d\varphi$$

$$= \delta i\,(\overline{K} - \overline{V}) - \delta i \int_0^1 \partial_x V(x(t))\,x(t)\,d\varphi = \delta i\,(\overline{K} - \overline{V}) + \delta i \int_0^1 m\ddot{x}(t)\,x(t)\,d\varphi$$

and we find, integrating again by parts,

G. Gallavotti, *Nonequilibrium and Irreversibility*,
Theoretical and Mathematical Physics, DOI: 10.1007/978-3-319-06758-2,
© Springer International Publishing Switzerland 2014

$$\delta \mathcal{A} = \delta i \, (\overline{K} - \overline{V}) - \delta i \int_0^1 m \, \dot{x}(t)^2 d\varphi = -\delta i \, (\overline{K} + \overline{V}) \qquad (A.3)$$

(always to first order in the variations); equating the *r.h.s* of Eq. (A.3) with $\delta \mathcal{A} \equiv \delta(i(\overline{K} - \overline{V})) \equiv (\overline{K} - \overline{V})\delta i + i \, \delta(\overline{K} - \overline{V})$ this is

$$\delta \overline{V} = \delta \overline{K} + 2\overline{K}\delta \log i \qquad (A.4)$$

hence proving Eq. (1.4.3) in the case in which the potential does not change.

When, instead, the potential V of the forces varies by $\delta\widetilde{V}$ the Eq. (A.3) has to be modified by simply adding $-i\delta\overline{\widetilde{V}}$, as $\delta\widetilde{V}$ is infinitesimal of first order. The quantity $\delta\overline{W} = -i\delta\overline{\widetilde{V}}$ has the interpretation of work done by the system.

From the above generalization of the least action principle it follows that the variation of the total energy of the system, $\overline{U} = \overline{K} + \overline{V}$, between two close motions, is

$$\delta\overline{U} + \delta\overline{W} = 2(\delta\overline{K} + \overline{K}\delta \log i) \equiv 2\overline{K} \, \delta \log(i\overline{K}) \qquad (A.5)$$

This variation must be interpreted as δQ, heat absorbed by the system. Hence setting $\overline{K} = cT$, with c an arbitrary constant, we find:

$$\frac{d\overline{U} + d\overline{W}}{T} = \frac{dQ}{T} = \frac{1}{c}d \log(c \, T \, i) \overset{def}{=} dS \qquad (A.6)$$

The Eq. (A.6) will be referred here as the *heat theorem* (abridging the original diction *second main theorem of the mechanical theory of heat*, [1, #2], [2]).

The above analysis admits an extension to Keplerian motions, discussed in [3, #39], provided one considers only motions with a fixed eccentricity.

Appendix B: Aperiodic Motions as Periodic with Infinite Period

The famous and criticized statement on the *infinite period of aperiodic motions*, [1, #2], is the heart of the application of the heat theorem to a gas in a box. Imagine, [4], the box containing the gas to be covered by a piston of section A and located to the right of the origin at distance L, so that the box volume is $V = AL$.

The microscopic model for the piston will be a potential $\overline{\varphi}(L - \xi)$ if $x = (\xi, \eta, \zeta)$ are the coordinates of a particle. The function $\overline{\varphi}(r)$ will vanish for $r > r_0$, for some $r_0 < L$, and diverge to $+\infty$ at $r = 0$. Thus r_0 is the width of the layer near the piston where the force of the wall is felt by the particles that happen to roam there.

Noticing that the potential energy due to the walls is $\varphi = \sum_j \overline{\varphi}(L - \xi_j)$ and that $\partial_V \varphi = A^{-1}\partial_L \varphi$ we must evaluate the time average of

$$\partial_L \varphi(x) = - \sum_j \overline{\varphi}'(L - \xi_j).$$ (B.1)

As time evolves the particles with ξ_j in the layer within r_0 of the wall will feel the force exercised by the wall and bounce back. Fixing the attention on one particle in the layer we see that it will contribute to the average of $\partial_L \varphi(x)$ the amount

$$\frac{1}{\text{total time}} 2 \int_{t_0}^{t_1} -\overline{\varphi}'(L - \xi_j) dt$$ (B.2)

if t_0 is the first instant when the point j enters the layer and t_1 is the instant when the ξ-component of the velocity vanishes "against the wall". Since $-\overline{\varphi}'(L - \xi_j)$ is the ξ-component of the force, the integral is $2m|\dot{\xi}_j|$ (by Newton's law), provided $\dot{\xi}_j > 0$ of course. One assumes that the density is low enough so that no collisions between particles occur while the particles travel within the range of the potential of the wall: i.e. the mean free path is much greater than the range of the potential $\overline{\varphi}$ defining the wall.

The number of such contributions to the average per unit time is therefore given by $\varrho_{wall} A \int_{v>0} 2mv \, f(v) \, v \, dv$ if ϱ_{wall} is the (average) density of the gas near the wall and $f(v)$ is the fraction of particles with velocity between v and $v + dv$. Using the ergodic hypothesis (i.e. the microcanonical ensemble) and the equivalence of the ensembles to evaluate $f(v)$ (as $\frac{e^{-\frac{\beta}{2}mv^2}}{\sqrt{2\pi\beta^{-1}}}$) it follows that:

$$p \overset{def}{=} -\langle \partial_V \varphi \rangle = \varrho_{wall} \beta^{-1}$$ (B.3)

where $\beta^{-1} = k_B T$ with T the absolute temperature and k_B Boltzmann's constant. Hence we see that Eq. (B.3) yields the correct value of the pressure, [4, Eq. (9.A3.3)], [5]; in fact it is often even taken as the microscopic definition of the pressure, [5].

On the other hand we have seen in Eq. (A.1) that if all motions are periodic the quantity p in Eq. (B.3) is the right quantity that would make the heat theorem work. Hence regarding all trajectories as periodic leads to the heat theorem with p, U, V, T having the *right physical interpretation*. And Boltzmann thought, since the beginning of his work, that trajectories confined into a finite region of phase space could be regarded as periodic *possibly with infinite period*, [1, p. 30], see p. 139.

Appendix C: The Heat Theorem Without Dynamics

The assumption of periodicity can be defended mathematically only in a discrete conception of space and time: furthermore the Loschmidt paradox had to be discussed in terms of *numbers of different initial states which determines their probability, which perhaps leads to an interesting method to calculate thermal equilibria*, [3, #39] and Sect. 6.11.

The statement, admittedly somewhat vague, had to be made precise: the subsequent paper, [6, #42], deals with this problem and adds major new insights into the matter.

It is shown that it is possible to forget completely the details of the underlying dynamics, except for the tacit ergodic hypothesis in the form that all microscopic states compatible with the constraints (of volume and energy) are visited by the motions. The discreteness assumption about the microscopic states is for the first time not only made very explicit but it is used to obtain in a completely new way, once more, that the equilibrium distribution is equivalently a canonical or a microcanonical one. The method is simply a combinatorial analysis on the number of particles that can occupy a given energy level. The analysis is performed first in the one dimensional case (i.e. assuming that the energy levels are ordered into an arithmetic progression) and then in the three dimensional case (i.e. allowing for degeneracy by labeling the energy levels with three integers). The combinatorial analysis is the one that has become familiar in the theory of the Bose–Einstein gases: if the levels are labeled by a label i a microscopic configuration is identified with the occupation numbers (n_i) of the levels ε_i under the restrictions

$$\sum_i n_i = N, \qquad \sum_i n_i \varepsilon_i = U \qquad (C.1)$$

fixing the total number of particles N and the total energy U. The calculations in [6, #42] amount at selecting among the N particles the n_i in the i-level with the energy restriction in Eq. (C.1): forgetting the latter the number of microscopic states would be $\frac{N!}{\prod_i n_i!}$ and the imposition of the energy value would lead, as by now usual by the Lagrange's multipliers method, to an occupation number per level

$$n_i = N \frac{e^{\mu - \beta \varepsilon_i}}{Z(\mu, \beta)} \qquad (C.2)$$

if Z is a normalization constant.

The Eq. (C.2) is the canonical distribution which implies that $\frac{dU + p\,dV}{T}$ is an exact differential if U is the average energy, p the average mechanical force due to the collisions with the walls, T is the average kinetic energy per particle, [4, Chaps. 1, 2]: it is apparently not necessary to invoke dynamical properties of the motions.

Clearly this is not a proof that the equilibria are described by the microcanonical ensemble. However it shows that for most systems, independently of the number of degrees of freedom, one can define a *mechanical model of thermodynamics*. The reason we observe approach to equilibrium over time scales far shorter than the recurrence times is due to the property that on most of the energy surface the actual values of the observables whose averages yield the pressure and temperature assume the same value. This implies that this value coincides with the average and therefore satisfies the *heat theorem*, i.e. the statement that $(dU + p\,dV)/T$ is an exact differential if p is the pressure (defined as the average momentum transfer to the walls per unit time and unit surface) and T is proportional to the average kinetic energy.

Appendix D: Keplerian Motion and Heat Theorem

It is convenient to use polar coordinates (ϱ, θ), so that if $A = \varrho^2\dot\theta$, $E = \frac12 m\dot{\mathbf{x}}^2 - \frac{gm}{\varrho}$, m being the mass and g the gravity attraction strength $(g = kM$ if k is the gravitational constant and M is the central mass) then

$$E = \frac{1}{2}m\dot\varrho^2 + \frac{mA^2}{2\varrho^2} - \frac{mg}{\varrho}, \qquad \varphi(\varrho) = -\frac{gm}{\varrho} \tag{D.1}$$

and, from the elementary theory of the two body problem, if e is the eccentricity and a is the major semiaxis of the ellipse

$$\dot\varrho^2 = \frac{2}{m}\left(E - \frac{mA^2}{2\varrho^2} + \frac{mg}{\varrho}\right) \overset{def}{=} A^2\left(\frac{1}{\varrho} - \frac{1}{\varrho_+}\right)\left(\frac{1}{\varrho_-} - \frac{1}{\varrho}\right)$$

$$\frac{1}{\varrho_+\varrho_-} = \frac{-2E}{mA^2}, \quad \frac{1}{\varrho_+} + \frac{1}{\varrho_-} = \frac{2g}{A^2}, \quad \frac{\varrho_+ + \varrho_-}{2} \overset{def}{=} a = \frac{mg}{-2E}$$

$$\sqrt{\varrho_+\varrho_-} \overset{def}{=} a\sqrt{1 - e^2}, \quad \sqrt{1 - e^2} = \frac{A}{\sqrt{ag}}. \tag{D.2}$$

Furthermore if a motion with parameters (E, A, g) is periodic (hence $E < 0$) and if $\langle\cdot\rangle$ denotes a time average over a period then

$$E = -\frac{mg}{2a}, \qquad \langle\varphi\rangle = -\frac{mg}{a}, \qquad \left\langle\frac{1}{\varrho^2}\right\rangle = \frac{1}{a^2\sqrt{1 - e^2}}$$

$$\langle K\rangle = \frac{mg}{2a} = -E, \qquad T = \frac{mg}{2a} \equiv \langle K\rangle \tag{D.3}$$

Hence if (E, A, g) are regarded as state parameters then

$$\frac{dE - \langle\frac{\partial_g\varphi(r)}{\partial g}\rangle dg}{T} = \frac{dE - 2E\frac{dg}{g}}{-E} = d\log\frac{g^2}{-E} \overset{def}{=} dS \tag{D.4}$$

Note that the equations $pg = 2T$ and $E = -T$ can be interpreted as, respectively, analogues of the "equation of state" and the "ideal specific heat" laws (with the "volume" being g, the "gas constant" being $R = 2$ and the "specific heat" $C_V = 1$).

If the potential is more general, for instance if it is increased by $\frac{b}{2r^2}$, the analogy fails, as shown by Boltzmann, [7, #19], see Sect. 6.11. Hence there may be cases in which the integrating factor of the differential form which should represent the heat production might not necessarily be the average kinetic energy: essentially all cases in which the energy is not the only constant of motion. Much more physically interesting examples arise in quantum mechanics: there in the simplest equilibrium statistics (free Bose–Einstein or free Fermi–Dirac gases) the average kinetic energy and the temperature do not coincide, [4].

Appendix E: Gauss' Least Constraint Principle

Let $\varphi(\dot{\mathbf{x}}, \mathbf{x}) = 0$, $(\dot{\mathbf{x}}, \mathbf{x}) = \{\dot{\mathbf{x}}_j, \mathbf{x}_j\}$ be a constraint and let $\mathbf{R}(\dot{\mathbf{x}}, \mathbf{x})$ be the constraint reaction and $\mathbf{F}(\dot{\mathbf{x}}, \mathbf{x})$ the active force.

Consider all the possible accelerations $\mathbf{a}$ compatible with the constraints and a given initial state $\dot{\mathbf{x}}$, $\mathbf{x}$. Then $\mathbf{R}$ is *ideal* or *satisfies the principle of minimal constraint* if the actual accelerations $\mathbf{a}_i = \frac{1}{m_i}(\mathbf{F}_i + \mathbf{R}_i)$ minimize the *effort* defined by:

$$\sum_{i=1}^{N} \frac{1}{m_i}(\mathbf{F}_i - m_i \mathbf{a}_i)^2 \leftrightarrow \sum_{i=1}^{N}(\mathbf{F}_i - m_i \mathbf{a}_i) \cdot \delta \mathbf{a}_i = 0 \qquad (E.1)$$

for all possible variations $\delta \mathbf{a}_i$ compatible with the constraint φ. Since all possible accelerations following $\dot{\mathbf{x}}$, $\mathbf{x}$ are such that $\sum_{i=1}^{N} \partial_{\dot{\mathbf{x}}_i} \varphi(\dot{\mathbf{x}}, \mathbf{x}) \cdot \delta \mathbf{a}_i = 0$ we can write

$$\mathbf{F}_i - m_i \mathbf{a}_i - \alpha \, \partial_{\dot{\mathbf{x}}_i} \varphi(\dot{\mathbf{x}}, \mathbf{x}) = \mathbf{0} \qquad (E.2)$$

with α such that

$$\frac{d}{dt}\varphi(\dot{\mathbf{x}}, \mathbf{x}) = 0, \qquad (E.3)$$

i.e.

$$\alpha = \frac{\sum_i (\dot{\mathbf{x}}_i \cdot \partial_{\mathbf{x}_i}\varphi + \frac{1}{m_i}\mathbf{F}_i \cdot \partial_{\dot{\mathbf{x}}_i}\varphi)}{\sum_i m_i^{-1}(\partial_{\dot{\mathbf{x}}_i}\varphi)^2} \qquad (E.4)$$

which is the analytic expression of the Gauss' principle, see [8] and [4, Appendix 9.A4].

Note that if the constraint is even in the $\dot{\mathbf{x}}_i$ then α is odd in the velocities: therefore if the constraint is imposed on a system with Hamiltonian $H = K + V$, with K quadratic in the velocities and V depending only on the positions, and if other purely positional forces (conservative or not) act on the system then the resulting equations of motion are reversible if time reversal is simply defined as velocity reversal.

The Gauss' principle has been somewhat overlooked in the Physics literature in statistical mechanics: its importance has again only recently been brought to the attention of researchers, see the review [9]. A notable, though by now ancient, exception is a paper of Gibbs, [10], which develops variational formulas which he relates to Gauss' principle of least constraint.

Conceptually this principle should be regarded as a *definition* of *ideal non holo-nomic constraint*, much as D'Alembert's principle or the least action principle are regarded as the definition of *ideal holonomic constraint*.

Appendix F: Non Smoothness of Stable/Unstable Manifolds

A simple argument to understand why even in analytic Anosov systems the stable and unstable manifolds might just be Hölder continuous can be given.

Let T^2 be the two dimensional torus $[0, 2\pi]^2$, and let S_0 be the Arnold's cat Anosov map on T^2; $(x', y') = S_0(x, y)$ and $S(x, y) = S_0(x, y) + \varepsilon f(x)$ be an analytic perturbation of S_0:

$$S_0 \to \begin{cases} x' = 2x + y \mod 2\pi \\ y' = x + y \mod 2\pi \end{cases}, \quad S(x, y) = S_0(x, y) + \varepsilon f(x, y) \quad \text{(F.1)}$$

with $f(x, y)$ periodic. Let v_+^0, v_-^0 be the eigenvectors of the matrix $\partial S_0 = \begin{pmatrix} 2 & 1 \\ 1 & 1 \end{pmatrix}$: which are the stable and unstable tangent vectors at all points of T^2; and let λ_+, λ_- be the corresponding eigenvalues.

Abridging the pair (x, y) into ξ, try to define a change of coordinates map $h(\xi)$, analytic in ε near $\varepsilon = 0$, which transforms S back into S_0; namely h such that

$$h(S_0\xi) = S(h(\xi)) + \varepsilon f(h(\xi)) \quad \text{(F.2)}$$

and which is *at least* mildly continuous, e.g. Hölder continuous with $|h(\xi) - h(\xi')| \leq B|\xi - \xi'|^\beta$ for some $B, \beta > 0$ (with $|\xi - \xi'|$ being the distance on T^2 of ξ from ξ'). If such a map exists it will transform stable or unstable manifolds for S_0 into corresponding stable and unstable manifolds of S.

It will now be shown that the map h cannot be expected to be differentiable but only Hölder continuous with an exponent β that can be prefixed as close as wished to 1 (at the expense of the coefficient B) but *not* $= 1$. Write $h(x) = x + \eta(x)$, with $\eta(x) = \sum_{k=0}^\infty \varepsilon^k \eta^{[k]}(x)$; then

$$\eta(S_0 x) = S_0 \eta(x) + \varepsilon f(x + \eta(x)) \quad \text{(F.3)}$$

$$\eta^{[k]}(S_0 x) - S_0 \eta^{[k]}(x) = \left(\varepsilon f(x + \eta(x))\right)^{[k]}$$

where the last term is the k-th order coefficient of the expansion in ε. If $h_\pm(\xi)$ are scalars so defined that $h(\xi) = h_+(\xi) v_+^0 + h_-(\xi) v_-^0$ it will be

$$h_+^{[k]}(\xi) = \lambda_+^{-1}\eta_+^{[k]}(S_0\xi) - \left(\varepsilon f_+(\xi + h(\xi))\right)^{[k]}$$

$$h_-^{[k]}(\xi) = \lambda_-\eta_+^{[k]}(S_0^{-1}\xi) + \left(\varepsilon f_+(S_0^{-1}\xi + h(S_0^{-1}\xi))\right)^{[k]} \qquad (F.4)$$

Hence for $k = 1$ it is

$$h_+^{[1]}(\xi) = -\sum_{n=0}^{\infty} \lambda_+^{-n} f_+(S_0^n\xi), \qquad h_-^{[1]}(\xi) = \sum_{n=0}^{\infty} \lambda_-^n f_-(S_0^{-n-1}\xi) \qquad (F.5)$$

the r.h.s. is a convergent series of differentiable functions as $\lambda_+^{-1} = |\lambda_-| < 1$.

However if differentiated term by term the n-th term will be given by $(\partial S_0)^n(\partial^n f_+)$ $(S_0^n\xi)$ or, respectively, $(\partial S_0)^{-n}(\partial^n f_-)(S_0^{-1-n}\xi)$ and these functions will grow as λ_+^n or as $|\lambda_-|^n$ for $n \to \infty$, respectively, unless $f_+ = 0$ in the first case or $f_- = 0$ in the second, so one of the two series cannot be expected to converge, i.e. the $h^{[1]}$ cannot be proved to be differentiable. Nevertheless if $\beta < 1$ the differentiability of f implies $|f(S_0^{\pm n}\xi) - f(S_0^{\pm n}\xi')| \le B|S_0^{\pm n}(\xi - \xi')|^\beta$ for some $B > 0$, hence

$$|h_+^{[1]}(\xi) - h_+^{[1]}(\eta)| \le \sum_{n=0}^{\infty} \lambda_+^{-n} B\lambda_+^{n\beta}|\xi - \eta|^\beta \le \frac{B}{1 - \lambda_+^{1-\beta}}|\xi - \eta|^\beta \qquad (F.6)$$

because $|S_0^{\pm n}(\xi - \eta)| \le \lambda_+^n|\xi - \eta|$ for all $n \ge 0$, and therefore $h^{[1]}$ is Hölder continuous.

The above argument can be extended to all order in ε to prove rigorously that $h(\xi)$ is analytic in ε near $\varepsilon = 0$ and Hölder continuous in ξ, for details see [11, Sect. 10.1]. It can also lead to show that locally the stable or unstable manifolds (which are the h-images of those of S_0) are infinitely differentiable surfaces (actually lines, in this 2-dimensional case), [11, Sect. 10.1].

Appendix G: Markovian Partitions Construction

This section is devoted to a mathematical proof of existence of Markovian partitions of phase space for Anosov maps on a 2-dimensional manifold, which at the same time provides an algorithm for their construction.[1] Although the idea can be extended to Anosov maps in dimension >2 the (different) original construction in arbitrary dimension is easily found in the literature, [11–13].

Let S be a smooth (analytic) map defined on a smooth compact manifold Ξ: suppose that S is hyperbolic, transitive. Let $\delta > 0$ be fixed, with the aim of constructing a Markovian partition with rectangles of diameter $\le \delta$.

---

[1] It follows an idea of M. Campanino.

Fig. G.1 An incomplete
rectangle: $\overline{C}$ enters it but does
not cover fully the stable axis
(completed by the *dashed
portion* in the second figure)

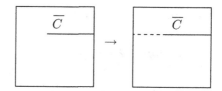

Suppose for simplicity that the map has a fixed point x_0. Let τ be so large that
$\overline{C} = S^{-\tau} W_\delta^s(x_0)$ and $\overline{D} = S^\tau W_\delta^u(x_0)$ (the notation used in Sect. 3.3 is $W_\delta(x) \overset{def}{=}$
connected part containing x of the set $W(x) \cap B_\gamma(x))$ fill $\varXi$ so that no point $y \in \varXi$
is at a distance $> \frac{1}{2}\delta$ from $\overline{C}$ and from $\overline{D}$.

The surfaces $\overline{C}, \overline{D}$ will repeatedly intersect forming, quite generally, rectangles
Q: however there will be cases of rectangles Q inside which part of the boundary
of $\overline{C}$ or of $\overline{D}$ will end up without *fully* overlapping with an axis of Q (i.e. the stable
axis w^s in Q containing $\overline{C} \cap Q$ or, respectively, the unstable axis w^u in Q containing
$\overline{D} \cap Q$) thus leaving an incomplete rectangle as in Fig. G.1.

In such cases $\overline{C}$ will be extended to contain w^s or $\overline{D}$ will be extended to contain
w^u: this means extending $W_\delta^s(x_0)$ to $W_\delta^s(x_0)'$ and $W_\delta^u(x_0)$ to $W_\delta^u(x_0)'$ by shifting
their boundaries by at most $\delta L^{-1} e^{-\lambda\tau}$ and the surfaces $\overline{C}' = S^{-\tau} W_\delta^s(x_0)'$ and
$\overline{D}' = S^\tau W_\delta^u(x_0)'$ will partition $\varXi$ into complete rectangles and there will be no
more incomplete ones.

By construction the rectangles $\mathcal{E}_0 = (E_1, \ldots, E_n)$ delimited by $\overline{C}', \overline{D}'$ will all
have diameter $< \delta$ and no pair of rectangles will have interior points in common.
Furthermore the boundaries of the rectangles will be smooth. Since by construction
$S^\tau \overline{C}' \subset \overline{C}'$ and $S^{-\tau} \overline{D}' \subset \overline{D}'$ if $L^{-1} e^{-\lambda\tau} < 1$, as it will be supposed, the main
property Eq. (3.4.1) will be satisfied with S^τ instead of S and $\mathcal{E}_0$ will be a Markovian
pavement for S^τ, if δ is small enough.

But if $\mathcal{E} = (E_1, \ldots, E_n), \mathcal{E}' = (E_1', \ldots, E_m')$ are Markovian pavements also
$\mathcal{E}'' = \mathcal{E} \vee \mathcal{E}'$, the pavement whose elements are the sets $E_i \cap E_j'$ with non empty
interior is Markovian and $\mathcal{E} \overset{def}{=} \vee_{i=0}^{\tau-1} S^i \mathcal{E}_0$ is checked to be Markovian for S (i.e. the
property in Eq. (3.4.1) holds not only for S^τ but also for S.

Anosov maps may fail to admit a fixed point: however they certainly have periodic
points(actually a dense set of periodic points).[2] If x_0 is a periodic point of period n
it is a fixed point for the map S^n. The iterates of an Anosov map are again Anosov
maps: hence there is an Markov pavement $\mathcal{E}_0$ for S^n: therefore $\mathcal{E} \overset{def}{=} \vee_{i=0}^{n-1} S^i \mathcal{E}_0$ is a
Markovian pavement for S.

---

[2] Let E be a rectangle: then if m is large enough $S^m E$ intersects E in a rectangle δ_1, and the image
$S^m \delta_1$ intersects δ_1 in a rectangle δ_2 and so on: hence there is an unstable axis δ_∞ of E with the
property $S^m \delta_\infty \supset \delta_\infty$. Therefore $S^{-km} \delta_\infty \subset \delta_\infty$ for all $k > 0$ hence $\cap_k S^{-km} \delta_\infty$ is a point x of
period m inside E.

Appendix H: Axiom C

H.1 A Simple Example

As an example consider an Anosov map S_* acting on $\mathcal{A}$ and on the identical set $\mathcal{A}'$; suppose that S_* admits a time reversal symmetry I^*: for instance $\mathcal{A}$ could be the torus T^2 and S the map in the example in Sect. 3.5. If x is a point in $\mathcal{A}$ the generic point of the phase space Ξ will be determined by a pair (x, z) where $x \in \mathcal{A}$ and z is a set of transverse coordinates that tell us how far we are from the attractor. The coordinate z takes two well defined values on $\mathcal{A}$ and $\mathcal{A}'$ that we can denote z_+ and z_- respectively.

The coordinate x identifies a point on the compact manifold $\mathcal{A}$ on which a reversible transitive Anosov map S_* acts (see [14]). And the map S on phase space Ξ is defined by:

$$S(x, z) = (S_* x, \widetilde{S} z) \tag{H.1}$$

where $\widetilde{S}$ is a map acting on the z coordinate (identifying a point on a compact manifold) which is an evolution leading from an unstable fixed point z_+ to a stable fixed point z_-.

To fix the ideas z could be a point on a circle with angular coordinate θ and $\widetilde{S}$ could be the time 1 evolution of θ defined by evaluating at integer times the solution of $\dot{\theta} = \sin \theta$. Such evolution sends θ to $z_- = \pi$ or $z_+ = 0$ as $t \to \pm \infty$ and $z_\pm$ are non marginal fixed points for $\widetilde{S}$.[3]

The map $\widetilde{I}$ acting on z by changing θ into $\pi - \theta$ is a time reversal for $\widetilde{S}$.

Thus if we set $S(x, z) = (S_* x, \widetilde{S} z)$ we see that our system is hyperbolic on the sets $\mathcal{A} \times \{z_-\}$ and $\mathcal{A}' \times \{z_+\}$ and the attracting set $\mathcal{A}$ can be identified with the set of points (x, z_+) with $x \in \mathcal{A}$ while the repelling set consists of points (x, z_-) with $x \in \mathcal{A}'$.

Furthermore the map $I(x, z) = (I^* x, \widetilde{I} z)$ is a time reversal for S. This is illustrated in the Fig. H.1:

Clearly $\mathcal{A}, \mathcal{A}'$ are mapped into each other by the map: $I(x, z_\pm) = (I^* x, z_\mp)$. But on each attractor a "local time reversal" acts: namely the map $(x, z_\pm) \leftrightarrow (I^* x, z_\pm)$ that, on $\mathcal{A}$ or $\mathcal{A}'$, can be thought as the composition of I and I', where $I'(x, z_\pm) = (x, z_\mp)$ is the correspondence defined by the lines of intersection between stable and unstable manifolds, see Fig. H.1.

The system is "chaotic" as it has an Axiom A attracting set consisting of the points having the form (x, z_+), for the motion towards the future; and a different Axiom A attracting set consisting of the points having the form (x, z_-), for the motion towards the past. In fact the dynamical systems $(\mathcal{A}, S)$ and $(\mathcal{A}', S)$ obtained by restricting S to the future or past attracting sets have been supposed Anosov systems (because they are regular manifolds).

---

[3] This is essentially the same example discussed later in Appendix L, see footnote 7.

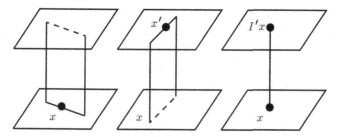

Fig. H.1 Each of the three *lower* surfaces represent the attractor $\mathcal{A}$ and the *upper* the repellers $\mathcal{A}'$. Motion S on $\mathcal{A}$ and $\mathcal{A}'$ is chaotic (Anosov). The system is reversible under the symmetry I but motion on $\mathcal{A}$ and $\mathcal{A}'$ is not reversible because $I\mathcal{A} = \mathcal{A}' \neq \mathcal{A}$. The stable manifold of the points on $\mathcal{A}$ sticks out of $\mathcal{A}$ as its tangent space contains the not only the plane tangent to $\mathcal{A}$ but also the plane determined by the directions on which contraction towards $\mathcal{A}$ occurs: this generates surfaces transversal to $\mathcal{A}$ represented vertical rectangle in the first drawing. Likewise the unstable manifolds of points in $\mathcal{A}'$ sticks out of $\mathcal{A}'$ forming a surface represented by the *vertical rectangle* in the intermediate figure. The axiom-C property requires that the two vertical surfaces extend to cross transversally both $\mathcal{A}$, $\mathcal{A}'$ as in the figure. Finally the pair of vertical surfaces intersect on a *line* crossing transversally both surfaces as in the third figure: thus establishing a one-to-one correspondence $x \leftrightarrow I'x$ between attractor and repeller. The composed map $I\,I'$ leaves $\mathcal{A}$ (and $\mathcal{A}'$) *invariant* and is a time-reversal for the restriction of the evolution S to $\mathcal{A}$ and $\mathcal{A}'$ (trivially equal to I^* in this example)

We may think that in generic reversible systems satisfying the chaotic hypothesis the situation is the above: namely there is an "irrelevant" set of coordinates z that describes the departure from the future and past attractors. The future and past attractors are copies (via the global time reversal I) of each other and on each of them is defined a map I^* which inverts the time arrow, *leaving the attractor invariant*: such map will be naturally called the *local time reversal*.

In the above case the map I^* and the coordinates (x, z) are "obvious". The problem is to see whether they are defined quite generally under the only assumption that the system is reversible and has a unique future and a unique past attractor that verify the Axiom A. This is a problem that is naturally solved in general when the system verifies the Axiom C, see Appendix H.

Finally the example given here is an example in which the pairing rule does not hold: it would be interesting to exhibit an example satisfying the pairing rule so that the idea leading to Eq. (4.4.2) could be tested.

H.2 Formal Definition

Definition Definition A smooth system $(\mathcal{C}, S)$ verifies Axiom C if it admits a unique attracting set $\mathcal{A}$ on which S is Anosov and a possibly distinct repelling set $\mathcal{A}'$ and:

(1) for every $x \in C$ the tangent space T_x admits a Hölder-continuous[4] decomposition as a direct sum of three subspaces T_x^u, T_x^s, T_x^m such that if $\delta(x) = \min(d(x, \mathcal{A}), d(x, \mathcal{A}'))$:

    (a) $dS\, T_x^\alpha \quad = T_{Sx}^\alpha \qquad\qquad \alpha = u, s, m$
    (b) $|dS^n w| \quad \le Ce^{-\lambda n}|w|, \qquad w \in T_x^s,\ n \ge 0$
    (c) $|dS^{-n}w| \le Ce^{-\lambda n}|w|, \qquad w \in T_x^u,\ n \ge 0$
    (d) $|dS^n w| \quad \le C\delta(x)^{-1}e^{-\lambda|n|}|w|,\ w \in T_x^m,\ \forall n$

where the dimensions of T_x^u, T_x^s, T_x^m are >0.
(2) if x is on the attracting set $\mathcal{A}$ then $T_x^s \oplus T_x^m$ is tangent to the stable manifold in x; viceversa if x is on the repelling set $\mathcal{A}'$ then $T_x^u \oplus T_x^m$ is tangent to the unstable manifold in x.

Although T_x^u and T_x^s are not uniquely determined the planes $T_x^s \oplus T_x^m$ and $T_x^u \oplus T_x^m$ are uniquely determined for $x \in \mathcal{A}$ and, respectively, $x \in \mathcal{A}$.

H.3 Geometrical and Dynamical Meaning

An Axiom C system is a system satisfying the chaotic hypothesis in a form which gives also properties of the motions away from the attracting set: i.e. it has a stronger, and more global, hyperbolicity property.

Namely, if $\mathcal{A}$ and $\mathcal{A}'$ are the attracting and repelling sets the stable manifold of a periodic point $p \in \mathcal{A}$ and the unstable manifold of a periodic point $q \in \mathcal{A}'$ not only have a point of transverse intersection, but they intersect transversely *all the way* on a manifold connecting $\mathcal{A}$ to $\mathcal{A}'$; the unstable manifold of a point in $\mathcal{A}'$ will accumulate on $\mathcal{A}$ *without winding around it*.

It will be helpful to continue referring to Fig. H.1 to help intuition. The definition implies that given $p \in \mathcal{A}$ the stable and unstable manifolds of x intersect in a line which intersects $\mathcal{A}'$ and establish a correspondence I' between $\mathcal{A}$ and $\mathcal{A}'$.

If there the map S satisfies the definition above and it has also a time reversal symmetry I then the map $I^* = I\,I'$ on $\mathcal{A}$ (and on $\mathcal{A}'$) has the property of a time reversal for the restriction of S to $\mathcal{A}$ (or $\mathcal{A}'$).

For more details see [17]. Axiom C systems can also be shown to have the property of structural stability: they are Ω-stable in the sense of [16, p. 749], see [17].

---

[4] One might prefer to require real smoothness, e.g. C^p with $1 \le p \le \infty$: but this would be too much for rather trivial reasons, like the ones examined in Appendix G. On the other hand Hölder continuity might be equivalent to simple C^0—continuity as in the case of Anosov systems, see [15, 16].

Appendix I: Pairing Theory

Consider the equations of motion, for an evolution in continuous time in $2Nd$ dimensions, for $X = (\boldsymbol{\xi}, \pi)$

$$\dot{\boldsymbol{\xi}} = \pi, \quad \dot{\pi} = -\partial\varphi(\boldsymbol{\xi}) + \frac{\partial\varphi(\boldsymbol{\xi}) \cdot \pi}{\pi^2}\pi \qquad (I.1)$$

which correspond to a thermostat fixing the temperature T as $dNk_BT = \pi^2$ in a system of N particles in a space of dimension d interacting via a potential φ which contains the internal and external forces. The position variables $\boldsymbol{\xi} = (\xi_1, \ldots, \xi_N)$ vary in R^d (in which case the external forces contain "wall potential barriers" which confine the system in space, possibly to non simply connected regions) or in a square box with periodic boundaries, T^d. The potential φ could possibly be not single valued, although its gradient is required to be single valued. The momentum variables $\pi = (\pi_1, \ldots, \pi_N)$ are constrained to keep π^2 constant because of the thermostat action $-\alpha(\boldsymbol{\xi}, \pi)\pi$ with $\alpha(\boldsymbol{\xi}, \pi) = -\frac{\partial\varphi(\boldsymbol{\xi}) \cdot \pi}{\pi^2}$.

Let $J(X)$ be the $2Nd \times 2Nd$ matrix of the derivatives of Eq. (I.1) so that the Jacobian matrix $W(X, t) = \partial_X S_t(X)$ of Eq. (I.1) is the solution of $\dot{W}(X, t) = J(S_t(X))W(X, t)$. The matrix $J(X)$ can be computed yielding

$$J(X) = \begin{pmatrix} 0 & \delta_{ij} \\ M_{ik}(\delta_{k,j} - \frac{\pi_k\pi_j}{\pi^2}) & -\alpha\delta_{ij} + \frac{\pi_i\partial_j\varphi}{\pi^2} - 2\frac{\partial_k\varphi\,\pi_k}{\pi^4}\pi_i\pi_j \end{pmatrix} \qquad (I.2)$$

with $M_{ik} \overset{def}{=} \partial_{ik}\varphi$; which shows explicitly the properties that an infinitesimal vector $\begin{pmatrix} \varepsilon\pi \\ 0 \end{pmatrix}$ joining two initial data $(\boldsymbol{\xi}, \pi)$ and $(\boldsymbol{\xi} + \varepsilon\pi, \pi)$ is eigenvector of $J(X)$ with eigenvalue 0.

This means that $W(X, t)$ maps $\begin{pmatrix} \varepsilon\pi \\ 0 \end{pmatrix}$ into $\begin{pmatrix} \varepsilon\pi_t \\ 0 \end{pmatrix}$ if $S_t(\boldsymbol{\xi}, \pi) = (\boldsymbol{\xi}_t, \pi_t)$ and that the time derivative $J\begin{pmatrix} 0 \\ \pi \end{pmatrix} = \begin{pmatrix} \pi \\ 0 \end{pmatrix}$ of $\begin{pmatrix} 0 \\ \pi \end{pmatrix}$ is orthogonal to $\begin{pmatrix} 0 \\ \pi \end{pmatrix}$ so that π^2 is a constant of motion, as directly consequence of Eq. (I.1). In other words the Jacobian $W(X, t)$ will have two vectors on which it acts expanding or contracting them less than exponentially: *hence the map S_t will have two zero Lyapunov exponents.*

The latter remark shows that it will be convenient two use a system of coordinates which, in an infinitesimal neighborhood of $X = (\boldsymbol{\xi}, \pi)$, is orthogonal and describes infinitesimal $2Nd$-dimensional phase space vectors $(d\boldsymbol{\xi}, d\pi)$ in terms of the Nd-dimensional components of $d\boldsymbol{\xi}$ and $d\pi$ in a reference frame with origin in $\boldsymbol{\xi}$ and with axes $\mathbf{e}_0, \ldots, \mathbf{e}_{Nd-1}$ with $\mathbf{e}_0 \equiv \frac{\pi}{|\pi|}$.

Calling J_π the matrix J in the new basis, it has the form

$$
\begin{pmatrix}
0 & \cdot & \cdot & & \cdot & 1 & \cdot & \cdot & \cdot & 0 \\
0 & \cdot & \cdot & & \cdot & 0 & 1 & \cdot & \cdot & 0 \\
0 & \cdot & \cdot & & \cdot & 0 & \cdot & \cdot & \cdot & 0 \\
0 & \cdot & \cdot & & \cdot & 0 & \cdot & \cdot & 1 & 0 \\
0 & \cdot & \cdot & & \cdot & 0 & \cdot & \cdot & \cdot & 1 \\
0 & M_{01} & M_{02} & & \cdot & 0 & \frac{\partial_1\varphi}{|\pi|} & \frac{\partial_2\varphi}{|\pi|} & \cdot & \cdot \\
0 & M_{11} & M_{12} & & \cdot & 0 & -\alpha & \cdot & \cdot & 0 \\
0 & M_{21} & M_{22} & & \cdot & 0 & 0 & -\alpha & \cdot & 0 \\
0 & \cdot & \cdot & & \cdot & 0 & \cdot & \cdot & \cdot & 0 \\
0 & \cdot & \cdot & M_{Nd,Nd} & 0 & \cdot & \cdot & \cdot & \cdot & -\alpha
\end{pmatrix}
\tag{I.3}
$$

where $M_{ij} \overset{def}{=} -\partial_{ij}\varphi(\boldsymbol{\xi})$. Therefore if $(x, \mathbf{u}, y, \mathbf{v})$ is a $2Nd$ column vector ($x, y \in R$, $\mathbf{u}, \mathbf{v} \in R^{Nd-1}$) and P is the projection which sets $x = y = 0$ the matrix J_π becomes $J_\pi = P J_\pi P + (1-P)J_\pi P + (1-P)J_\pi(1-P)$ (since $P J_\pi(1-P) = 0$) and, setting $Q \overset{def}{=} (1-P)$,

$$
J_\pi^{2n} = (P J_\pi P)^{2n} + (Q J_\pi Q)^{2n},
\tag{I.4}
$$
$$
J_\pi^{2n+1} = (P J_\pi P)^{2n+1} + (Q J_\pi Q)^{2n+1} + (Q J_\pi Q)^{2n} Q J_\pi P
$$

The key remark, [18], is that the $2(Nd-1) \overset{def}{=} 2D$ dimensional matrix $P J_\pi P$ has the form

$$
\begin{pmatrix} 0 & 1 \\ M' & -\alpha \end{pmatrix} = \begin{pmatrix} \frac{\alpha}{2} & 1 \\ M' & -\frac{\alpha}{2} \end{pmatrix} - \frac{1}{2}\alpha
\tag{I.5}
$$

with M' a symmetric matrix and the first matrix in the r.h.s. is an infinitesimal symplectic matrix:[5] therefore the the matrix $W_\pi(X, t)$ solution of the equation

$$
\dot{W}_\pi(X, t) = J_\pi(S_t X) W_\pi(X, t), \qquad W_\pi(X, 0) = 1,
\tag{I.6}
$$

will have the form $W_\pi(X, t) = W_{\pi,0}(X, t) e^{-\frac{1}{2}\int_0^t \alpha(S_t X)dt}$ with $W_{\pi,0}$ a symplectic matrix.

Ordering in decreasing order its $2D = 2(Nd-1)$ eigenvalues $\lambda_j(X, t)$ of the product $\frac{1}{2t}\log(W_\pi(X, t)^T W_\pi(X, t))$ it is

$$
\lambda_j(X, t) + \lambda_{2D-j}(X, t) = \int_0^\tau \alpha(S_t X)dt, \quad j = 0, \ldots D
\tag{I.7}
$$

that can be called "local pairing rule".

---

[5] It can be thought as the Jacobian matrix for the equations of motion with Hamiltonian $H(p, q) = \frac{1}{2}p^2 - \frac{1}{2}(q, Mq) + \frac{1}{2}apq$.

Since $(QJQ)^n = 0$ for $n \geq 3$ the 2×2 matrix QJQ will give 2 extra exponents which are 0.

Going back to the original basis let $R_0(X)$ a rotation which brings the axis 1 to π if $X = (\xi, \pi)$; and let $\widetilde{R}_0(X)$ be the $2Nd \times 2Nd$ matrix formed by two diagonal blocks equal to $R_0(X)$; then the Jacobian matrix $\partial S_t(X)$ can be written in the original basis as

$$\partial S_t(X) = \widetilde{R}_t(X)^T W_\pi(X, t) \widetilde{R}_0(X) \tag{I.8}$$

and we see that the eigenvalues of $\partial S_t(X)^T \partial S_t(X)$ and of $W_\pi(X, t)^T W_\pi(X, t)$ coincide and the pairing rule holds, [19] (after discarding the two 0 exponents). This will mean $\lambda_i + \lambda_{2D-j} = D\langle\alpha\rangle$ or if σ_+ is the average phase space contraction

$$\lambda_j + \lambda_{2(Nd-1)-j} = \langle\alpha\rangle(Nd - 1) = \sigma_+ \tag{I.9}$$

If α is replaced by a constant the argument leading to Eq. (I.7) can be adapted and gives a local pairing: a case discovered earlier in [18].

Appendix J: Gaussian Fluid Equations

The classic Euler equation for an inviscid fluid in a container Ω is a Hamiltonian equation for some Hamiltonian function H. To see this consider the Lagrangian density for a fluid:

$$\mathcal{L}_0(\dot{\delta}, \delta) = \frac{1}{2} \int_\Omega \dot{\delta}(\mathbf{x})^2 d\mathbf{x} \tag{J.1}$$

defined on the space $\mathcal{D}$ of the diffeomorphisms $\mathbf{x} \to \delta(\mathbf{x})$ *of the box Ω.* Impose on the mechanical system defined by the above Lagrangian an *ideal incompressibility constraint*:

$$\det J(\delta)(\mathbf{x}) \overset{def}{=} \det \frac{\partial\delta}{\partial\mathbf{x}}(\mathbf{x}) = \partial\delta_1(\mathbf{x}) \wedge \partial\delta_2(\mathbf{x}) \cdot \partial\delta_3(\mathbf{x}) \equiv 1 \tag{J.2}$$

Consider δ as labeled by $\mathbf{x}$, $\mathbf{x} \in \Omega$ and $i = 1, 2, 3$. The partial derivatives with respect to $\delta_i(\mathbf{x})$ will be, correspondingly, functional derivatives; we shall "ignore" this because a "formally proper" analysis is easy and leads to the same results. By "formal" we do not mean rigorous, but *only* rigorous if the considered functions have suitably strong smoothness properties: a fully rigorous treatment is of course impossible, at least in 3 dimensions for want of reasonable existence, uniqueness and regularity theorems for the Euler's (and later Navier–Stokes') equations.

If Q is a Lagrange multiplier, the stationarity condition corresponding to the Lagrangian density:

$$\mathcal{L}(\dot{\delta}, \delta) = \frac{1}{2} \int \dot{\delta}^2 d\mathbf{x} + \int Q(\mathbf{x})(\det J(\delta)(\mathbf{x}) - 1) d\mathbf{x} \tag{J.3}$$

which leads, after taking into account several cancellations, to:

$$\ddot{\delta} = -(\det J)(J^{-1}\partial Q) = -\det J \frac{\partial \mathbf{x}}{\partial \underset{\sim}{\delta}} \cdot \underset{\sim}{\partial} Q = -(\det J)\frac{\partial p(\delta)}{\partial \delta} \tag{J.4}$$

where $p(\delta(\mathbf{x})) \overset{def}{=} Q(\mathbf{x})$. So that setting $\mathbf{u}(\delta(\mathbf{x})) \overset{def}{=} \dot{\delta}(\mathbf{x})$, we see that:

$$\frac{d\mathbf{u}}{dt} \equiv \frac{\partial \mathbf{u}}{\partial t} + \underset{\sim}{u} \cdot \underset{\sim}{\partial} \mathbf{u} = -\partial p \tag{J.5}$$

which are the Euler equations. And the multiplier $Q(\mathbf{x})$ can be computed as:

$$Q(\mathbf{x}) = p(\delta(\mathbf{x})) = -\left[\Delta^{-1}(\partial \underset{\sim}{u} \cdot \underset{\sim}{\partial} \mathbf{u})\right]_{\delta(\mathbf{x})} \tag{J.6}$$

where the functions in square brackets are regarded as functions of the variables δ and the differential operators also operate over such variable; after the computation the variable δ has to be set equal to $\delta(\mathbf{x})$.

Therefore by using the Lagrangian:

$$\mathcal{L}_i(\dot{\delta}, \delta) = \int \left(\frac{\dot{\delta}(\mathbf{x})^2}{2} - \left(\left[\Delta^{-1}(\partial \underset{\sim}{u} \cdot \underset{\sim}{\partial} \mathbf{u})\right]_{\delta(\mathbf{x})}\right)(\det J(\delta)|_{\mathbf{x}} - 1)\right) d\mathbf{x} \tag{J.7}$$

Lagrangian equations are defined for which the "surface" Σ of the *incompressible diffeomorphisms* in the space $\mathcal{D}$ is *invariant*: these are the diffeomorphisms $\mathbf{x} \to \delta(\mathbf{x})$ such that $J(\delta) = \partial \delta_1 \wedge \partial \delta_2 \cdot \partial \delta_3 \equiv 1$ at every point $\mathbf{x} \in \Omega$.

The above is a rephrasing of the well known idea of Arnold, [15, 20], which implies that the flow generated by the Euler equations can be considered as the *geodesic flow* on the surface Σ defined by the ideal holonomic constraint $\det J(\delta)|_{\mathbf{x}} - 1 = 0$ on the free flow, on the space of the diffeomorphisms, generated by the unconstrained Lagrangian density in Eq. (J.1).

Then Σ is invariant in the sense that the solution to the Lagrangian equations with initial data "on Σ", i.e. such that $\delta \in \Sigma$ and $\partial \cdot \dot{\delta}(\mathbf{x}) = 0$, evolve remaining "on Σ".

The Hamiltonian for the Lagrangian Eq. (J.7) is obtained by computing the canonical momentum $\mathbf{p}(\mathbf{x})$ and the Hamiltonian as:

$$\mathbf{p}(\mathbf{x}) = \frac{\delta \mathcal{L}_i}{\delta \dot{\delta}(\mathbf{x})} = \dot{\delta}(\mathbf{x}) + \cdots$$

$$H(\mathbf{p}, \mathbf{q}) = \frac{1}{2}(G(\mathbf{q})\mathbf{p}, \mathbf{p}) \tag{J.8}$$

where $G(\mathbf{q})$ is a suitable quadratic form that can be read directly from Eq. (J.7) (but it has a somewhat involved expression of no interest here), and the ... (that can also be read from Eq. (J.7)) *are terms that vanish if* $\delta \in \Sigma$ *and* $\partial \cdot \dot{\delta} = 0$, i.e. they vanish on the incompressible motions.

Modifying the Euler equations by the addition of a force $\mathbf{f}(\mathbf{x})$ such that *locally* $\mathbf{f}(\mathbf{x}) = -\partial\,\Phi(\mathbf{x})$ means modifying the equations into:

$$\frac{d\mathbf{u}}{dt} = -\partial p - \partial_{\mathbf{x}}\Phi \tag{J.9}$$

which can be derived from a Lagrangian:

$$\mathcal{L}_i^{\Phi}(\dot{\delta}, \delta) = \mathcal{L}_i(\dot{\delta}, \delta) - \int \Phi(\delta(\mathbf{x}))\,d\mathbf{x} \tag{J.10}$$

which leads to the equations:

$$\dot{\mathbf{u}}(\delta(\mathbf{x})) = -\frac{1}{\varrho}\partial_{\delta}p(\delta(\mathbf{x})) + \partial_{\delta}\Phi(\delta(\mathbf{x})) \tag{J.11}$$

Adding as a further constraint via Gauss's least constraint principle, Eq. (E.4), that the total energy $\mathcal{E} = \int(\dot{\delta}(x))^2 = const$ or the dissipation (per unit time) $\mathcal{D} = \int(\partial\dot{\delta}(x))^2 = const$ should be constant, new equations are obtained that will be called Euler dissipative equations. They have the form:

$$\dot{\mathbf{q}} = \partial_{\mathbf{p}}H \tag{J.12}$$
$$\dot{\mathbf{p}} = -\partial_{\mathbf{q}}H - \partial\Phi - \alpha(\mathbf{p})\mathbf{p}$$

where $\alpha(\mathbf{u})$ is

$$\alpha(\mathbf{u}) = \frac{\int \partial\Phi(x) \cdot \mathbf{u}(x)\,dx}{\int \mathbf{u}(x)^2\,dx}, \quad \text{if } \mathcal{E} \stackrel{def}{=} \int \mathbf{u}^2 dx = const$$

$$\alpha(\mathbf{u}) = -\begin{cases} \dfrac{\int \left(\partial\mathbf{u}\cdot(\partial(\underline{\mathbf{u}}\cdot\underline{\partial})\mathbf{u}) + \Delta\mathbf{u}\cdot\partial\Phi\right)dx}{\int(\Delta\mathbf{u}(x))^2\,dx}, \\[4pt] \text{if } \mathcal{D} \stackrel{def}{=} \int(\partial\mathbf{u})^2 dx = const \end{cases} \tag{J.13}$$

as far as the motions which have an incompressible initial datum are concerned.

The equations can be written in more familiar notation. for instance in the first case, as

$$\dot{\delta}(x) = \mathbf{u}(\delta(x))$$

$$\dot{\mathbf{u}}(x) = -\partial p + \partial\Phi(x) - \alpha(\mathbf{u})\mathbf{u}(x), \quad \alpha(\mathbf{u}) = \frac{\int \partial\Phi(x) \cdot \mathbf{u}(x)dx}{\int \mathbf{u}(x)^2\,dx} \tag{J.14}$$

Since $\int \mathbf{p}^2 \equiv \int \mathbf{u}^2$ is the motion energy the pairing proof in Appendix I applies formally and the Lyapunov exponents are paired in the sense of Sect. 4.4. Actually the pairing occurs also locally, as in the case of Appendix I.

If $\alpha(\mathbf{p})$ is replaced by a constant χ the pairing remains true, as it follows from Appendix I.

Clearly the equations in Eulerian form have "half the number of degrees of freedom", as they involve only the velocities. This means that a pairing rule does not apply: however it might be that the exponents of the Euler equations bear a trace of the pairing rule, as discussed in [21].

In the second case the equations become $\dot{\delta}(x) = \mathbf{u}(\delta(x))$ with $\partial \cdot \mathbf{u} = 0$ and (using $\dot{\mathbf{u}} = \partial_t \mathbf{u} + (\partial \cdot \underline{\mathbf{u}})\mathbf{u}$)

$$\dot{\mathbf{u}}(x) = -\partial p + \partial \Phi(x) + \alpha(\mathbf{u})\Delta \mathbf{u}(x),$$

$$\alpha(\mathbf{u}) = -\frac{\int (\widehat{\partial}\mathbf{u} \cdot (\widehat{\partial}(\underline{\mathbf{u}} \cdot \partial)\mathbf{u}) + \Delta \mathbf{u} \cdot \mathbf{g})\, dx}{\int (\Delta \mathbf{u}(x))^2 dx}, \tag{J.15}$$

with $\partial \cdot \mathbf{u} = 0$ which can be called the "Gaussian Navier–Stokes equations", [21].

Appendix K: Jarzinsky's Formula

An immediate consequence of the fluctuation theorem is

$$\langle e^{-\int_0^\tau \varepsilon(S_t x)\, dt}\rangle_{SRB} = e^{O(1)} \tag{K.1}$$

i.e. $\langle e^{-\int_0^\tau \varepsilon(S_t x)\, dt}\rangle_{SRB}$ stays bounded as $\tau \to \infty$. This is a relation that I call *Bonetto's formula* (private communication, [22, Eq. (16)]), see [4, Eq. (9.10.4)]; it can be also written, somewhat imprecisely and for mnemonic purposes, [22],

$$\langle e^{-\int_0^\tau \varepsilon(S_t x)\, dt}\rangle_{SRB} \xrightarrow[\tau \to \infty]{} 1 \tag{K.2}$$

which *would be exact* if the fluctuation theorem in the form Eq. (4.6.1) held without the $O(1)$ corrections for finite τ (rather than in the limit as $\tau \to \infty$).

This relation bears resemblance to *Jarzynski's formula*, [23], which deals with a canonical Gibbs distribution (in a finite volume) corresponding to a Hamiltonian $H_0(p, q)$ and temperature $T = (k_B \beta)^{-1}$, and with a time dependent family of Hamiltonians $H(p, q, t)$ which interpolates between H_0 and a second Hamiltonian H_1 as t grows from 0 to 1 (in suitable units) which is called *a protocol*.

Imagine to extract samples (p, q) with a canonical probability distribution $\mu_0(dpdq) = Z_0^{-1} e^{-\beta H_0(p,q)} dpdq$, with Z_0 being the canonical partition function, and let $S_{0,t}(p, q)$ be the solution of the Hamiltonian *time dependent* equations

$\dot{p} = -\partial_q H(p, q, t), \dot{q} = \partial_p H(p, q, t)$ for $0 \leq t \leq 1$. Then [23, 24], establish an identity as follows.

Let $(p', q') \overset{def}{=} S_{0,1}(p, q)$ and let $W(p', q') \overset{def}{=} H_1(p', q') - H_0(p, q)$, then the distribution $Z_1^{-1} e^{-\beta H_1(p', q')} dp' dq'$ is exactly equal to $\frac{Z_0}{Z_1} e^{-\beta W(p', q')} \mu_0(dpdq)$. Hence (as $dp'dq' \equiv dpdq$ by Liouville's theorem for Hamiltonian flows):

$$\langle e^{-\beta W} \rangle_{\mu_0} = \frac{Z_1}{Z_0} = e^{-\beta \Delta F(\beta)} \quad \text{or equivalently} \tag{K.3}$$

$$\langle e^{\beta(\Delta F - W)} \rangle = 1$$

where the average is with respect to the Gibbs distribution μ_0 and ΔF is the free energy variation between the equilibrium states with Hamiltonians H_1 and H_0 respectively.

Remarks (i) The reader will recognize in this exact identity an instance of the Monte Carlo method (analogically implemented rather than in a simulation). Its interest lies in the fact that it can be implemented *without actually knowing* neither H_0 nor H_1 nor the *protocol* $H(p, q, t)$. It has to be stressed that the protocol, i.e. the process of varying the Hamiltonian, has an arbitrarily prefixed duration which has *nothing to do* with the time that the system will need to reach the equilibrium state with Hamiltonian H_1 of which we want to evaluate the free energy variation.

(ii) If one wants to evaluate the difference in free energy between two equilibrium states at the same temperature in a system that one can construct in a laboratory then "all one has to do" is

   (a) Fix a protocol, i.e. a procedure to transform the forces acting on the system along a well defined *fixed once and for all* path from the initial values to the final values in a fixed time interval ($t = 1$ in some units), and
   (b) Measure the energy variation W generated by the "machines" implementing the protocol. This is a really measurable quantity at least in the cases in which W can be interpreted as work done on the system, or related to it.
   (c) Then average of the exponential of $-\beta W$ with respect to a large number of repetition of the protocol. This can be useful even, and perhaps mainly, in biological experiments.

(iii) If the "protocol" conserves energy (like a Joule expansion of a gas) or if the difference $W = H_1(p', q') - H_0(p, q)$ has zero average in the equilibrium state μ_0 we get, by Jensen's inequality (i.e. by the convexity of the exponential function: $\langle e^A \rangle \geq e^{\langle A \rangle}$), that $\Delta F \leq 0$ as expected from Thermodynamics.

(iv) The measurability of W is a difficult question, to be discussed on a case by case basis. It is often possible to identify it with the "work done by the machines implementing the protocol".

The two formulae Eqs. (K.1) and (K.3) are however very different:

(1) the $\int_0^T \sigma(S_t x)\, dt$ is an entropy production in a non equilibrium stationary state rather than $\Delta F - W$ in a *not stationary process* lasting a prefixed time i.e. two completely different situations. In Sect. 4.8 the relation between Eq. (K.1) and the Green-Kubo formula is discussed.
(2) the average is over the SRB distribution of a stationary state, in general out of equilibrium, rather than on a canonical equilibrium state.
(3) the Eq. (K.1), says that $\langle e^{-\int_0^T \varepsilon(S_t x)\, dt} \rangle_{SRB}$ is bounded as $\tau \to \infty$ rather than being 1 exactly, [24].

The Eq. (K.3) has proved useful in various equilibrium problems (to evaluate the free energy variation when an equilibrium state with Hamiltonian H_0 is compared to one with Hamiltonian H_1); hence it has some interest to investigate whether Eq. (K.2) can have some consequences.

If a system is in a steady state and produces entropy at rate ε_+ (e.g. a living organism feeding on a background) the fluctuation theorem Eq. (4.6.1) and its consequence, Eq. (K.2), gives us informations on the fluctuations of entropy production, i.e. of heat produced, and Eq. (K.2) *could be useful*, for instance, to check that all relevant heat transfers have been properly taken into account.

Appendix L: Evans–Searles' Formula

It has been remarked that time reversal I puts some constraints on fluctuations in systems that evolve *towards non equilibrium* starting *from an equilibrium state* μ_0 or, more generally from a state μ_0 which is proportional to the volume measure on phase space and $\mu_0(IE) \equiv \mu_0(E)$ (but not necessarily stationary).

For instance if the equations of motion are $\dot{x} = f(x)$ and $-\sigma(x) =$ divergence of f, i.e. $\sigma(x) = -\partial \cdot f(x)$, where $\dot{x} = f_0(x) + Eg(x)$ with $\dot{x} = f_0(x)$ a volume preserving evolution and E a parameter. It is supposed that $\sigma(Ix) = -\sigma(x)$ and $\mu_0(IE) \equiv \mu_0(E)$. Then one could pose the question, [25],

Which is the probability that in time t the volume contracts by the amount e^A with $A = \int_0^t \sigma(S_t x)dt$, compared to that of the opposite event $-A$?

If $\mathcal{E}_A =$ set of points whose neighborhoods contract with contraction A in time t, then the set $\mathcal{E}_A$ at time t becomes (by definition) the set $S_t \mathcal{E}_A$ with $\mu_0(S_t \mathcal{E}_A) = e^{-A}\mu_0(\mathcal{E}_A)$, $A = \int_0^t \sigma(S_\tau x)d\tau$.

However $\mathcal{E}_A^- \overset{def}{=} I S_t \mathcal{E}_A$ is the set of points $\mathcal{E}_{-A}$ which contract by $-A$ as:

$$e^{-\int_0^T \sigma(S_\tau I S_t x)d\tau} \equiv e^{-\int_0^t \sigma(S_\tau S_{-t} Ix)d\tau} \equiv e^{-\int_0^t \sigma(I S_{-\tau} S_t x)d\tau}$$

$$\equiv e^{+\int_0^T \sigma(S_{t-\tau}x)d\tau} \equiv e^{+\int_0^T \sigma(S_\tau x)d\tau} \equiv e^A \qquad (L.1)$$

In other words the set $\mathcal{E}_A$ of points which contract by A in time t becomes the set of points whose time reversed images is the set $\mathcal{E}_A^{-} \overset{def}{=} I S_t \mathcal{E}_A$ which contract by A. The measures of such sets are $\mu_0(\mathcal{E}_A)$ and $\mu_0(I S_t \mathcal{E}_A) \equiv \mu_0(\mathcal{E}_A) e^{-A} \equiv \mu_0(\mathcal{E}_A^{-})$ (recall that I is measure preserving), hence

$$\frac{\mu_0(\mathcal{E}_A)}{\mu_0(\mathcal{E}_A^{-})} \equiv e^A \tag{L.2}$$

for any A (as long as it is "possible", [25]).[6]

This has been called *transient fluctuation theorem*. It is extremely general and does not depend on any chaoticity assumption. Just reversibility and time reversal symmetry and the evolution of an initial distribution μ_0 which is invariant under time reversal (independently of the dynamics that evolves it in time). It says nothing about the SRB distribution (which is singular with respect to the Liouville distribution).

Some claims that occasionally can be found in the literature that the above relation is equivalent to the fluctuation theorem rely on further assumptions.

The similarity with the conceptually completely different expression of the fluctuation theorem Eq. (4.6.5) explains, perhaps, why this is very often confused with the fluctuation theorem.

It is easy to exhibit examples of time reversible maps or flows, with as many Lyapunov exponents, positive and negative, for which the transient fluctuation theorem holds but the fluctuation relation fails because the chaotic hypothesis fails (i.e. the fluctuation theorem cannot be applied). The example in [26, Eq. (4)] has an attracting set which is not chaotic, yet it proves that it could not be claimed that the fluctuation theorem is a consequence of the above transient theorem (in absence of further assumptions): furthermore it is as easy to give also examples with chaotic systems.[7]

Relations of the kind of the transient fluctuation theorem have appeared in the literature quite early in the development of non equilibrium theories, perhaps the first have been [27, 28].

---

[6] For instance in the Hamiltonian case $A \neq 0$ would be impossible.

[7] As an example (from F. Bonetto) let $x = (\varphi, \psi, \xi)$ with φ and ψ on the sphere and ξ a point on a manifold $\mathcal{A}$ of arbitrarily prefixed dimension on which a reversible Anosov map S_0 acts with time reversal map I_0; let $\overline{S}$ be a map of the sphere which has the north pole as a repelling fixed point and the south pole as an attractive fixed point driving any other point exponentially fast to the south pole (and exponentially fast away from the north pole). Define $S(\varphi, \psi, \xi) \overset{def}{=} (\overline{S}\varphi, \overline{S}^{-1}\psi, S_0\xi)$ and let $I(\varphi, \psi, \xi) \overset{def}{=} (\psi, \varphi, I_0\xi)$: it is $IS = S^{-1}I$, the motions are chaotic, but the system is not Anosov and obviously the fluctuation relation does not hold if the initial data are sampled, for instance, with the distribution $\frac{d\varphi d\psi}{(4\pi)^2} \times d\xi$, $d\xi$ being the normalized volume measure on $\mathcal{A}$; *however the transient fluctuation theorem holds*, of course.

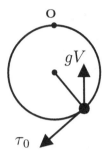

Fig. M.1 Pendulum with inertia J, gravity $2Vg$ ("directed up"), torque τ_0, subject to damping ξ and white noise w (not represented) the "equilibrium position" at $\tau_0 = 0$ is O

Appendix M: Forced Pendulum with Noise

The analysis of a non equilibrium problem will be, as an example, a pendulum subject to a torque τ_0, friction ξ and white noise $\sqrt{\frac{2\xi}{\beta}}\,\dot{w}$ ("Langevin stochastic thermostat") at temperature β^{-1}, related to the model in Eq. (5.8.1).

Appendices M, N, O, P describe the work in [29]. The equation of motion is the stochastic equation on $T^1 \times R$:

$$\dot{q} = \frac{p}{J}, \qquad \dot{p} = -\partial_q U - \tau_0 - \frac{\xi}{J}p + \sqrt{\frac{2\xi}{\beta}}\,\dot{w}, \qquad U(q) \overset{def}{=} 2\,V_0 \cos q \qquad \text{(M.1)}$$

where J is the pendulum inertia, ξ the friction, $\dot{w}$ a standard white noise with increments $dw = w(t+dt) - w(t)$ of variance dt, so that $\sqrt{\frac{2\xi}{\beta}}\,\dot{w}$ is a Langevin random force at inverse temperature β; the gravity constant will be $-2V_0$ (Fig. M.1).

If gravity $V_0 = 0$ or torque $\tau_0 = 0$ the stationary distribution is simply, respectively, given by $F(q, p) = \dfrac{e^{-\frac{\beta}{2}(p + \frac{\tau_0 J}{\xi})^2}}{\sqrt{2\pi\beta^{-1}}}$ or by $F(q, p) = \dfrac{e^{-\frac{\beta}{2}(p^2 + U(q))}}{\sqrt{2\pi\beta^{-1}}}$.

The stationary state of this system can be shown to exist and to be described by a smooth function $0 \le F(p, q) \in L_1(dpdp) \cap L_2(dpdq)$ on phase space: it solves the differential equation, [30]:

$$\mathcal{L}^* F \overset{def}{=} -\left\{ \left(\frac{p}{J}\partial_q F(q, p) - (\partial_q U(q) + \tau_0)\partial_p F(q, p) \right) \right.$$
$$\left. - \xi \left(\beta^{-1}\partial_p^2 F(q, p) + \frac{1}{J}\partial_p(p\,F(q, p)) \right) \right\} = 0 \qquad \text{(M.2)}$$

Consider only cases in which τ_0, V_0 are small; there are two qualitatively different regimes: if $\tau_0 > 0$, $V_0 = gV$ then for g small ($\tau_0 \gg gV$) the pendulum will in the average rotate on a time scale of order $\frac{J\tau_0}{\xi}$; if, instead, $V_0 > 0, \tau_0 = g\tau$ the pendulum will oscillate, very rarely performing full rotations.

Here $\tau_0 = g\tau$, $V_0 = gV$ will be chosen with τ, V fixed and g a dimensionless strength parameter.

The solution of $\mathcal{L}^* F = 0$ will be searched within the class of probability distributions satisfying:

(H1) *The function* $F(p, q)$ *is smooth and admits an expansion in Hermite's polynomials (or "Wick's monomials")* H_n *of the form*[8]:

$$F(q, p) = G_\beta(p) \sum_a \varrho_a(q) : p^a :, \qquad G_\beta(p) = \frac{e^{-\frac{\beta}{2J}p^2}}{\sqrt{2\pi J \beta^{-1}}}$$

$$: p^n : \overset{def}{=} \left(2J\beta^{-1}\right)^{\frac{n}{2}} H_n\left(\frac{p}{\sqrt{2J\beta^{-1}}}\right) \tag{M.3}$$

where $a \geq 0$ *are integers; so that* $\int : p^n :: p^m : G_\beta(p)\,dp = \delta_{nm} n! (J\beta^{-1})^n$.

(H2) *The coefficients* $\varrho_n(q)$ *are* C^∞-*differentiable in* q, g *and the* p, q, g-*derivatives of* F *can be computed by term by term differentiation, obtaining asymptotic series.*

It is known that the equation $\mathcal{L}^* F(p, q) = 0$ admits a unique smooth and positive solution in $L_1(dpdp) \cap L_2(dpdq)$ (cf. [30]), with $\int F dpdq = 1$. However whether they satisfy the (H1),(H2) does not seem to have been established mathematically.

Consider the expansion for the cefficients $\varrho_a(q)$ in Eq. (M.3) (asymptotic by assumption (H2)):

$$\varrho_n(q) = \sum_{r \geq 0} \varrho_n^{[r]}(q) g^r. \tag{M.4}$$

The properties (H1), (H2) allow us to perform the algebra needed to turn the stationarity condition $\mathcal{L}^* F = 0$ into a hierarchy of equations for the coefficients $\varrho_n(q)$, $\forall n \geq 0$. After some algebraic calculations it is found:

$$n\beta^{-1}\partial\varrho_n(q) + \left[\frac{1}{J}\partial\varrho_{n-2}(q) + \frac{\xi}{b}J(\partial U(q) + \tau)\varrho_{n-2}(q)\right.$$

$$\left. + (n-1)\frac{\xi}{J}\varrho_{n-1}(q)\right] = 0 \tag{M.5}$$

where $\varrho_{-1}, \varrho_{-2}$ are to be set $= 0$.

The main result is about a formal solution in powers of g of Eq. (M.5): it is possible to exhibit an asymptotic expansion $\varrho_n(q) = \sum_{r \geq 0} \varrho_n^{[r]}(q) g^r$ which solves the Eq. (M.5) *formally* (i.e. order by order) with coefficients $\varrho_n^{[r]}(q)$ which are well defined and such that the series $\sum_{n=0}^\infty \varrho_n^{[r]}(q) : p^n : \overset{def}{=} \varrho^{[r]}(p, q)$ is convergent for all $r \geq 0$ so that:

---

[8] The normalization of H_n here is 2^{-n} the one in [31], so that the leading coefficient of $: p^n :$ is 1.

Theorem *For all orders $r \geq 0$ the derivatives $\varrho_n^{[r]}(q) \overset{def}{=} \partial_g^r \varrho_n(q)|_{g=0}$ have Fourier transforms $\sum_{k=-\infty}^{\infty} \varrho_{n,k}^{[r]} e^{ikq}$ and*

(1) $\varrho_{n,k}^{[r]}$ can be determined by a constructive algorithm

(2) the coefficients $\varrho_{n,k}^{[r]}$ vanish for $|k| > r$ and satisfy the bounds

$$\xi^n |\varrho_{n,k}^{[r]}| \leq \mathcal{A}_r \frac{r^{2n}}{n!} \delta_{|k| \leq r}, \qquad \forall r, k \tag{M.6}$$

for $\mathcal{A}_r$ suitably chosen.

Remarks (1) Adapting [30] it can be seen that $\mathcal{L}^* F(p, q) = 0$ admits a unique solution smooth in p, q. However its analyticity in g and the properties of its representation in the form in Eq. (M.3), if possible, are not solved by Theorem 1 as it only yields Taylor coefficients of a *formal expansion* of $F(p, q)$, Eq. (M.3), or (equivalently) of $\varrho(p, q)$ in powers of g around $g = 0$.

(2) The result is not really satisfactory because convergence or summation rules conditions for the series are not determined; hence the "solution" remains a formal one in the above sense. This is a very interesting problem: if τ_0 is taken much smaller than gV, e.g. $\tau_0 = g^2 \tau$, or much larger $\tau_0 = g\tau_0$, $U = g^2 \tau_0$ the problem does not look simpler: and of course the transition between the two regimes (if any) is a kind of phase transition (this explains, perhaps, why the problem seems still open).

(3) Therefore, by remark (2), analyticity in g, for g small is not to be expected. The same method of proof yields formal expansions in powers of g if $V_0 = gV$, $\tau_0 = g^2 \tau$ or if $V_0 = g^2 V$, $\tau_0 = g\tau$ (or even if τ_0 is fixed and $V_0 = gV$, [29]): in this case analyticity at small g could be expected. But the estimates that could be derived, by the methods of in the following Appendices, in the corresponding versions of theorem 1 seem to be essentially the same.

(4) If $\bar{\varrho}_n \overset{def}{=} \int \varrho_n(q) \frac{dq}{2\pi}$ and $\tilde{\varrho}_n(q) \overset{def}{=} \varrho_n(q) - \bar{\varrho}_n$ Eq. (M.6) yields the identities:

$$\int \varrho_0(q) \frac{dq}{2\pi} = 1, \quad \tilde{\varrho}_1 = 0 \tag{M.7}$$

Equation (M.4) for the functions $n! \xi^n \varrho_n(q)$ has dimensionless form, for $n \geq 1$,

$$\partial \tilde{\sigma}_n = - \eta(n-1)(\partial \tilde{\sigma}_{n-2} + \beta \widetilde{\partial U \tilde{\sigma}}_{n-2} + \beta \partial U \bar{\sigma}_{n-2} \tag{M.8}$$
$$+ \beta \tau_0 \tilde{\sigma}_{n-2} + \tilde{\sigma}_{n-1})$$
$$\bar{\sigma}_n = - \left(\overline{\beta \partial U \tilde{\sigma}_{n-1}} + \beta \tau_0 \bar{\sigma}_{n-1} \right)$$

where σ's with negative labels are intended to be 0.

The equation is conveniently written for $\sigma_{n,k} \overset{def}{=} \frac{1}{2\pi} \int_0^{2\pi} e^{-ikq} \sigma_n(q) dq$, $k = 0, 1, \ldots$ ($\sigma_{n,-k} \equiv \sigma_{n,k}^{c.c.}$). After defining $\mathbf{S}_{n,k} \overset{def}{=} \begin{pmatrix} \widetilde{\sigma}_{n,k} \\ \widetilde{\sigma}_{n-1,k} \end{pmatrix}$ for $n \geq 1$, it is natural to introduce g, τ–*independent* 2×2 *matrices* $M_{n,k}$

$$M_{n+1,k} \overset{def}{=} \begin{pmatrix} i\frac{n}{k}\eta & -n\eta \\ 1 & 0 \end{pmatrix} \tag{M.9}$$

so that the Eq. (M.8) can be written more concisely, for $n \geq 0$,

$$\mathbf{S}_{n+1,k} = M_{n+1,k}\left(\mathbf{S}_{n,k} + \mathbf{X}_{n+1,k}\right), \quad \mathbf{X}_{n+1,k} \overset{def}{=} \begin{pmatrix} 0 \\ x_{n+1,k} \end{pmatrix},$$

$$x_{n+1,k} \overset{def}{=} \beta g V \left(\delta_{|k|=1}\overline{\sigma}_{n-1} + \sum_{k'=\pm 1} \frac{k'}{k}\widetilde{\sigma}_{n-1,k-k'}\right) + \frac{\beta g \tau}{ik}\widetilde{\sigma}_{n-1,k},$$

$$\overline{\sigma}_{n+1} = -(\overline{\beta \partial U \widetilde{\sigma}_n} + \beta g \tau \overline{\sigma}_n) \overset{def}{=} v_{n+1}, \quad x_{1,k} \overset{def}{=} 0. \tag{M.10}$$

Expanding the latter equation in powers of g the recursion can be reduced to an iterative determination of $x_{n,k}^{[r]}$, $\overline{\sigma}_n^{[r]}$ starting from $r = 1$, as the case $r = 0$ can be evaluated as $x_{n,k}^{[0]} = 0$, $\overline{\sigma}_n^{[0]} = (-\beta\tau)^n$, $\mathbf{S}_n^{[0]} = 0$, since $\widetilde{\sigma}_n^{[0]} \equiv 0$.

Setting $\mathbf{S}_{0,k}^{[r]} \overset{def}{=} \begin{pmatrix} y_k^{[r]} \\ 0 \end{pmatrix}$, $\mathbf{S}_{0,k}^{[r]} \overset{def}{=} \begin{pmatrix} 0 \\ y_k^{[r]} \end{pmatrix}$, in agreement with Eq. (M.7), for $r \geq 1$ it is

$$\mathbf{S}_{2,k}^{[r]} = \begin{pmatrix} -\eta\left(y_k^{[r]} + x_{2,k}^{[r]}\right) \\ 0 \end{pmatrix} = \begin{pmatrix} \widetilde{\sigma}_{2,k}^{[r]} \\ 0 \end{pmatrix}, \quad \overline{\sigma}_0^{[r]} = 0. \tag{M.11}$$

The Eq. (M.10), for $r \geq 1$, is related to the general, r-independent, equations for $n \geq 2$ conveniently written computing the inverse matrix $M_{n,k}^{-1}$

$$M_{n+1,k}^{-1} = \begin{pmatrix} 0 & 1 \\ -\frac{1}{n\eta} & -\frac{1}{ik} \end{pmatrix}$$

$$\mathbf{S}_{n,k}^{[r]} = M_{n+1,k}^{-1}\mathbf{S}_{n+1,k}^{[r]} - \mathbf{X}_{n+1,k}^{[r]}, \quad \mathbf{S}_{2,k}^{[r]} = \widetilde{\sigma}_{2,k}^{[r]}\begin{pmatrix} 1 \\ 0 \end{pmatrix}$$

$$\mathbf{X}_{n,k}^{[r]} = x_{n,k}^{[r]}\begin{pmatrix} 0 \\ 1 \end{pmatrix}, \quad x_{0,k}^{[r]} = x_{1,k}^{[r]} = 0$$

$$\overline{\sigma}_n^{[r]} = v_n^{[r]}, \quad \overline{\sigma}_0^{[r]} = 0, \tag{M.12}$$

For a given pair (r, k), these are inhomogeneous equations in the unknowns $(\mathbf{S}_n, \overline{\sigma}_n)_{n \geq 2}$ imagining $\mathbf{S}_2, \mathbf{X}_n, v_n$ as known inhomogeneous quantities as prescribed by Eq. (M.10).

Let $(M_p^{-1})^{*s} \overset{def}{=} M_p^{-1} \cdots M_{p+s-1}^{-1}$ for $s \geq 1$, $(M_p^{-1})^{*0} \overset{def}{=} 1$. Define:

$$\boldsymbol{\xi}_n \overset{def}{=} -\sum_{h=n}^{\infty} (M_{n+1}^{-1})^{*(h-n)} \mathbf{X}_{h+1}, \qquad \overline{\sigma}_n \overset{def}{=} w, \qquad n \geq 2 \qquad (\text{M.13})$$

then $M_{n+1}^{-1} \boldsymbol{\xi}_{n+1} = \boldsymbol{\xi}_n + \mathbf{X}_{n+1}$, if the series converges. Hence the $\mathbf{S}_2^{[r]}$ is determined simply by the conditions

(a) convergence of the series in Eq. (M.13) and
(b) the second component of $\mathbf{S}_{2,k}^{[r]}$ vanishes.

The $x_{n,k}^{[r]}$, $v_n^{[r]}$ are determined, for $r \geq 1$, in terms of the lower order quantities: the first of Eq. (M.11) determines $y_k^{[r]} + x_{2,k}^{[r]}$, and $x_{2,k}^{[r]}$, $\mathbf{S}_0^{[r]}$ are derived from Eq. (M.10) for $n = 1$ determines while $\overline{\sigma}_n^{[r]}$ is determined by Eq. (M.10).

It remains to see if the convergence and vanishing conditions, (a) and (b), on the series in Eq. (M.13) can be met recursively.

The iteration involves considering products of the matrices $M_{k,n}^{-1}$, hence leads to a problem on continued fractions. The estimates are somewhat long but standard and the theorem follows: more details are in Appendices N, O, P.

A simpler problem is the so called overdamped pendulum, which can be solved exactly and which shows, nevertheless, surprising properties, [32]: it corresponds to the equation $\dot{q} = -\frac{1}{\xi}(\partial U + \tau) + \sqrt{\frac{2}{\beta\xi}}\,\dot{w}$.

Appendix N: Solution of Eq. (M.10)

With reference to Eq. (M.13) define $|0\rangle \equiv |\downarrow\rangle \equiv \binom{0}{1}$, $|1\rangle \equiv |\uparrow\rangle \equiv \binom{1}{0}$ and:

$$\langle\nu|(M_{n+1}^{-1})^{*(h-n)}|\nu'\rangle \overset{def}{=} \frac{\Lambda(n+\nu, h-\nu')}{(-\eta(h-1))^{\nu'}}, \qquad \nu, \nu' = 0, 1, \quad n \leq h \qquad (\text{N.1})$$

which, *provided* $\Lambda(n, h) \neq 0$, implies the identities

$$\zeta(n, h) \overset{def}{=} \binom{\zeta(n,h)_1}{1}, \quad \zeta(n, h)_1 = \frac{\Lambda(n+1, h)}{\Lambda(n, h)}, \qquad 2 \leq n \leq h \qquad (\text{N.2})$$

$$(M_{n'+1}^{-1})^{*(n-n')}\zeta(n, n') = \frac{\Lambda(n', N)}{\Lambda(n, N)}\zeta(n', N), \quad n+1 > n'$$

(interpret $\zeta(n, n)_1$ as 0), where the second relation will be called the *eigenvector property* of the $\zeta(n, h)$. It also implies the recurrence

$$\varphi(n, h) \overset{def}{=} -\frac{\zeta(n, h)_1}{ik} = \frac{1}{1 + \frac{z}{n}\varphi(n+1, h)} \tag{N.3}$$

$$= \frac{1}{1 + \frac{z}{n}\cfrac{1}{1+\frac{z}{n+1}} \cdots \cfrac{1}{1+\frac{z}{n-2}}}, \quad h - 2 \geq n, \quad z \overset{def}{=} \frac{k^2}{\eta} > 0$$

and $\varphi(n - 1, n) = 1$, $\varphi(n, n) = 0$, representing the ζ's as continued fractions and showing that $\zeta(n, h)$ and, as $h \to \infty$, the limits $\zeta(n, \infty)$ are analytic in z for $|z| < \frac{1}{4}$, [33, p. 45].

The continued fraction is the S-fraction $\frac{n-1}{z} \mathbf{K}_{m=n-1}^{\infty}(\frac{z/m}{1})$, following [33, p. 35], and defines a holomorphic function of z in the complex plane cut along the negative real axis, see [33, p. 47, (A)]. The $\varphi(n, h)$ is also a (truncated) S-fraction obtained by setting $m = \infty$ for $m \geq h$ in the previous continued fraction. Hence, by [33, p. 47, (B)], $\varphi(n, h)$, $\varphi(n, \infty)$ are holomorphic for $|z| < \frac{1}{4}$, continuous and bounded by $\frac{1}{2}$ in $|z| \leq \frac{1}{4}$, [33, p. 45].

The definitions imply $\boldsymbol{\xi}_n \equiv -\sum_{h=n}^{\infty} x_{h+1} \Lambda(n, h)\zeta(n, h)$, if the series converges. Furthermore, if the limits $\lim_{N \to \infty} \frac{\Lambda(n,N)}{\Lambda(2,N)}$ exist, symbolically denoted $\frac{\Lambda(n,\infty)}{\Lambda(2,\infty)}$, then

$$\mathbf{T}_n^0 \overset{def}{=} \frac{\Lambda(n, \infty)}{\Lambda(2, \infty)}\zeta(n), \quad \zeta(n) \overset{def}{=} \zeta(n, \infty) \tag{N.4}$$

is a solution of $M_{n+1}^{-1}\boldsymbol{\xi}_{n+1} = \boldsymbol{\xi}_n + \mathbf{X}_{n+1}$, with $\mathbf{X} = 0$ and some initial data for $n = 2$. A solution to the r-th order equations will thus have the form

$$\mathbf{S}_n = \boldsymbol{\xi}_n + \lambda \mathbf{T}_n^0, \tag{N.5}$$

where the constant λ will be fixed to match the data at $n = 2$ (i.e. to have a vanishing second component of $\mathbf{S}_n$). In the case of the r-th order equation, the initial data of interest are $\bar{\sigma}_2^{[r]}$ and $x_{2,k}^{[r]}$. Furthermore $\mathbf{X}_{n,k}^{[r]}$, $v_n^{[r]}$ are given by Eq. (M.10), in terms of quantities of order $r - 1$.

This means that the (unique) solution to the recursion with the initial data $\mathbf{S}_2^{[r]} = s| \uparrow \rangle$ has necessarily the form

$$\mathbf{S}_{2,k}^{[r]} = -\sum_{h=2}^{\infty} x_{h+1,k}^{[r]} \Lambda(2, h)(\zeta(2, h) - \zeta(2)), \tag{N.6}$$

which is proportional to $| \uparrow \rangle$, because the second components of $\zeta(2, h)$ and $\zeta(2)$ are identically 1, by definition, and it implies $\lambda = \sum_{h=2}^{\infty} x_{h+1,k}^{[r]} \Lambda(2, h)$.

Proceeding formally, the $\mathbf{S}_n^{[r-1]}$ will be given, for $n > 2$, by applying the recursion; since $\boldsymbol{\xi}_n$ is a formal solution and $\zeta(n)$ has the eigenvector property Eq. (N.2) it is:

$$\mathbf{S}_{n,k}^{[r]} = \sum_{h=2}^{n-1} x_{h+1,k}^{[r]} \Lambda(2,h) \frac{\Lambda(n,\infty)}{\Lambda(2,\infty)} \zeta(n) \tag{N.7}$$

$$- \sum_{h=n}^{\infty} x_{h+1,k}^{[r]} \left(\Lambda(n,h)\zeta(n,h) - \Lambda(2,h) \frac{\Lambda(n,\infty)}{\Lambda(2,\infty)} \zeta(n) \right).$$

It should be stressed that the series in Eq. (N.3) might diverge and, nevertheless, in Eq. (N.7) cancellations may (and will) occur so that it would still be a solution if the series in Eq. (N.7) converges (as it can be checked by inserting it in the Eq. (N.2)).

To compute the first component of $\mathbf{S}_{n,k}^{[r]}$, we left multiply Eq. (N.7) by $\langle \uparrow \mid$ considering that $\zeta_k(n,m)_1 = \frac{\Lambda(n+1,m)}{\Lambda(n,m)}$ by the first of the (N.2) and that $\Lambda(n+1,n) = \zeta_k(n,n)_1 \Lambda(n,n) \equiv 0$. Setting $\Lambda(n,m) = 0, \forall m < n$, we obtain (after patient algebra):

$$\tilde{\sigma}_{n,k}^{[r]} = \sum_{m=2}^{n} x_{m+1,k}^{[r]} \Lambda(2,m) \frac{\Lambda(n+1,\infty)}{\Lambda(2,\infty)} \tag{N.8}$$

$$- \sum_{m=n+1}^{\infty} x_{m+1,k}^{[r]} \Lambda(n+1,m) \left(1 - \frac{\Lambda(2,m)}{\Lambda(n+1,m)} \frac{\Lambda(n,\infty)}{\Lambda(2,\infty)} \right)$$

$$\equiv \sum_{m=2}^{n} x_{m+1,k}^{[r]} \left(\prod_{j=2}^{m-1} \frac{\zeta(j,\infty)_1}{\zeta(j,m)_1} \right) \left(\prod_{j=m}^{n} \zeta(j,\infty)_1 \right)$$

$$- \sum_{m=n+1}^{\infty} x_{m+1,k}^{[r]} \left(\prod_{j=n+1}^{m-1} \frac{1}{\zeta(j,m)_1} \right) \left(1 - \prod_{j=2}^{n} \frac{\zeta(j,\infty)_1}{\zeta(j,m)_1} \right),$$

for $n \geq 2$. From $\tilde{\sigma}_{2,k}^{[r]}$, and from $x_{2,k}^{[r]}$ derived from Eq. (M.10), and using also $\tilde{\sigma}_{1,k}^{[r]} = 0$ (see Eq. N.1) the "main unknown" $\tilde{\sigma}_{0,k}^{[r]}$, is computed.

As stressed above, Eqs. (N.6) and (N.8) are acceptable if the series converge. For $r = 1$ the series in Eqs. (N.3) and (N.8) are identically 0 because $x_{n+1,k}^{[1]} = -\beta V \delta_{n=1} \delta_{|k|=1}$. So it will be possible to try an iterative construction, $\forall \eta, \beta\tau, r > 0$.

Remarks Hence $\mathbf{S}_n^{[1]} = 0, \forall n \geq 2$, and consequently $\tilde{\sigma}_{0,k}^{[1]} = -x_{2,k}^{[1]} \equiv \beta V \delta_{|k|=1}$.

Appendix O: Iteration for Eq. (M.10)

If $x_{n,k}^{[r']}, \bar{\sigma}_n^{[r']}$ are known for $r' < r$, it is possible to compute $\tilde{\sigma}_{n,k}^{[r]}$ from Eq. (N.8) and using them to define implicitly the kernels $\theta_k(n,m)$:

$$\tilde{\sigma}_{n,k}^{[r]} = \sum_{m=2}^{\infty} \theta_k(n,m) x_{m+1,k}^{[r]}, \quad n \geq 2$$

$$\tilde{\sigma}_1^{[r]} = 0$$

$$\tilde{\sigma}_{0,k}^{[r]} = \frac{-1}{\eta} \left(\sum_{m=2}^{\infty} \theta_k(2,m) x_{m+1,k}^{[r]} + \eta \beta V \sum_{|k'|=\pm 1} \frac{k'}{k} \tilde{\sigma}_{0,k-k'}^{[r-1]} + \eta \beta \tau \frac{\tilde{\sigma}_{0,k}^{[r-1]}}{ik} \right) \quad (O.1)$$

for all $r > 1$ and, also for all $r > 1$:

$$\overline{\sigma}_n^{[r]} = -\beta V \sum_{k'=\pm 1} ik' \tilde{\sigma}_{n-1,-k'}^{[r-1]} - \beta \tau \overline{\sigma}_{n-1}^{[r-1]}, \qquad n \geq 1$$

$$\overline{\sigma}_0^{[r]} = 0$$

$$x_{n+1,k}^{[r]} = \beta V \left(\overline{\sigma}_{n-1}^{[r-1]} \delta_{|k|=1} + \sum_{k'=\pm 1} \frac{k'}{k} \tilde{\sigma}_{n-1,k-k'}^{[r-1]} \right) + \frac{\beta \tau}{ik} \tilde{\sigma}_{n-1,k}^{[r-1]}, \quad n \geq 2 \quad (O.2)$$

where $\overline{\sigma}_0^{[r-1]} = \delta_{r=1}$.

To proceed it is convenient to introduce the operator ϑ operating on the sequences of two components vectors $\sigma_{n,k}^{[r]}$, $\alpha = 1, 2$, with

$$\sigma_{n,k}^{[r]} = \begin{pmatrix} \sigma_{n,k,1}^{[r]} \\ \sigma_{n,k,2}^{[r]} \end{pmatrix} \overset{def}{=} \begin{pmatrix} \tilde{\sigma}_{n,k}^{[r]} \\ \overline{\sigma}_n^{[r]} \end{pmatrix}, \qquad n = 0, 1, \ldots, \ k = 1, 2 \ldots \quad (O.3)$$

to abridge the Eqs. (O.1), (O.2) into the form: $\sigma^{[r]} = \vartheta \sigma^{[r-1]}$ with

$$(\vartheta \sigma)_{n,k;1,1} \overset{def}{=} \sum_{m,k'} \theta_k(n,m) \left(\frac{\beta V}{ik} ik' \tilde{\sigma}_{m-1,k-k'} \delta_{|k'|=1} + \frac{\beta \tau}{ik} \tilde{\sigma}_{m-1,k} \delta_{k=k'} \right)$$

$$(\vartheta \sigma)_{n,k;1,2} \overset{def}{=} \sum_m \theta_k(n,m) \beta V \overline{\sigma}_{m-1}$$

$$(\vartheta \sigma)_{n;1,2} \overset{def}{=} -\beta V \sum_m \delta_{n=m} \sum_{k'=\pm 1} ik' \tilde{\sigma}_{m-1,-k'}$$

$$(\vartheta \sigma)_{n;2,2} \overset{def}{=} -\beta \tau \sum_m \delta_{n=m} \overline{\sigma}_{m-1} \quad (O.4)$$

for $n \geq 2$ The kernels $\vartheta_k(n,m)$ are defined by

$$\theta_k(n,m) \overset{def}{=} \left(\prod_{j=2}^{m-1} \frac{\zeta_k(j,\infty)}{\zeta_k(j,m)} \right) \left(\prod_{j=m}^{n} \zeta_k(j,\infty) \right), \quad 2 \le m \le n,$$

$$\theta_k(n,m) \overset{def}{=} \left(\prod_{j=n+1}^{m-1} \frac{1}{\zeta_k(j,m)} \right) \left(\prod_{j=2}^{n} \frac{\zeta_k(j,\infty)}{\zeta_k(j,m)} - 1 \right), \quad 2 \le n < m,$$

$$\theta_k(0,m) \overset{def}{=} -\frac{\theta_k(2;m)}{\eta} \delta_{m \ge 2} - \delta_{m,1}, \tag{O.5}$$

where the undefined elements $\theta_k(n;m)$ are set $= 0$, and products over an empty set of labels is interpreted as 1.

Remarks An interesting consistency check is that if $\tau = 0$ the recursion gives $\overline{\sigma}_n^{[r]} \equiv 0$, $\forall n \ge 1$, $r \ge 0$ and $\widetilde{\sigma}_{2,k}^{[r]} = 0$, $\forall r > 0$ and this leads, as expected, to $\sigma_n \equiv 0$, $\forall n > 0$ and $\sigma_0(q) = Z^{-1} e^{-g\beta 2V \cos(q)}$, after some algebra and after summation of the series in g, and $\overline{\sigma}_0 = 1$.

Appendix P: Bounds for the Theorem in Appendix M

Let $z = \frac{k^2}{\eta}$. From the theory of the continued fraction $\varphi(j,m)$ the following inequalities can be derived from the inequality in [33, p. 138] for $j < m$, $m \le n$:

$$\prod_{j=2}^{m-1} \frac{\varphi(j,\infty)}{\varphi(j,m)} \prod_{j=m}^{n} \varphi(j,\infty) \le 1 \quad 2 \le m \le n$$

$$|\theta_k(n,m)| \le k^{(n-m+1)} \left(\frac{(z\, e^{\frac{1}{3}z})^{m-n-1} e^z\, e^{e^{2z}}}{(m-n-1)!} \right)^{\delta_{n<m}}, \quad n,m \ge 2$$

$$|\theta_k(0,m)| \le \delta_{m=1} + \frac{k\,\delta_{m=2}}{\eta} + \frac{(z\, e^{\frac{1}{3}z})^{m-3} e^z\, e^{e^{2z}}}{k^{m-3}(m-3)!} \frac{\delta_{m>2}}{\eta}, \quad m \ge 1$$

$$b(z,\eta) \overset{def}{=} \sum_{m>n} |\theta_k(n,m)| \le 1 + \frac{\sqrt{\eta z}}{\eta} + \frac{e^{\frac{\sqrt{z}}{\sqrt{\eta}}} e^{\frac{4}{3}z} e^{e^{2z}}}{\eta}, \quad \forall n \tag{P.1}$$

with $k = \sqrt{\eta z}$.

The above bounds imply that $\widetilde{\sigma}^{[1]}$ is well defined because $\widetilde{\sigma}_{n,k}^{[0]} = 0$, $\overline{\sigma}_n^{[0]} = \delta_{n=0}$ imply $\widetilde{\sigma}_{n,k}^{[1]} = \theta_k(n,1)\beta V \delta_{|k|=1}$ and $\overline{\sigma}_n^{[1]} = -\beta\tau\delta_{n=1}$.

Rewriting Eq. (O.3), for $r > 1$, as

$$\sigma_{n,k,\alpha}^{[r]} = \sum_{n,m;k,k';\alpha,\alpha'} T_{n,m;k,k';\alpha,\alpha'} \sigma_{m,k',\alpha'}^{[r-1]} \tag{P.2}$$

it is possible to write the general $\sigma_{n,k,\alpha}^{[r]}$ and bound it by

$$\sum_{\{n_i\},\{k_i\},\{\alpha_i\}} \delta_{|k_{r-1}|=1} \prod_{i=1}^{r-1} |T_{n_{i-1},n_i;k_{i-1},k_i;\alpha_{i-1},\alpha_i}| \, |\tilde{\sigma}_{n_r-1,k_{r-1}}^{[1]}| \tag{P.3}$$

with $n_0 = n$, $k_0 = k$, $\alpha_0 = \alpha$, and $|k_j| \leq r - j$.

Taking into account that the summation over the labels k_i involves at most 3^r choices (as there are only three choices for $k_i - k_{i+1}$ in Eq. (P.3), while the summation over the labels α_i involves 2^r choices (due to the two possibilities for the labels α_i) and using the bounds in Eq. (P.1) and summing a few elementary series (geometric and exponential) it is found (for $r > 1$):

$$A_r = \frac{e^{z_r} e^{2^{2z_r}}}{(z_r e^{\frac{1}{3} z_r})^3}, \qquad B_r = r \qquad C_r = \sum_{p=0}^{\infty} \left[r^{-p} + \frac{A_r (r \, B_r)^p}{p!} \right] \tag{P.4}$$

with $z_r = \frac{r^2}{\eta}$. This implies, suitably defining $\mathcal{A}_r$, $|\sigma_{n,k}^{[r]}| \leq (\beta V + \beta \tau)^r \mathcal{A}_r r^n$ and, therefore, the theorem is proved.

Remarks The bounds above *are far from optimal* and can be improved: but the coefficient A_r does not seem to become good enough to sum over r. The bounds are sufficient to control the sum over n and yield the $\varrho^{[r]}(q)$, as in the statement of the theorem.

Appendix Q: Hard Spheres, BBGKY Hierarchy

As a second example an attempt is discussed to study a non equilibrium stationary problem which is Hamiltonian: the heat conduction in a (rarefied) gas. The system will be a gas of mass 1 particles elastically interacting via a hard core potential of radius $\frac{1}{2} r$, with centers confined in container with smooth elastic walls.

As mentioned in Sect. 2.1 the container must reach infinity, where the temperature will be fixed. So the simplest geometry is the one illustrated in Fig. H.1, with the container reaching infinity in two regions (symbolically $\pm\infty$) which *are not* connected through infinity. In this way temperature at $\pm\infty$ could be assigned with different values.

However finite containers Ω will also be considered and we look for stationary states (if any). The first question is to determine the equations of motion for the evolution of a probability distribution which assigns to a configuration $\mathbf{p}_n, \mathbf{q}_n$, of exactly n hard balls of unit mass in a *finite* container Ω, a probability to be found in $d\mathbf{p}_n d\mathbf{q}_n$ given by

$$D_n(\mathbf{p}_n, \mathbf{q}_n) \frac{d\mathbf{p}_n d\mathbf{q}_n}{n!} \tag{Q.1}$$

here $n \leq N_\Omega$ if N_Ω is the maximum number of hard balls of radius $\frac{1}{2}r$ which can fit inside Ω, and D_n are given functions, symmetric for permutations of pairs (p_i, q_i). It will be supposed that the D_n are smooth with their derivatives for $|q_i - q_j| > r, \forall i \neq j$, and that they admit limits at particles contacts.

Remarks A special case is $D_n \equiv 0$ for all $n \neq N$ with $N < N_\Omega$: it will be referred to as the microcanonical case.

Define the correlation functions as

$$\varrho(\mathbf{p}_n, \mathbf{q}_n) \stackrel{def}{=} \sum_{m \geq 0} \int D_{n+m}(\mathbf{p}_n, \mathbf{q}_n, \mathbf{p}'_m, \mathbf{q}'_m) \frac{d\mathbf{p}'_m \, d\mathbf{q}'_m}{m!} \qquad (Q.2)$$

so that $\frac{1}{n!}\varrho(\mathbf{p}_n, \mathbf{q}_n)d\mathbf{p}_n \, d\mathbf{q}_n$ is the probability of finding n particles in $d\mathbf{p}_n \, d\mathbf{q}_n$.

Let $|q_i - q_j| > r, i, j = 1, \dots, n$; then the dynamics yields $\partial_t D_n(\mathbf{p}_n, \mathbf{q}_n) = \sum_{j=1}^n p_j \partial_{q_j} D_n(\mathbf{p}_n, \mathbf{q}_n)$. Therefore:

$$\partial_t \varrho(\mathbf{p}_n, \mathbf{q}_n) = \int \sum_{i=1}^n p_i \partial_{q_i} D_{n+m}(\mathbf{p}_n, \mathbf{q}_n, \mathbf{p}'_m, \mathbf{q}'_m) \frac{d\mathbf{p}'_m \, d\mathbf{q}'_m}{m!}$$

$$+ \int \sum_{j=1}^m p'_j \partial_{q'_j} D_{n+m}(\mathbf{p}_n, \mathbf{q}_n, \mathbf{p}'_m, \mathbf{q}'_m) \frac{d\mathbf{p}'_m \, d\mathbf{q}'_m}{m!} \qquad (Q.3)$$

the integrals being over $R^3 \times \Omega$ in the coordinates of each ball. The first sum becomes

$$\sum_{i=1}^n p_i \left(\partial_{q_i} \varrho(\mathbf{p}_n, \mathbf{q}_n) - \int D(\mathbf{p}_n, \mathbf{q}_n, \mathbf{p}'_m, \mathbf{q}'_m) \sum_{j=0}^m \partial_{q_i} \chi(|q_i - q'_j) \frac{d\mathbf{p}'_m \, d\mathbf{q}'_m}{m!} \right)$$

$$(Q.4)$$

$$= \sum_{i=1}^n p_i \left(\partial_{q_i} \varrho(\mathbf{p}_n, \mathbf{q}_n) + \int_{s(q_i)} w\varrho(p_n, q_n, p', q, q_i + r\omega) \, dp' \, d\sigma_\omega \right)$$

where $s(q)$ is the sphere of radius r and center q and ω is the external normal to $s(q)$. The second term in Eq. (Q.3) is integrated by parts and for each j becomes an integral over the boundaries of the $n + m - 1$ balls and over the surface of Ω:

$$- \sum_{i=1}^n \int_{s(q_i)} \omega \cdot p' \varrho(\mathbf{p}_n, \mathbf{q}_n, p', q_i + r\omega) d\sigma_\omega dp' + X \qquad (Q.5)$$

where X is defined by

$$X \stackrel{def}{=} \frac{1}{2} \int\limits_{s(q')} \omega \cdot (p' - p'') \varrho(\mathbf{p}_n, \mathbf{q}_n, p', q', p'', q' + r\omega) dp' dp'' dq' d\sigma_\omega$$

$$+ \int\limits_{\partial\Omega} p' \cdot n_{in} \, \varrho(\mathbf{p}_n, \mathbf{q}_n, p', q) \, dp' \, d\sigma_q \qquad (Q.6)$$

where n_{in} is the internal normal to $\partial\Omega$ at the surface element $d\sigma_q \subset \partial\Omega$.

Therefore the time derivative of ϱ at $t = 0$ will be the following *BBGKY hierarchy*:

$$\partial_t \varrho(\mathbf{p}_n, \mathbf{q}_n) = \sum_i p_i \partial_{q_i} \varrho(p_n, \mathbf{q}_n)$$

$$+ \int\limits_{s(q_i)} \omega \cdot (p_i - p') \varrho(\mathbf{p}_n, \mathbf{q}_n, p', q_i + r\omega) d\sigma_\omega \, dp' + X \qquad (Q.7)$$

Having determined the equation of motion or, better, the time derivative of the correlations at $t = 0$, under the mentioned smoothness assumption, for the correlations we look for stationary solutions.

Given a function $\beta(q) \equiv (1 + \varepsilon(q))\beta_0$, $\beta_0 > 0, 0 < \varepsilon(q) \le \varepsilon_0$, imagine a distribution over the positions $\mathbf{q}_n$ with correlations

$$\varrho_0(\mathbf{q}_n) = \sum_{m \ge 0} \int D_{n+m}(\mathbf{q}_n, \mathbf{q}'_m) \frac{d\mathbf{q}'_m}{m!} \qquad (Q.8)$$

with $D_k(\mathbf{q}_k) = \delta_{k=N} z_0^k \prod_{i=1}^k \frac{\beta(q_i)}{\beta_0}$; its derivatives will be:

$$\partial_{q_i} \varrho_\emptyset(\mathbf{q}_n) = \sum_{m \ge 0} \int \partial_{q_i} \left(D_{n+m}(\mathbf{q}_n, \mathbf{q}'_m) \prod_j \chi(|q_i - q'_j| > r) \right) \frac{d\mathbf{q}'_m}{m!}$$

$$= \frac{\partial_{q_i} \beta(q_i)}{\beta(q_i)} \varrho_\emptyset(\mathbf{q}_n) - \int\limits_{s(q_i)} \omega_i \varrho_\emptyset(\mathbf{q}_n q_i + r\omega) d\sigma_\omega \qquad (Q.9)$$

It follows that setting $G_\beta(p) \stackrel{def}{=} \frac{\exp(-\frac{\beta}{2} p^2)}{(2\pi \beta^{-1})^{\frac{1}{2}}}$

$$\varrho(\mathbf{p}_n, \mathbf{q}_n) \stackrel{def}{=} \varrho_\emptyset(\mathbf{q}_n) \prod_{i=1}^n \left(\frac{\beta_0}{\beta(q_i)} \left(G_{\beta_0}(p_i) + \left(\frac{\beta(q_i)}{\beta_0} - 1 \right) \delta(p) \right) \right) \qquad (Q.10)$$

the $\varrho(\mathbf{p}_n, \mathbf{q}_n)$ satisfy identically the Eq. (Q.7) with $\partial_t \varrho = 0$, $X = 0$, i.e. the Eq. (Q.10) is a *formal* solution of the stationary BBGKY hierarchy:

$$\sum_i \left(p_i \partial_{q_i} \varrho(p_n, \mathbf{q}_n) + \int\limits_{s(q_i)} \omega \cdot (p_i - p') \varrho(\mathbf{p}_n, \mathbf{q}_n, p', q_i + r\omega) d\sigma_\omega \, dp' \right) = 0 \qquad (Q.11)$$

+∞

−∞

Fig. Q.1 Hyperboloid-like container $\Omega \subset R^3$ shape is symbolic: e.g. a cylinder of height H and area S, continued in two truncated cones is also "hyperboloid-like" for other examples see Fig. 3.5

it is formal because it is not smooth (as instead assumed in the derivation). This is also a solution of the hierarchy for infinite containers Ω, e.g. Fig. Q.1, provided the $\varrho_\emptyset$ is well defined: i.e. if Ω is finite and $N < N_\Omega$ or, for Ω is infinite, if $z_0 \frac{\beta(q)}{\beta_0}$ is small so that the Mayer series converges.

Appendix R: Interpretation of BBGKY Equations

Let $\varrho(\mathbf{p}_n, \mathbf{q}_n)$ at time t be defined for $|q_i - q_j| = r$ as a limit of the values as $|q_i - q_j| \to r^+$ at the time t: a key question arises considering two configurations with $n+2$ particles $x_n, q, p, q+r\omega, \pi$ and $x_n, q, p', q+r\omega, \pi'$ in which $q, p, q+r\omega, \pi$ is an incoming collision in the direction of the unit vector ω which is changed into the outgoing collision $q, p', q+r\omega, \pi'$.

The microscopic dynamics is described by the elastic collisions between pairs of particles. This means that if $q' = q + r\omega$, *i.e.* if particles with momenta p, π collide at a point in the cone $d\omega$ (hence $(p - \pi) \cdot \omega > 0$) cutting the surface $d\sigma_\omega$ on the sphere $s(q)$ of radius r centered at q in the direction of the unit vector ω, see Fig. R.1, Question: should it be supposed that

$$\varrho(x_n, q, p, q+r\omega, \pi) = \varrho(x_n, q, p', q+r\omega, \pi') \, ? \tag{R.1}$$

Obviously the immediate answer is *no!*: because it is possible to imagine initial distributions which do not enjoy of this property, which will be called here *transport continuity*, [34].

Yet it is true that if Eq. (R.1) holds at time 0 then it is preserved by the evolution to any *finite t*, [34], at least if the dynamics is well defined and randomly selected initial configurations, out of an equilibrium state in Ω has, with probability 1, an evolution in which only pair collisions take place. This property is known if $\Omega = R^d$, $\forall d$, [35], or if Ω is finite, [36].

While if it does not hold initially, the evolution will take the difference between the two sides of Eq. (R.1) traveling as a discontinuity in phase space, making the singularity points of the distribution density denser and denser and smoothness at

$$p' = p - \omega \cdot (p - \pi)\,\omega$$
$$\pi' = \pi + \omega \cdot (p - \pi)\,\omega \qquad \omega \cdot (p - \pi) > 0$$

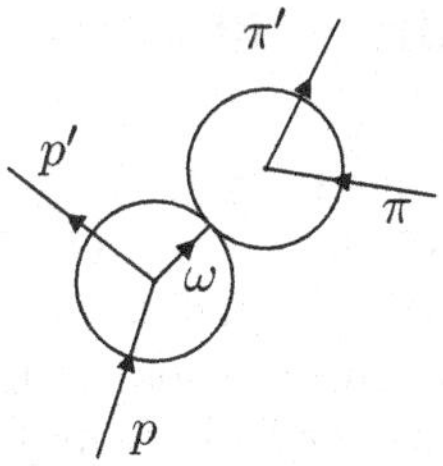

Fig. R.1 A pair collision representation

finite time cannot even be supposed: in this case the Eq. (Q.5) makes sense only at time $t = 0$ if the initial distribution is smooth.

Therefore the real question is whether Eq. (R.1) *holds with* $X = 0$, $\partial_t \varrho = 0$ in the limit of $t \to +\infty$ for the stationary distributions, once supposed that the stationary correlations are smooth.

Even assuming that the only interesting distributions should be the limits as $t \to \infty$ of the solutions of smooth distributions satisfying initially (hence forever) Eq. (R.1) it is not clear that the limits will still satisfy it.

In fact Eq. (R.1) leads to a strong and remarkable simplification of Eq. (Q.5) into $X \equiv 0$ and

$$\partial_t \varrho(\mathbf{p}_n, \mathbf{q}_n) = \sum_{i=1}^{n} (-p_i \cdot \partial_i \varrho(\mathbf{p}_n, \mathbf{q}_n)) + \int\limits_{\omega \cdot (p_i - \pi) > 0} d\sigma_\omega \, d\pi \, |\omega \cdot (\pi - p_i)| \quad \text{(R.2)}$$
$$\cdot (\varrho(\mathbf{p}_n', \mathbf{q}_n, \pi', q_i + r\omega) - \varrho(\mathbf{p}_n, \mathbf{q}_n, \pi, q_i + r\omega))$$

where $\mathbf{p}_n', \mathbf{q}_n, \pi', q_i + r\omega$ is the configuration obtained from $\mathbf{p}_n, \mathbf{q}_n, \pi, q_i + r\omega$ after the collision between p_i and π within the solid angle ω (hence $\omega \cdot (\mathbf{p}_i - \pi) > 0$).

This is the "usual" version of the BBGKY hierarchy, [37], which has been often used, [37, Eq. (2.14)], [38, p. 86], particularly in the derivation of Boltzmann's equation from the hierarchy, (although not always, see [34]).

It should be stressed that the transport continuity question also arises in Boltzmann's equation itself: although derived supposing the transport continuity, the final equation is obtained after a suitable limit (with t fixed but vanishing density with finite mean free path, i.e. the *Grad limit*, [4]), and it has $\varrho(p, q)\varrho(\pi, q) - \varrho(p', q)\varrho(\pi', q))$ in the integral in Eq. (R.2) (for $n = 1$): thus Eq. (R.1) would imply that the latter quantity is 0 and, by the classical argument of Maxwell, that $\varrho(q, p)$ is a Gaussian in p, which of course can hardly be expected.

In conclusion the transport continuity seems to be an open question, and a rather important one.

Appendix S: BGGKY; An Exact Solution (?)

It is tempting to try to find a solution to the stationary BBGKY hierarchy in situations in which the system is close to an equilibrium state whose correlations can be computed via a virial or Mayer's expansion. If the system is in equilibrium and the momentum distribution is supposed Gaussian then it can be shown that in fact the BBGKY hierarchy is equivalent to the Kirkwood-Salsburg equations at least if the parameters are in the region of convergence of the Mayer's series, see [39] for the soft potentials case and [40] for the hard spheres case.[9]

The formal solution in Appendix Appendix Q: does not satisfy the collision continuity property. Furthermore it contains an arbitrary function $\beta(q)$ which is such that $\langle \frac{1}{2} p^2 \rangle (q) \overset{def}{=} \int \varrho(p, q) \frac{1}{2} p^2 dp = \frac{3}{2} \beta(q)^{-1}$, i.e. it sets a quantity that could be called the local temperature to $\beta(q)^{-1}$, an arbitrary value.

This indicates that the Eq. (Q.7) needs extra "boundary conditions". The collision continuity would be natural but, as remarked, it is not satisfied by Eq. (Q.10).

There are other "collision continuity" conditions that can be considered. For instance if $(\mathbf{p}, \pi) \to (\mathbf{p}', \pi')$ is a collision with centers of $\mathbf{p}, \pi$ away by $r\,\omega$, then in the paper of Maxwell, [43], the equation

$$\sum_\alpha \int p_\alpha Q(p) \partial_\alpha f(p, q) = \int_{\omega \cdot (\mathbf{p} - \pi) > 0} \omega \cdot (\mathbf{p} - \pi) \, d\mathbf{p} d\pi d\sigma_\omega$$
$$\cdot (Q(p') - Q(p)) f(p, q, \pi, q + r\omega) \qquad (S.1)$$

is the key stationary equation and is used[10] for $Q = 1$, p_α, p^2, $p_\alpha p^2$.[11]

Validity of Eq. (S.1) for $Q = p^2$ is implied, at least for all points of Ω farther from $\partial\Omega$, by $\geq 2r$ by the strong condition that $\beta(q)$ is harmonic.[12] This would be of some interest if it could be proved that the correlations define a probability measure[13] and if the measure is stationary.

(a) The ϱ are > 0 if $\frac{\beta(q)}{\beta_0} \geq 1$: a condition that would be interesting, for instance in the case of Fig. Q.1, if it could be satisfied by requiring $\beta(q)$ to be harmonic with Neumann's boundary conditions on $\partial\Omega$ and $\beta(q) = \beta_0$ at $q = -\infty$ and $\beta(q) = \beta_0(1 + \varepsilon_0)$, $\varepsilon_0 > 0$ at $q = +\infty$.

---

[9] The latter work also corrects an error on the third and higher orders of an expansion designed, in [41], to present a simplified account of the proof in [42] of the convergence of the virial series.

[10] But with the cross section of a potential proportional to r^{-4} instead of the hard balls cross section considered in Eq. (S.1), i.e. with $\omega \cdot (\mathbf{p} - \pi)$ replaced by a constant.

[11] If $f(p, q, \pi, q + r\omega)$ were a product $f(p, q) f(\pi, q)$, Eq. (S.1) would follow from the Boltzmann's equation for hard spheres.

[12] Simply insert the Eq. (Q.10) in Eq. (S.1) and remark that it is a consequence of the harmonic average theorem.

[13] If the $\varrho(\mathbf{p}_n, \mathbf{q}_n)$ are non negative then they actually are correlations of a probability measure: this follows from the fact that the $\varrho_\emptyset$ are correlations of a probability measure on the positions.

However the harmonicity of $\beta(q)$ implies, in the same conditions, that $Q(p) = p_\alpha$ satisfies Eq. (S.1) with the *r.h.s.* multiplied by the factor 2 (!). And Eq. (S.1) also fails for $Q(p) = p_\alpha p^2$ as well as for all other observables.

(b) Furthermore a rather strong argument in support of the lack of stationarity of the discussed exact solutions can be based on the ergodicity properties of the finite hard spheres systems, see Appendix T.

Appendix T: Comments on BGGKY and Stationarity

(1) The main question is: why "transport continuity", as Eq. (S.1) may be called, does not hold for $Q(p) \neq 1$, p^2? (in the cases in Appendix S) or perhaps: why should it hold at all? It might seem that if transport continuity for $Q = 1$, p_α, p^2 fails, also conservation of mass, momentum and energy fail: but this is not the case because the conservation of $Q(p)$ in a stationary state satisfying Eq. (Q.11) with $\partial_t \varrho = 0$ and $X = 0$, is simply obtained by multiplying both sides, with $n = 1$, by $Q(p)$ and integrating over p. Therefore there seems to be no obvious reason to require nor to impose transport continuity. Not even for $Q = p^2$. But then the function $\beta(q)$ remains quite arbitrary.

(2) A consequence of the analysis is that if $\beta = \beta(q) \neq \beta_0$ the correlations in Eq. (Q.10) yield a solution of the equilibrium equations in infinite volume *which is not the Gibbs state*, and which has, at positions q, average kinetic energy per particle $\frac{3}{2}\beta(q)^{-1}$ and activity $z = z_0 \frac{\beta(q)}{\beta_0}$. Actually the above comment seems to indicate that the stationary states defined by the considered exact solution Eq. (Q.10), could be considered as new equilibrium states rather than genuine non equilibrium states. Therefore it is important to understand whether the exact solution is really a stationary solution for the hard spheres dynamics.

(3) It is possible, as discussed below, to give an argument showing that the exact solution, Eq. (Q.10), *is not in general a stationary solution* for the hard sphere dynamics in a finite container. The question in infinite volume would be more difficult because there is no existence theorem for infinite hard spheres systems. However in the case of N balls in a finite container Ω with elastic reflecting walls Eq. (Q.10) is an exact property of the functions ϱ at time $t = 0$ and the dynamics is well defined almost everywhere o7n each energy surface (with respect to the area measure of the surface), [35].

The questions on whether the Eq. (Q.10) is or not an exact solution, or if $\beta(q)$ and the boundary conditions can be chosen so that it is a stationary solution, are well posed in the case of a finite container with N balls and have some interest.

Consider initial configurations in which some $k \leq N$ balls have *zero momentum*. Let $\mathbf{p} \overset{def}{=} (p_1, \ldots, p_N)$ and $\mathbf{q} \overset{def}{=} (q_1, \ldots, q_N)$ and $q_X \overset{def}{=} (q_{j_1}, \ldots, q_{j_k})$, $p_X \overset{def}{=} (p_{j_1}, \ldots, p_{j_k})$ if $X = (j_1, \ldots, j_k) \subset \{1, \ldots, N\}$, $X^c =$ complement of X in $\{1, \ldots, N\}$. Let $(\mathbf{p}, \mathbf{q})$ be chosen with the distribution proportional to

$$G_{\beta_0}(p_{X^c}) \prod_{j \in X} (\varepsilon(q_j)\delta(p_j)) \qquad (\text{T.1})$$

with $\varepsilon \geq 0$ as in Eq. (Q.10).

It seems reasonable to *conjecture* that, with probability 1, the above configurations $(\mathbf{p}, \mathbf{q})$ will be typical[14], see [44, 45], for the ergodic properties of the hard balls motions.[15]

Then, if the ergodicity conjecture holds for the finite systems of hard balls, the measure in Eq. (T.1) will evolve in time to a distribution proportional to $G_{\beta_0}(\mathbf{p}) \frac{\sigma_{N-k}}{\sigma_N} \mathbf{p}^{-3k}$, with σ_k the surface of the unit ball in $3k$ dimension times $(\sqrt{2\pi}\beta_0^{-1})^{3k}$.

Hence the Eq. (Q.10) will evolve to a distribution proportional to

$$z_0^N \Big(\prod_{j=1}^{N} \varepsilon(q_j)\delta(p_j) \Big) + z_0^N G(\mathbf{p}) \Big(\sum_{k=1}^{N} \binom{N}{k} e_k |\mathbf{p}|^{-3k} \Big(\sqrt{2\pi\beta_0^{-1}} \Big)^{3k} \Big) \qquad (\text{T.2})$$

where $e_k \overset{def}{=} \langle \prod_{j \in X} \varepsilon(q_j) \rangle$ if $|X| = k$ and $\langle \cdot \rangle$ denotes average with respect to the equilibrium distribution with activity $z(q) = z_0 \frac{\beta(q)}{\beta_0}$.

Since the distribution in Eq. (T.2), if $\beta(q)$ is not constant, is certainly different from the distribution which solves exactly the BBGKY hierarchy Eq. (Q.10), simply because it does not contain any partial product of delta functions, it follows that the Eq. (Q.10) will be different from its own time average and *therefore it is not a stationary distribution* even though at time 0 it has formally zero time derivative.

Of course it might be that phase space points with one or more particles standing with 0 momentum are not typical on their energy surface: therefore the above argument has a heuristic nature.

In conclusion the Eq. (Q.11) says that the time derivative of the distribution Eq. (Q.10) is 0 at time 0: however the equation is not an ordinary differential equation and hence this does not imply that it remains stationary. And if it is not stationary then it will become discontinuous at $t > 0$ and it will not even obey the BBGKY hierarchy in the form, Eq. (Q.7), in which it has been derived under smoothness hypotheses on the correlations.

---

[14] Typical means that the time averages of smooth observables on the trajectory generated by $(\mathbf{p}, \mathbf{q})$ will exist and be given by the average over the surface of energy $\frac{1}{2}\mathbf{p}^2 \overset{def}{=} E$ with respect to the normalized area measure.

[15] The results on ergodicity of the motions on the energy surfaces for hard balls systems are still not complete, [46].

References

1. Boltzmann, L.: Über die mechanische Bedeutung des zweiten Hauptsatzes der Wärmetheorie, vol. 1, #2 of Wissenschaftliche Abhandlungen, ed. F. Hasenöhrl. Chelsea, New York (1968)
2. Clausius, R.: Ueber die Zurückführung des zweites Hauptsatzes der mechanischen Wärmetheorie und allgemeine mechanische Prinzipien. Annalen der Physik 142, 433–461 (1871)
3. Boltzmann, L.: Bemerkungen über einige Probleme der mechanischen Wärmetheorie, vol. 2, #39 of Wissenschaftliche Abhandlungen, ed. F. Hasenöhrl. Chelsea, New York (1877)
4. Gallavotti, G.: Statistical Mechanics. A Short Treatise. Springer, Berlin (2000)
5. Marchioro, C., Presutti, E.: Thermodynamics of particle systems in presence of external macroscopic fields. Communications in Mathematical Physics 27, 146–154 (1972)
6. Boltzmann, L.: Über die Beziehung zwischen dem zweiten Hauptsatze der mechanischen Wärmetheorie und der Wahrscheinlichkeitsrechnung, respektive den Sätzen über das Wärmegleichgewicht, vol. 2, #42 of Wissenschaftliche Abhandlungen, ed. F. Hasenöhrl. Chelsea, New York (1968)
7. Boltzmann, L.: Analytischer Beweis des zweiten Hauptsatzes der mechanischen Wärmetheorie aus den Sätzen über das Gleichgewicht des lebendigen Kraft, vol. 1, #20 of Wissenschaftliche Abhandlungen, ed. F. Hasenöhrl. Chelsea, New York (1968)
8. Whittaker, E.T.: A Treatise on the Analytic Dynamics of Particles and Rigid Bodies. Cambridge University Press, Cambridge (1917) (reprinted 1989)
9. Holian, B.D., Hoover, W.G., Posch, H.A.W.: Resolution of loschmidt paradox: the origin of irreversible behavior in reversible atomistic dynamics. Phys. Rev. Lett. 59, 10–13 (1987)
10. Gibbs, J.: Elementary Principles in Statistical Mechanics. Schribner, Cambridge (1902)
11. Gallavotti, G., Bonetto, F., Gentile, G.: Aspects of the Ergodic, Qualitative and Statistical Theory of Motion. Springer, Berlin (2004)
12. Sinai, Y.G.: Markov partitions and C-diffeomorphisms. Funct. Anal. Appl. 2(1), 64–89 (1968)
13. Bowen, R.: Markov partitions for axiom A diffeomorphisms. Am. J. Math. 92, 725–747 (1970)
14. Gallavotti, G.: Reversible Anosov diffeomorphisms and large deviations. Math. Phys. Elect. J. 1, 1–12 (1995)
15. Arnold, V., Avez, A.: Ergodic Problems of Classical Mechanics. Benjamin, New York (1966)
16. Smale, S.: Differentiable dynamical systems. Bull. Am. Math. Soc. 73, 747–818 (1967)
17. Bonetto, F., Gallavotti, G.: Reversibility, coarse graining and the chaoticity principle. Commun. Math. Phys. 189, 263–276 (1997)
18. Dressler, U.: Symmetry property of the lyapunov exponents of a class of dissipative dynamical systems with viscous damping. Phys. Rev. A 38, 2103–2109 (1988)
19. Dettman, C., Morriss, G.: Proof of conjugate pairing for an isokinetic thermostat. Phys. Rev. E 53, 5545–5549 (1996)
20. Arnold, V.: Méthodes Mathématiques de la Mécanique classique. MIR, Moscow (1974)
21. Gallavotti, G.: Dynamical ensembles equivalence in fluid mechanics. Physica D 105, 163–184 (1997)
22. Gallavotti, G.: Chaotic dynamics, fluctuations, non-equilibrium ensembles. Chaos 8, 384–392 (1998)
23. Jarzynski, C.: Nonequilibrium equality for free energy difference. Phys. Rev. Lett. 78, 2690–2693 (1997)
24. Jarzynski, C.: Hamiltonian derivation of a detailed fluctuation theorem. J. Stat. Phys. 98, 77–102 (1999)
25. Evans, D.J., Searles, D.J.: Equilibriun microstates which generate second law violating steady states. Phys. Rev. E 50, 1645–1648 (1994)
26. Cohen, E.G.D., Gallavotti, G.: Note on two theorems in nonequilibrium statistical mechanics. J. Stat. Phys. 96, 1343–1349 (1999)
27. Bochkov, G.N., Kuzovlev, YuE: Nonlinear fluctuation-dissipation relations and stochastic models in nonequilibrium thermodynamics: I. Generalized fluctuation-dissipation theorem. Physica A 106, 443–479 (1981)

28. Bochkov, G.N., Kuzovlev, YuE: Nonlinear fluctuation-dissipation relations and stochastic models in nonequilibrium thermodynamics: II. Kinetic potential and variational principles for nonlinear irreversible processes. Physica A **106**, 480–520 (1981)

29. Gallavotti, G., Iacobucci, A., Olla, S.: Nonequilibrium stationary state for a damped pendulum. arXiv:1310.5379, pp. 1–18 (2013)

30. Mattingly, J.C., Stuart, A.M.: Geometric ergodicity of some hypo-elliptic diffusions for particle motions. Markov Process. Rel. Fields **8**, 199–214 (2002)

31. Gradshtein, I.S., Ryzhik, I.M.: Table of Integrals, Series, and Products. Academic Press, New York (1965)

32. Faggionato, A., Gabrielli, G.: A representation formula for large deviations rate functionals of invariant measures on the one dimensional torus. Annales de l' Institut H. Poincaré, (Probabilité et Statistique), **48**, 212–234 (2012)

33. Cuyt, A., Petersen, V., Verdonk, B., Waadeland, H., Jones, W.: Handbook of Continued Fractions for Special Functions. Springer, Berlin (2004)

34. Spohn, H.: On the integrated form of the BBGKY hierarchy for hard spheres. arxiv: math-ph/0605068, pp. 1–19 (2006)

35. Marchioro, C., Pellegrinotti, A., Presutti, E.: Existence of time evolution for ν dimensional statistical mechanics. Commun. Math. Phys. **40**, 175–185 (1975)

36. Marchioro, C., Pellegrinotti, A., Presutti, E., Pulvirenti, M.: On the dynamics of particles in a bounded region: a measure theoretical approach. J. Math. Phys. **17**, 647–652 (1976)

37. Cercignani, C.: The Boltzmann Equation and Its Applications. Applied Mathematical Sciences, vol. 67. Springer, Heidelberg (1988)

38. Lanford, O.: Time evolution of large classical systems. Moser, J. (ed.) Dynamical Systems, Theory and Applications, vol. 38, pp. 1–111. Lecture Notes in Physics, Springer, Berlin (1974)

39. Gallavotti, G.: On the mechanical equilibrium equations. Il Nuovo Cimento **57 B**, 208–211 (1968)

40. Genovese, G., Simonella, S.: On the stationary BBGKY hierarchy for equilibrium states. J. Stat. Phys. (in print), arXiv:1205.2788 1–27 (2012)

41. Gallavotti, G., Verboven, E.: On the classical KMS boundary condition. Il Nuovo Cimento **28 B**, 274–286 (1975)

42. Morrey, C.B.: On the derivation of the equations of hydrodynamics from statistical mechanics. Commun. Pure Appl. Math. **8**, 279–326 (1955)

43. Maxwell, J.C.: On the dynamical theory of gases. In: Niven, W.D. (ed.) The Scientific Papers of J.C. Maxwell, vol. 2. Cambridge University Press, Cambridge (1964)

44. Sinai, Y.: Dynamical systems with elastic reflections. Russ. Math. Surv. **25**, 137–189 (1970)

45. Bunimovich, L.A.: On ergodic properties of certain billiards. Funct. Anal. Appl. **8**, 254–255 (1979)

46. Szasz, D.: Some challenges in the theory of (semi)-dispersing billiards. Nonlinearity **21**, T187–T193 (2008)

47. Gallavotti, G.: Entropy, Thermostats and Chaotic Hypothesis. Chaos **16**, 043114 (+6) (2006)

Author Index

G. Gallavotti, *Nonequilibrium and Irreversibility*,
Theoretical and Mathematical Physics, DOI: 10.1007/978-3-319-06758-2,
© Springer International Publishing Switzerland 2014

Subject Index

G. Gallavotti, *Nonequilibrium and Irreversibility*,
Theoretical and Mathematical Physics, DOI: 10.1007/978-3-319-06758-2,
© Springer International Publishing Switzerland 2014

Printed in the United States
By Bookmasters